PUNCTUATED EQUILIBRIUM

Stephen Jay Gould

PUNCTUATED EQUILIBRIUM

THE BELKNAP PRESS OF
HARVARD UNIVERSITY PRESS
CAMBRIDGE, MASSACHUSETTS
LONDON, ENGLAND
2007

Printed in the United States of America

First paperback edition published with permission of Turbo, Inc.

Material in this book has been reprinted from Chapters 1 and 9 of *The Structure of Evolutionary Theory* by Stephen Jay Gould, The Belknap Press of Harvard University Press, 2002 (abbreviated in the text as *SET*). Cross references to chapter numbers in the text refer to chapters in the present volume, unless otherwise noted.

Library of Congress Cataloging-in-Publication Data

Gould, Stephen Jay.
Punctuated equilibrium / Stephen Jay Gould.—1st pbk. ed.
p. cm.
Includes bibliographical references and index.
ISBN-13: 978-0-674-02444-1
ISBN-10: 0-674-02444-3
1. Punctuated equilibrium (Evolution) I. Title.

QH398.G68 2007
576.8′2—dc22 2007001439

CONTENTS

PUNCTUATED
EQUILIBRIUM

INTRODUCTION

I admire my friend Oliver Sacks extravagantly as a writer, and I could never hope to match him in general quality or human compassion. He once said something that touched me deeply, despite my continuing firm disagreement with his claim (while acknowledging the validity of the single statement relevant to the present context). Oliver said that he envied me because, although we had both staked out a large and generous subject for our writing (he on the human mind, me on evolution), I had enjoyed the privilege of devising and developing a general theory that allowed me to coordinate all my work into a coherent and distinctive body, whereas he had only written descriptively and aimlessly, albeit with some insight, because no similar central focus underlay his work. I replied that he had surely sold himself short, because he had been beguiled by conventional views about the nature and limits of what may legitimately be called a central scientific theory—and that he certainly held such an organizing concept in his attempt to reintroduce the venerable "case study method" of attention to irreducible peculiarities of individual patients in the practice of cure and healing in medicine. Thus, I argued, he held a central theory about the importance of individuality and contingency in general medical theory, just as I and others had stressed the centrality of historical contingency in any theoretical analysis and understanding of evolution and its actual results.

Oliver saw the theory of punctuated equilibrium itself, which I developed with Niles Eldredge and discuss at inordinate length in this book, as my coordinating centerpiece, and I would not deny this statement. But punctuated equilibrium stands for a larger and coherent set of mostly iconoclastic concerns, and I must present some intellectual autobiography to explain the reasons and the comings together, as best I understand them myself.

As my first two scientific commitments, I fell in love with paleontology when I met *Tyrannosaurus* in the Museum of Natural History at age five, and with evolution at age 11, when I read G. G. Simpson's *The Meaning of Evolution,* with great excitement but minimal comprehension, after my parents, as members of a book club for folks with intellectual interests but little economic opportunity or formal credentials, forgot to send back the "we don't want anything this month" card, and received the book they would never have ordered (but that I begged them to keep because I saw the little stick figures of dinosaurs on the dust jacket). Thus, from day one, my developing professional interests united paleontology and evolution. For some reason still unclear to me, I always found the theory of how evolution works more fascinating than the realized pageant of its paleontological results, and my major interest therefore always focused upon principles of macroevolution.[1] I did come to understand the vague feelings of dissatisfaction (despite Simpson's attempt to resolve them in an orthodox way by incorporating paleontology within the Modern Synthesis) that some paleontologists have always felt with the Darwinian premise that microevolutionary mechanics could construct their entire show just by accumulating incremental results through geological immensity.

As I began my professional preparation for a career in paleontology, this vague dissatisfaction coagulated into two operational foci of discontent. First (and with Niles Eldredge, for we worried this subject virtually to death as graduate students), I became deeply troubled by the Darwinian convention that attributed all non-gradualistic literal appearances to imperfections of the geological record. This traditional argument contained no logical holes, but the practical consequences struck me as unacceptable (especially at the outset of a career, full of enthusiasm for empirical work, and trained in statistical techniques that would permit the discernment of small evolutionary changes). For, by the conventional rationale, the study of microevolution became virtually nonoperational in paleontology—as one almost never found this anticipated form of gradual change up geological sections, and one therefore had to interpret the vastly predominant signal of stasis and geologically abrupt appearance as a sign of the record's imperfection, and therefore as no empirical guide to the nature of evolution. Second, I became increasingly disturbed that, at the higher level of evolutionary trends within clades, the majority of well documented examples (reduction of stipe number in graptolites, increasing symmetry of crinoidal cups, growing complexity of ammonoid sutures, for example) had never been adequately explained in the terms demanded by Darwinian convention—that is, as adaptive improvements of constituent organisms in anagenetic sequences. Most so-called explanations amounted to little more than what Lewontin and I, following Kipling, would later call "just-so stories," or plausible claims without tested evidence, whereas other prominent trends couldn't even generate a plausible story in adaptationist terms at all.

As Eldredge and I devised punctuated equilibrium, I did use the theory to resolve these two puzzles to my satisfaction, and each resolution, when finally generalized and further developed, led to my two major critiques of the first two branches of the essential triad of Darwinian central logic—so Oliver Sacks's identification of punctuated equilibrium as central to my theoretical world holds, although more as a starting point than as a coordinating focus. By accepting the geologically abrupt appearance and subsequent extended stasis of species as a fair description of an evolutionary reality, and not only as a sign of the poverty of paleontological data, we soon recognized that species met all criteria for definition and operation as genuine Darwinian individuals in the higher-level domain of macroevolution—and this insight (by complex routes discussed in the chapters of this book) led us to concepts of species selection in particular and, eventually, to the full hierarchical model of selection as an interesting theoretical challenge and contrast to Darwinian convictions about the exclusivity of organismal selection. In this way, punctuated equilibrium led to the reformulation proposed herein for the first branch of essential Darwinian logic.

Meanwhile, in trying to understand the nature of stasis, we initially focused (largely in error, I now believe) upon internal constraints, as vaguely represented by various concepts of "homeostasis," and as exemplified in the model of Galton's polyhedron (see *The Structure of Evolutionary Theory* (hereafter *SET*), Chapter 4). These thoughts led me to extend my doubts about adaptation and the sufficiency of functionalist mechanisms in general—especially in conjunction with my old worries about paleontological failures to explain cladal trends along traditional adaptationist lines. Thus, these aspects of punctuated equilibrium strongly contributed to my developing critiques of adaptationism and purely functional mechanics on the second branch of essential Darwinian logic (although other arguments struck me as even more important).

Nonetheless, and despite the centrality of punctuated equilibrium in developing a broader critique of conventional Darwinism, my sources extended outward into a diverse and quirky network of concerns that seemed, to me and at first, isolated and uncoordinated, and that only later congealed into a coherent critique. For this curious, almost paradoxical, reason, I have become even more convinced that the elements of my overall critique hang together, for I never sensed the connections when I initially identified the components as, individually, the most challenging and intriguing items I had encountered in my study of evolution. When one accumulates a set of things only for their independent appeals, with no inkling that any common intellectual ground underlies the apparent miscellany, then one can only gain confidence in the "reality" of a conceptual basis discerned only later for the cohesion. I would never argue that this critique of strict Darwinism gains any higher probability of truth value for initially infecting me in such an uncoordinated and mindless way. But I would assert that a genuinely coherent and general alternative formulation must exist "out there" in the philosophical universe of intellectual possibilities—whatever its empirical validity—if its isolated components could coagulate, and be discerned and selected, so unconsciously.

If I may make a somewhat far-fetched analogy to my favorite Victorian novel, *Daniel Deronda* (the last effort of Darwin's friend George Eliot), the hero of this story, a Jew raised in a Christian family with no knowledge of his ethnic origins, becomes, as an adult, drawn to a set of apparently independent activities with no coordinating theme beyond their relationship, entirely unknown to Deronda at the time of his initial fascination, to Jewish history and customs. Eventually, he recognizes the unifying theme behind such apparent diversity, and learns the truth of his own genetic background. (I forgive Eliot for this basically silly fable of genealogical determinism because her philosemitic motives, however naïve and a bit condescending, shine forth so clearly in the surrounding antisemitic darkness of her times.) But I do feel, to complete the analogy, rather like a modern, if only culturally or psychologically predisposed, Deronda who gathered the elements of a coherent critique solely because he loved each item individually—and only later sensed an underlying unity, which therefore cannot be chimaerical, but may claim some logical existence prior to any conscious formulation on my part.

In fact, the case for an external and objective coherence of this alternative view of evolution seems even stronger to me because I gathered the independent items not only in ignorance of their coordination, but also at a time when I held a conscious and conventional view of Darwinian evolution that would have actively denied their critical unity and meaning. I fledged in science as a firm adaptationist, utterly beguiled by the absolutist beauty (no doubt, my own simplistic reading of a more subtle, albeit truly hardened, Modern Synthesis) of asserting, à la Cain and other ecological geneticists of the British school, that all aspects of organismal phenotypes, even the most trivial nuances, could be fully explained as adaptations built by natural selection.

I remember two incidents of juvenilia with profound embarrassment today: First, an undergraduate evening bull session with the smartest physics major at Antioch College, as his skepticism evoked my stronger insistence that our science matched his in reductionistic rigor because "we" now knew for certain that natural selection built everything for optimal advantage, thus making evolution as quantifiable and predictive as classical physics. Second, as a somewhat more sophisticated, but still beguiled, assistant professor, I remember my profound feeling of sadness and disappointment, nearly amounting to an emotional sense of betrayal, upon learning that an anthropological colleague favored drift as the probable reason for apparently trivial genetic differences among isolated groups of Papua-New Guinea peoples. I remember remonstrating with him as follows: Of course your argument conforms to logic and empirical possibility, and I admit that we have no proof either way. But your results are also consistent with selection—and our panselectionist paradigm has forged a theory of such beauty and elegant simplicity that one should never favor exceptions for their mere plausibility, but only for documented necessity. (I recall this discussion with special force because my emotional feelings were so strong, and my disappointment in his "unnecessary apostasy" so keen, even though I knew that neither of us had the empirical "goods.") Finally, if I could, in a species of Devil's bargain, wipe any of my publications off the face of the earth and out of all memory, I would gladly nominate my unfortunately rather popular review article on "Evolutionary paleontology and the science of form" (Gould, 1970a)—a ringing paean to selectionist absolutism, buttressed by the literary barbarism that a "quantifunctional" paleontology, combining the best of biometric and mechanical analyses, could prove panadaptationism even for fossils that could not be run through the hoops of actual experiments.

Against this orthodox background—or, rather, within it and quite unconsciously for many years—I worked piecemeal, producing a set of separate and continually accreting revisionary items along each of the branches of

Darwinian central logic, until I realized that a "Platonic" something "up there" in ideological space could coordinate all these critiques and fascinations into a revised general theory with a retained Darwinian base.

The first branch of levels in selection proceeded rather directly and linearly because the generality flowed so clearly from punctuated equilibrium itself, once Eldredge and I finally worked through the implications and extensions of our own formulations (Eldredge and Gould, 1972). Steve Stanley (1975) and Elisabeth Vrba (1980) helped to show us what we had missed in ramifications leading from the phenomenology of stasis and geologically abrupt appearance, to recognizing species as genuine Darwinian individuals, to designating species as, therefore and potentially, the basic individuals of macroevolution (comparable with the role of the organism in microevolution), to the validity of species selection, and eventually to the full hierarchical model and its profound departure from the exclusively organismal accounts of conventional Darwinism (or the even more reduced and equally monistic genic versions of Williams and Dawkins)—see Vrba and Gould, 1986. Finally, by adopting the interactor rather than the replicator approach to defining selection, and by recognizing emergent fitness, rather than emergent characters, as the proper criterion for identifying higher-level selection (Lloyd and Gould, 1993; Gould and Lloyd, 1999), I think that we finally reached, by a circuitous route around many stumbling blocks of my previous stupidity, a consistent and truly operational theory of hierarchical selection (see *SET,* Chapter 8).

I must also confess to some preconditioning beyond punctuated equilibrium. I had admired Wynne-Edwards's pluck (1962) from the start, even though I agreed with Williams's (1966) trenchant criticisms of his particular defenses for group selection, rooted in the ability of populations to regulate their own numbers in the interests of group advantage. Still, I felt, for no reason beyond vague intuition, that group selection made logical sense and might well find other domains and formulations of greater validity—a feeling that has now been cashed out by modern reformulations of evolutionary theory (see especially Wilson and Sober, 1998; *SET,* Chapter 8; and the arguments herein).

My odyssey on the second branch of balancing internal constraint with external adaptation in understanding the patterning and creative population of novel places in evolutionary morphospace followed a much more complex, meandering and diverse set of pathways. As an undergraduate, I loved D'Arcy Thompson's *Growth and Form* (1917; see Gould, 1971a, for my first "literary" paper), and wrote a senior thesis on his theory of morphology. But I thought that I admired the book only for its incomparable prose, and I attacked the anti-Darwinian (and structuralist) components of his theory unmercifully. I then took up allometry for my first empirical studies, somehow fascinated by structural constraint and correlation of growth, but thinking all the while that my task must center on a restoration of adaptationist themes to this "holdout" bastion of formalist thought—particularly the achievement of biomechanical optima consistent with the Galilean principle of decreasing surface/volume ratios with increasing size in isometric forms. I remain proud of my first review article, dedicated to this subject (Gould, 1966), written when I was still a graduate student, but I am now embarrassed by the fervor of my adaptationist convictions.

I emphasized allometric analysis, now in a directly multivariate reformulation, in my first set of empirical studies on the Bermudian pulmonate snail *Poecilozonites* (see especially Gould, 1969—the published version of my Ph.D. dissertation). And yet, of all the long and largely adaptationist treatises in this series, and for some reason that I could not identify at the time, the conclusion that I reached with most satisfaction, and that I somehow regarded as most theoretically innovative (without knowing why), resided in a short, and otherwise insignificant, article that I wrote for a specialized paleontological journal on a case of convergence produced by structural necessity, given modes of coiling and allometry in this genus, rather than by selectionist honing (for some cases rested upon ecophenotypic expression, others on paedomorphosis, and still others on gradual change that could be read as conventionally adaptive): "Precise but fortuitous convergence in Pleistocene land snails" (Gould, 1971b).

Five disparate reasons underlie my more explicit recognition, during the 1970s and early 1980s, of the importance and theoretical interest (and iconoclasm versus Darwinian traditions) of nonadaptationist themes rooted in structural and historical constraint. First, I stood under the dome of San Marco during a meeting in Venice and then wrote a notorious paper with Dick Lewontin on the subject of spandrels, or nonadaptive sequelae of prior structural decisions (Gould and Lewontin, 1979—see *SET,* Chapter 11, pp. 1246–1258). Second, I recognized, with Elisabeth Vrba, that the lexicon of evolutionary biology possessed no term for the evidently important phenomenon of structures coopted for utility from different sources of origin (including nonadaptive spandrels), and not directly built as adaptations for their current function. We therefore devised the term "exaptation" (Gould and Vrba, 1982) and explored its implications for structuralist revisions to pure Darwinian functionalism. Third, I worked with a group of paleontological colleagues (Raup et al., 1973; Raup and Gould, 1974; Gould et al., 1977) to develop more rigorous criteria for identifying the signals that required selectionist, rather than stochastic, explanation of apparent order in phyletic patterns. This work left me humbled by the insight that our brains seek pattern, while our cultures favor particular kinds of stories for explaining these patterns—thus imposing a powerful bias for ascribing conventional deterministic causes, particularly adaptationist scenarios in our Darwinian traditions, to patterns well within the range of expected outcomes in purely stochastic systems. This work sobered me against such *a priori* preferences for adaptationist solutions, so often based upon plausible stories about results, rather than rigorous documentation of mechanisms.

Fourth, and most importantly, I read the great European structuralist literatures in writing my book on *Ontogeny and Phylogeny* (Gould, 1977). I don't see how anyone could read, from Goethe and Geoffroy down through Severtzov, Remane and Riedl, without developing some appreciation for the plausibility, or at least for the sheer intellectual power, of morphological explanations outside the domain of Darwinian functionalism—although my resulting book, for the last time in my career, stuck closely to selectionist orthodoxy, while describing these alternatives in an accurate and sympathetic manner. Fifth, my growing unhappiness with the speculative character of many adaptationist scenarios increased when, starting in the mid 1970s, the growing vernacular (and some of the technical) literature on sociobiology touted conclusions that struck me as implausible, and that also (in some cases) ran counter to my political and social beliefs as well.

Personal distaste, needless to say, bears no necessary relationship to scientific validity. After all, what could be more unpleasant, but also more factually undeniable, than personal mortality? But when distasteful conclusions gain popularity by appealing to supposedly scientific support, and when this "support" rests upon little more than favored speculation in an orthodox mode of increasingly dubious status, then popular misuse can legitimately sharpen a scientist's sense of unhappiness with the flawed theoretical basis behind a particular misuse. In any case, I trust that this compendium of reasons will dispel Cain's (1979) hurtful assertion that Lewontin, I, and other evolutionists who questioned early forms of sociobiology by developing a general critique of adaptationism, had acted cynically, and even antiscientifically, in opposing biological theories that we knew to be true because we disliked their political implications for explaining human behavior. My own growing doubts about adaptationism arose from several roots, mostly paleontological, with any displeasure about sociobiology serving as a late and minor spur to further examination and synthesis.

I then tried to apply my general critique of pure Darwinian functionalism, and my conviction that important and positive constraints could be actively identified by quantitative morphometric study (and not merely passively inferred from failures of adaptationist scenarios) in my work on "covariance sets" in the growth, variation, and evolution of the West Indian pulmonate *Cerion* (Gould, 1984a and b), a snail that encompasses its maximal diversity in overt form among populations within a constraining set of pervasive allometries in growth. I discuss some of this work in my text on the empirical validation of positive constraint (see *SET,* Chapter 10, pages 1045–1051).

My doubts on the third branch of extrapolationism and uniformity began even earlier, and in a more inchoate way, but then gained expression in my efforts in the history of science, and not so much in my direct empirical work—hence, in part, the reduced attention devoted to this theme (*SET,* Chapters 6 and 12) compared with the first two branches of selection's agency and efficacy. On a fieldtrip in my freshman geology course, my professor took us to a travertine mound and argued that the deposit must be about 11,000 years old because he had measured the current rate of accumulation and then extrapolated back to a beginning. When I asked how he could assume such constancy of rate, he replied that the fundamental rule of geological inference, something called "the principle of uniformitarianism" permitted such inferences because we must regard the laws of nature as constant if we wish to reach any scientific conclusions about the past. This argument struck me as logically incorrect, and I pledged myself to making a rigorous analysis of the reasons.

As a joint major in geology and philosophy, I studied this issue throughout my undergraduate years, producing a paper entitled "Hume and uniformitarianism" that eventually transmogrified into my first publication (Gould, 1965), "Is uniformitarianism necessary?" (Norman Newell, my graduate advisor, urged me to send the paper to *Science* where, as I learned to my amusement much later, my future "boss" at Harvard, the senior paleontology professor Bernie Kummel, rejected it roundly as a reviewer. Properly humbled—although I still regard his reasons as ill founded—I then sent the paper to a specialty journal in geology.)

May I share one shameful memory of this otherwise iconoclastic first paper, from which I still draw some pride? In my undergraduate work on this theme, I made a personal discovery (as others did independently) that became important in late 20th-century studies of the history of geology. I had been schooled in the conventional view that the catastrophists (*aka* "bad guys") had invoked supernatural sources of paroxysmal dynamics in order to compress the earth's history into the strictures of biblical chronology. I read and reread all the classical texts of late 18th and early 19th century catastrophism in their original languages—and I could find no claim for supernatural influences upon the history of the earth. In fact, the catastrophists seemed to be advancing the opposite claim that we should base our causal conclusions upon a literal reading of the empirical record, whereas the uniformitarians (*aka* "good guys") seemed to be arguing, in an opposing claim less congenial with the stereotypical empiricism of science, that we must make hypothetical inferences about the gradualistic mechanics that a woefully imperfect record does not permit us to observe directly.

But, although I had developed and presented an iconoclastic exegesis of Lyell, I simply lacked the courage to state so general a claim for inverting the standard view about uniformitarians and catastrophists. I assumed that I must be wrong, and that I must have misunderstood catastrophism because I had not read enough, or could not comprehend the subtleties at this fledgling state of a career. So I scoured the catastrophist literature again until I found a quote from William Buckland (both a leading divine and the first reader in geology at Oxford) that could be interpreted as a defense of supernaturalism. I cited the quotation (Gould, 1965, p. 223) and stuck to convention on this broader issue, while presenting an original analysis of multiple meanings—some valid (like the invariance of law) and some invalid (like my professor's claim for constancy in range of rates)—subsumed by Lyell under the singular description of "uniformity" in nature.

This work led me, partly from shame at my initial cowardice, and as others reassessed the scientific character of catastrophism, to a more general analysis of the potential validity of catastrophic claims, and particularly to an understanding of how assumptions of gradualism had so stymied and constrained our comprehension of the earth's much richer history. These ideas forced me to question the necessary basis for Darwin's key assumption that observable, small-scale processes of microevolution could, by extension through the immensity of geological time, explain all patterns in the history of life—namely, the Lyellian belief in uniformity of rate (one of the invalid meanings of the hybrid concept of uniformitarianism). This exegesis led to a technical book about concepts of time and direction in geology (Gould, 1987), to an enlarged view that encouraged the development of punctuated equilibrium, and to a position of cautious favor towards such truly catastrophic proposals as the Alvarez theory of mass extinction by extraterrestrial impact—a concept ridiculed by nearly all other paleontologists when first proposed (Alvarez et al., 1980), but now affirmed for the K-T event, and accepted as an empirical basis for expanding our range of scientifically legitimate hypotheses beyond the smooth extrapolationism demanded by this third branch of Darwinian central logic.

In addition to these disparate accretions of revisionism on the three branches of Darwinian central logic, one further domain—my studies in the history of evolutionary thought—served as a *sine qua non* for wresting a coherent critique from such an inchoate jumble of disparate items. Above all, if I had not studied Darwin's persona and social context so intensely, I doubt that I would ever have understood the motivations and consistencies—also the idiosyncrasies of time, place and manner—behind the abstract grandeur of his view of life. History must not be dismissed as a humanistic frill upon the adamantine solidity of "real" science, but must be embraced as the coordinating context for any broad view of the logic and reasoning behind a subject so close to the bone of human concern as the science of life's nature and structure. (Of the two greatest revolutions in scientific thought, Darwin surely trumps Copernicus in raw emotional impact, if only because the older transition spoke mainly of real estate, and the later of essence.)

Some of my historical writing appeared in the standard professional literature, particularly my thesis about the "hardening" of the Modern Synthesis (Gould, 1980c, 1982a, 1983), a trend (but also, in part, a drift) towards a stricter and less pluralistic Darwinism. Several full-time historians of science then affirmed this hypothesis (Provine, 1986; Beatty, 1988; Smocovitis, 1996). But much of the historical analysis behind the basic argument of this book had its roots (in my consciousness at least) in the 300 consecutive monthly essays that I wrote from 1974 to 2001 in the popular forum of *Natural History* magazine, where I tried to develop a distinctive style of "mini intellectual biography" in essay form—attempts to epitomize the key ideas of a professional career in a biographic context, and within the strictures of a few thousand words. By thus forcing myself to emphasize essentials and to discard peripherals (while always searching out the truly lovely details that best exemplify any abstraction), I think that I came to understand the major ideological contrasts between the defining features of Darwinian theory and the centerpieces of alternative views. In this format, I first studied such structuralist alternatives as Goethe's theory of the archetypal leaf, Geoffroy's hypothesis on the vertebral underpinning of all animals, and on dorso-ventral inversion of arthropods and vertebrates, and Owen's uncharacteristic English support for this continental view of life. I also developed immense sympathy for the beauty and raw intellectual power of various alternatives, even if I eventually found them wanting in empirical terms. And I came to understand the partial validity, and even the moral suasion, in certain proposals unfairly ridiculed by history's later victors—as in reconsidering the great hippocampus debate between Huxley and Owen, and recognizing how Owen used his (ultimately false) view in the service of racial egalitarianism, while Huxley misused his (ultimately correct) interpretation in a fallacious defense of traditional racial ranking.

Finally, my general love of history in the broadest sense spilled over into my empirical work as I began to explore the role of history's great theoretical theme in my empirical work as well—contingency, or the tendency of complex systems with substantial stochastic components, and intricate nonlinear interactions among components, to be unpredictable in principle from full knowledge of antecedent conditions, but fully explainable after time's actual unfoldings. This work led to two books on the pageant of life's history (Gould, 1989a; Gould, 1996). Although this book, by contrast, treats general theory and its broad results (pattern *vs.* pageant), rather than contingency and the explanation of life's particulars, the science of contingency must ultimately be integrated with the more conventional science of general theory—for we shall thus attain our best possible understanding of both pattern and pageant, and their different attributes and predictabilities.

When I ask myself how all these disparate thoughts and items fell together into one long argument, I can only cite—and I don't know how else to put this—my love of Darwin and the power of his genius. Only he could have presented such a fecund framework of a fully consistent theory, so radical in form, so complete in logic, and so expansive in implication. No other early evolutionary thinker ever developed such a rich and comprehensive starting point. From this inception, I only had to explicate the full original version, tease out the central elements and commitments, and discuss the subsequent history of debate and revision for these essential features, culminating in a consistent reformulation of the full corpus in a helpful way that leaves Darwin's foundation intact while constructing a larger edifice of interestingly different form thereupon. Clearly I do not honor Darwin by hagiography, if only because such obsequious efforts would make any honest character cringe (and would surely cause Darwin to spin in his grave, thus upsetting both the tourists in Westminster Abbey and the adjacent bones of Isaac Newton). I honor Darwin's struggles as much as his successes, and I focus on his few weaknesses as entry points for needed revision—his acknowledged failure to solve the "problem of diversity," or his special pleading for progress in the absence of any explicit rationale from the operation of his central mechanism of natural selection.

As a final comment, if this section has violated the norms of scientific discourse (at least in our contemporary world, although not in Darwin's age) by the liberty that I have taken in explicating personal motives, errors, and corrections, at least I have shown how we all grope upward from initial stupidity, and how we would never be able to climb without the help and collaboration of innumerable colleagues, all engaged in the intensely social enterprise called modern science. I experienced no eureka moment in developing the long argument contained in *The Structure of Evolutionary Theory.* I forged the chain link by link, from initial possession of a few separate items that I didn't even appreciate as pieces of a single chain, or of any chain at all. I made my linkages one by one, and then often cut the segments apart, in order to refashion the totality in a different order. So many people helped me along the way—from long-dead antecedents by their wise words to younger colleagues by their wisecracks—that I must view this outcome as a social project, even though I, the most arrogant of literati, insisted on writing every word. Perhaps I can best express my profound thanks to the members of such an intellectual collectivity by stating, in the most literal sense, that this book would not exist without their aid and sufferance. Scientists fight and squabble as all folks do (and I have scarcely avoided a substantial documentation thereof in this book). But we are, in general, a reasonably honorable lot, and we do embrace a tendency to help each other because we really do revel in the understanding of nature's facts and ways—and most of us will even trade some personal acclaim for the goal of faster and firmer learning. For all the tensions and unhappinesses in any life, I can at least say, with all my heart, that I chose to work in the best of all enterprises at the best of all possible times. May our contingent future only improve this matrix for my successors.

1 WHAT EVERY PALEONTOLOGIST KNOWS

An Introductory Example

If Hugh Falconer (1808–1865) had not died before writing his major and synthetic works, he might be remembered today as perhaps the greatest vertebrate paleontologist of the late 19th century. Falconer went to India in 1830 as a surgeon for the East India Company, but spent most of his time as a naturalist in two very different realms. In 1832, he became superintendent of the botanical garden at Saharanpur, at the base of the Siwaliks, a "foothill" range of the Himalayas. There he played a major role in fostering the cultivation of Indian tea, but he also collected and described one of the most famous and important of all fossil faunas, the Tertiary mammalian remains of the Siwalik Hills. Broken health forced a return to England in 1842, where he worked for several years on the collection of Indian fossils at the British Museum. He then returned to India, this time as professor of botany at Calcutta Medical College, but declining health forced his permanent repatriation to England in 1855. During the last decade of his life, Falconer studied the late Tertiary and Quaternary mammals of Europe and North America, particularly the history of fossil elephants.

Colleagues revered Falconer for his prodigious memory, his gargantuan capacity for work, and his inexhaustible attention to the minutest details. Darwin, as discussed in Chapter 1, pp. 1–6 of *The Structure of Evolutionary Theory* (Gould 2002; hereafter *SET*), held immense respect for Falconer, and invested much hope and trepidation in the prospect that such a master of detail might be persuaded about the probable truth of evolution.

Among all his observations and general conclusions, Falconer took greatest interest in the stability he observed in species of fossil vertebrates, often through long geological periods, and across such maximal changes of environment as the recent glacial ages. Falconer, of course, began with the usual assumption that such stability implied creation and permanence of species. Darwin included him among the great paleontologists who supported such a view. Noting the strength of this opposition to evolution, Darwin wrote (1859, p. 310): "We see this in the plainest manner by the fact that all the most eminent paleontologists, namely Cuvier, Agassiz, Barrande, Falconer, E. Forbes, *etc.* . . . have unanimously, often vehemently, maintained the immutability of species."

Darwin sent Falconer a copy of the first edition of the *Origin of Species,* preceded by the following note (letter of November 11, 1859): "Lord, how savage you will be, if you read it, and how you will long to crucify me alive! I fear it will produce no other effect on you; but if it should stagger you in ever so slight a degree, in this case, I am fully convinced that you will become, year after year, less fixed in your belief in the immutability of species. With this audacious and presumptuous conviction, I remain, my dear Falconer, Yours most truly, Charles Darwin." (Several years before, Darwin had chosen Falconer as one of the very few scientists to whom he confided his beliefs about evolution. Falconer had not, to say the least, reacted positively. In a letter to Hooker on October 13, 1858, Darwin had written of Falconer's jocular, but entirely serious, response: ". . . dear old Falconer, who some few years ago once told me that I should do more harm than any ten other naturalists would do good, [and] that I had half-spoiled you already!")

Falconer wrote to Darwin on June 23, 1861, expressing his great respect (and that of so many others) for the *Origin,* though not his agreement: "My dear Darwin, I have been rambling through the north of Italy, and Germany lately. Everywhere have I heard your views and your admirable essay canvassed—the views of course often dissented from, according to the special bias of the speaker—but the work, its honesty of purpose, grandeur of conception, felicity of illustration, and courageous exposition, always referred to in terms of the highest admiration. And among your warmest friends no one rejoiced more heartily in the just appreciation of Charles Darwin than did, Yours very truly, H. Falconer." Darwin, greatly relieved, replied the next day: "I shall keep your note amongst a very few precious letters. Your kindness has quite touched me."

Hugh Falconer did reassess his worldview, and did accept the principle of evolution (though not causality by natural selection)—but only within the context of the one overarching phenomenon that so strongly governed the nature of the fossil record according to his extensive and meticulous observations: the long-term stability of fossil species, even through major environmental changes. Falconer published his reassessment in an 1863 monograph entitled: "On the American fossil elephant of the regions bordering the Gulf of Mexico (*E. columbi,* Falc.); with general observations on the living and extinct species." But he first sent a copy of the manuscript to Darwin (on September 24, 1862), in eager anticipation of Darwin's reaction to his new views. In the first paragraph of his letter, Falconer reemphasized the stability of species through great climatic changes, arguing that any evolutionary account must deal with this primary fact of paleontology:

> Do not be frightened at the enclosure. I wish to set myself right by you before I go to press. I am bringing out a heavy memoir on elephants—an *omnium gatherum* affair, with observations on the fossil and recent species. One section is devoted to the persistence in time of the specific characters of the mammoth. I trace him from before the Glacial period, through it and after it, unchangeable and unchanged as far as the organs of digestion (teeth) and locomotion are concerned. Now, the Glacial period was no joke: it would have made ducks and drakes of your dear pigeons and doves.

Darwin, of course, was delighted. He wrote to Lyell on October 1, 1862: "I found here a short and very kind note of Falconer, with some pages of his 'Elephant Memoir,' which will be published, in which he treats admirably on long persistence of type. I thought he was going to make a good and crushing attack on me, but, to my great satisfaction, he ends by pointing out a loophole, and adds, '. . . The most rational view seems to be that they [Mammoths] are the modified descendants of earlier progenitors, etc.' This is capital. There will not be soon one good paleontologist who believes in immutability."

If we turn to the key section of Falconer's 1863 monograph, entitled "persistence in time of the distinctive characters of the European fossil elephants," we can trace the development of an important evolutionary argument (I am quoting from the posthumous two-volume 1868 collection of Falconer's complete works). Falconer begins with his basic claim about the constancy of species: "If there is one fact, which is impressed on the conviction of the observer with more force than any other, it is the persistence and uniformity of the characters of the molar teeth in the earliest known Mammoth, and his most modern successor" (p. 252). Falconer then extends his observations from this single species to the entire clade of European fossil elephants: "Taking the group of four European fossil species . . . do they show any signs, in the successive deposits of a transition from the one form into the other? Here again, the result of my observation, in so far as it has extended over the European area, is, that the specific characters of the molars are constant in each, within a moderate range of variation, and that we nowhere meet with intermediate forms" (p. 253).

Falconer finds this constancy all the more significant, given the extreme climatic variation of the glacial ages: "If we cast a glance back on the long vista of physical changes which our planet has undergone since the Neozoic Epoch, we can nowhere detect signs of a revolution more sudden and pronounced, or more important in its results, than the intercalation and subsequent disappearance of the Glacial period. Yet the 'dicyclotherian' Mammoth lived before it, and passed through the ordeal of all the hard extremities which it involved, bearing his organs of locomotion and digestion all but unchanged" (pp. 252–253).

But Falconer then declines to use these observations of stability and sudden geological appearance without intermediates as evidence for special creation. He proclaims himself satisfied with Darwin's basic evolutionary premise, and draws the obvious inference that new species of elephants did not evolve by transformation of older European species, but must have emerged from other stocks:

> The inferences which I draw from these facts are not opposed to one of the leading propositions of Darwin's theory. With him I have no faith in the opinion that the Mammoth and other extinct Elephants made their appearance suddenly, after the type in which their fossil remains are presented to us. The most rational view seems to be, that they are in some shape the modified descendants of earlier progenitors. But if the asserted facts be correct, they seem clearly to indicate that the older Elephants of Europe . . . were not the stocks from which the later species . . . sprung, and that we must look elsewhere for their origin (pp. 253–254).

Falconer thus anticipates a primary inference of punctuated equilibrium—that a local pattern of abrupt replacement does not signify macromutational transformation *in situ,* but an origin of the later species from an ancestral population living elsewhere, followed by migration into the local region. Falconer suggests that the ancestry of later European species may be sought among Miocene species in India: "The nearest affinity, and that a very close one . . . is with the Miocene . . . of India" (p. 254).

Falconer then summarizes the puzzles that such stability—of such long-lasting, widespread forms in such variable environments—raises for evolutionary theory: "The whole range of the Mammalia, fossil and recent, cannot furnish a species which has had a wider geographical distribution, and at the same time passed through a longer term of time, and through more extreme changes of climatal (*sic*) conditions, than the Mammoth. If species are so unstable, and so susceptible of mutation through such influences, why does that extinct form stand out so signally, a monument of stability?" (p. 254).

Darwin's reaction to these famous pages in the history of paleontology make fascinating reading, especially in the light of persistence (or reemergence) of all major issues in our modern debate about punctuated equilibrium. First, with his usual insight into the mechanics of his own theory, Darwin expresses special surprise that teeth should be so stable within species—for the same features vary so greatly among species. As many modern evolutionists have remarked—though Darwin did not use the same terminology—natural selection works by converting variation within populations to differences among populations: a primary expression of the extrapolationist principle in Darwinian logic. But the stasis of species challenges such continuationism. (Darwin included his remarks in a long letter to Falconer, written on October 1, 1862, as a response to the manuscript on elephants that Falconer had sent Darwin, and that would become the 1863 publication quoted above): "Your case seems the most striking one which I have met with of the persistence of specific characters. It is very much the more striking as it relates to the molar teeth, which differ so much in the species of the genus, and in which consequently I should have expected variation."

Darwin then searches for ways to mitigate the surprise of such stasis in the face of environmental changes that should have altered selective pressures. He suggests, first, that the global fluctuations of ice-age climates might not have seemed so extensive to elephants. Perhaps they migrated with a favored climatic belt, therefore experiencing little fluctuation, and perhaps no major selective pressures for change: "You speak of these animals as having being exposed to a vast range of climatal changes from before to after the Glacial period. I should have thought, from analogy of sea-shells, that by migration (or local extinction when migration is not possible) these animals might and would have kept under nearly the same climate."

Searching for another way to explain the absence of anticipated (and gradual) change, Darwin then argued that altering climates may generally imply evolutionary modification, but that groups in serious decline, including elephants, often become stalled in their capacity to vary, and especially to form new taxa: "A rather more important consideration, as it seems to me, is that the whole proboscidean group may, I presume, be looked at as verging towards extinction . . . Numerous considerations and facts have led me in the *Origin* to conclude that it is the flourishing or dominant members of each order which generally give rise to new races, sub-species, and species; and under this point of view I am not at all surprised at the constancy of your species." But if Darwin had not been surprised, or at least disturbed, why did he try so hard to reconcile this unexpected phenomenon with his general theory? Falconer, in any case, replied that elephants remained in vigor, and could not be considered as a group on the verge of elimination.

I recount this story at some length, as an introduction to punctuated equilibrium, both because Falconer and Darwin presage in such a striking manner, the main positions of supporters and opponents (respectively) of punctuated equilibrium in our generation, and because the tale itself illustrates the central fact of the fossil record so well—geologically abrupt origin and subsequent extended stasis of most species. Falconer, especially, illustrates the transition from too easy a false resolution under creationist premises, to recognizing a puzzle (and proposing some interesting solutions) within the new world of evolutionary explanation. Most importantly, this tale exemplifies what may be called the cardinal and dominant fact of the fossil record, something that professional paleontologists learned as soon as they developed tools for an adequate stratigraphic tracing of fossils through time: the great majority of species appear with geological abruptness in the fossil record and then persist in stasis until their extinction. Anatomy may fluctuate through time, but the last remnants of a species usually look pretty much like the first representatives. In proposing punctuated equilibrium, Eldredge and I did not discover, or even rediscover, this fundamental fact of the fossil record. Paleontologists have always recognized the longterm stability of most species, but we had become more than a bit ashamed by this strong and literal signal, for the dominant theory of our scientific culture told us to look for the opposite result of gradualism as the primary empirical expression of every biologist's favorite subject—evolution itself.

Testimonials to Common Knowledge

The common knowledge of a profession often goes unrecorded in technical literature for two reasons: one need not preach commonplaces to the initiated; and one should not attempt to inform the uninitiated in publications they do not read. The longterm stasis, following a geologically abrupt origin, of most fossil morphospecies, has always been recognized by professional paleontologists, as the previous story of Hugh Falconer testifies. This fact, as discussed on the next page, established a basis for bistratigraphic practice, the primary professional role for paleontology during most of its history.

But another reason, beyond tacitly shared knowledge, soon arose to drive stasis more actively into textual silence. Darwinian evolution became the great intellectual novelty of the later 19th century, and paleontology held the archives of life's history. Darwin proclaimed insensibly gradual transition as the canonical expectation for evolution's expression in the fossil record. He knew, of course, that the detailed histories of species rarely show such a pattern, so he explained the literal appearance of stasis and abrupt replacement as an artifact of a woefully imperfect fossil record. Thus, paleontologists could be good Darwinians and still acknowledge the primary fact of their profession—but only at the price of sheepishness or embarrassment. No one can take great comfort when the primary observation of their discipline becomes an artifact of limited evidence rather than an expression of nature's ways. Thus, once gradualism emerged as the expected pattern for documenting evolution—with an evident implication that the fossil record's dominant signal of stasis and abrupt replacement can only be a sign of evidentiary poverty—paleontologists became cowed or puzzled, and even less likely to showcase their primary datum.

But this puzzlement did sometimes break through to overt statement. For example, in 1903, H. F. Cleland, a paleontologist's paleontologist—that is, a respected expert on local minutiae, but not a general theorist—wrote of the famous Devonian Hamilton section in New York State (which has since become the "type" for an important potential extension of punctuated equilibrium to the integrated behavior of entire faunas, the hypothesis of "coordinated stasis"—see pp. 222–229):

> In a section such as that of the Hamilton formation at Cayuga Lake . . . if the statement *natura non facit saltum* is granted, one should, with some con-

> fidence, expect to find many—at least some—evidences of evolution. A careful examination of the fossils of all the zones, from the lowest to the highest, failed to reveal any evolutional changes, with the possible exception of *Ambocoelia praeumbona* [a brachiopod]. The species are as distinct or as variable in one portion of the section as in another. Species varied in shape, in size, and in surface markings, but these changes were not progressive. The conclusion must be that . . . the evolution of brachiopods, gastropods, and pelecypods either does not take place at all or takes place very seldom, and that it makes little difference how much time elapses so long as the conditions of environment remain unchanged (quoted in Brett, Ivany, and Schopf, 1995, p. 2).

But far better than such explicit testimonies—and following various gastronomical metaphors about the primacy of practice (knowing by fruits, proofs of the pudding, etc.)—the most persuasive testimony about dominant stasis and abrupt appearance inheres, without conscious intent or formulation, in methods developed by the people who use fossils in their daily, practical work. Evolutionary theory may be a wonderful intellectual frill, but workaday paleontology, until very recently, used fossils primarily in the immensely useful activity (in mining, mapping, finding oil, etc.) of dating rocks and determining their stratigraphic sequence. These practical paleontologists dared not be wrong in setting their criteria for designating ages and environments. They had to develop the most precise system that empirical recognition could supply for specifying the age of a stratum; they could not let theory dictate a fancy expectation unsupported by observation. Whom would you hire if you wanted to build a bridge across your local stream—the mason with a hundred spans to his credit, or the abstract geometer who has never left his ivory tower? When in doubt, trust the practitioner.

If most fossil species changed gradually during their geological lifetimes, biostratigraphers would have codified "stage of evolution" as the primary criterion for dating by fossils. In a world dominated by gradualism, maximal resolution would be obtained by specifying a precise stratigraphic position within a continuum of steady change, and much information would be lost by listing only the general name of a species rather than its immediate state within a smooth transition. But, in fact, biostratigraphers treat species as stable entities throughout their documented ranges—because the vast majority so appear in the empirical record. Finer resolution can then be obtained by two major strategies: first, by identifying species with unusually short durations, but wide geographic spread (so-called "index fossils"); and, second, by documenting the differing ranges of many species within a fauna and then using the principle of "overlapping range zones" to designate geological moments of joint occurrence for several taxa (see Fig. 1-1).

This peculiar situation of discordance between the knowledge of practical experts and the expectation of theorists impressed Eldredge and me deeply when we formulated punctuated equilibrium. We therefore made the following remarks in closing our first paper on the application of our model to biostratigraphy (Eldredge and Gould, 1977):

> [We] wondered why evolutionary paleontologists have continued to seek, for over a century and almost always in vain, the "insensibly graded series" that Darwin told us to find. Biostratigraphers have known for years that morphological stability, particularly in characters that allow us to recognize species-level taxa, is the rule, not the exception. It is time for evolutionary theory to catch up with empirical paleontology, to confront the phenomenon of evolutionary non-change, and to incorporate it into our theory, rather than simply explain it away . . . We believe that, unconsciously, biostratigraphic methodology has been evolutionarily based all along, since biostratigraphers have always treated their data as if species do not change

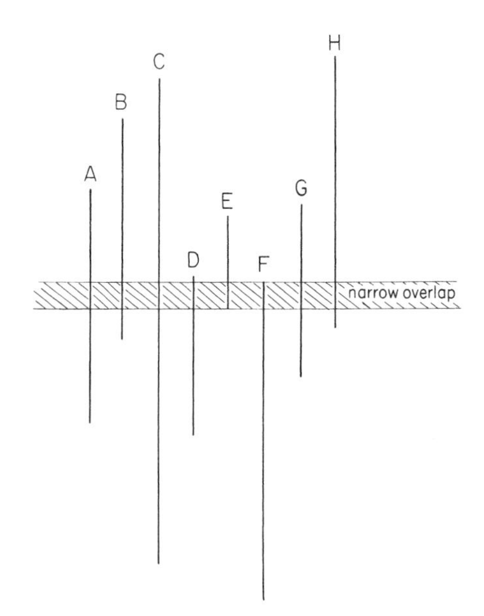

1-1. Knowledge by working biostratigraphers of stasis observed in the vast majority of fossil species led these field scientists, for practical and not for theoretical reasons, to use the criterion of "overlapping range zones" for maximal precision in stratigraphic correlation. If most species changed gradualistically within their geological lifetime, stage of evolution within individual species would provide a better criterion for correlation.

> much during their [residence in any local section], are tolerably distinguishable from their nearest relatives, and do not grade insensibly into their close relatives in adjacent stratigraphic horizons . . . Biostratigraphers, thankfully, have ignored theories of speciation, since the only one traditionally available to them has not made much sense. To date, evolutionary theory owes more to biostratigraphy than vice versa. Perhaps in the future evolutionary theory can begin to repay its debt.

Finally, the witness of experts engaged in a lifelong study of particular groups and times provides especially persuasive testimony because, as I have emphasized, natural history is a science of relative frequencies, not of unique cases, however well documented. We[1] have never doubted that examples of both gradualism and punctuation can be found in the history of almost any group. The debate about punctuated equilibrium rests upon our claim for a dominant relative frequency, not for mere occurrence. The summed experiences of long and distinguished careers therefore provide a good basis for proper assessment.

The paleontological literature, particularly in the "summing up" articles of dedicated specialists, abounds in testimony for predominant stasis, often viewed as surprising, anomalous, or even a bit embarrassing, because such experts had been trained to expect gradualism, particularly as the reward of diligent study. To choose some examples in just three prominent fossil groups representing the full span of conventional "complexity" in the invertebrate record, most microorganisms seem to show predominant stasis—despite the excellent documentation of a few "best cases" of gradualism in Cenozoic planktonic Foraminifera (see pp. 84–96). For example, MacGillavry (1968, p. 70) wrote from long practical experience: "During my work as an oil paleontologist, I had the opportunity to study sections meeting these rigid requirements [of continuous sedimentation and sufficient span of time]. As an ardent student of evolution, moreover, I was continually on the watch for evidence of evolutionary change . . . The great majority of species do not show any appreciable evolutionary change at all. These species appear in the section (first occurrence) without obvious ancestors in underlying beds, are stable once established, and disappear higher up without leaving any descendants."

Echoing the hopes and disappointments of many paleontologists (including both Eldredge and me), who trained themselves in statistical methods primarily to find the "subtle" cases of gradualism that had eluded traditional, subjective observation, Reyment (1975, p. 665) wrote: "The occurrences of long sequences within species are common in boreholes and it is possible to exploit the statistical properties of such sequences in detailed biostratigraphy. It is noteworthy that gradual, directed transitions from one species to another do not seem to exist in borehole samples of microorganisms."

Moving to a metazoan group generally regarded as relatively "simple" in form, and especially prominent in the fossil record, particularly in Paleozoic strata, Roberts (1981, p. 123) concluded from many years of studying Australian Carboniferous brachiopods: "There is no evidence of 'gradualistic' evolutionary processes affecting brachiopod species either within or between zones, and the succession of faunas can be regarded as 'punctuated.'"

Johnson (1975), inspired by Ziegler's (1966) documentation of one putatively gradualistic sequence in the brachiopod *Eocoelia,* decided to search for others—and found only examples of punctuation and stasis throughout the Paleozoic record. He wrote (1975, p. 657):

> After completion of Ziegler's paper we talked a number of times about the possibilities of duplicating his efforts with other fossils and in other times. It was a heady prospect . . . In subsequent years many workers have attempted to seek out and define lineages of brachiopod species and other megafossils in the lower and middle Paleozoic with little success. My conclusion, subjective in many ways, is that speciation of brachiopods in the mid-Paleozoic via a phyletic mode has been rare. Rather, it is probable that most new brachiopod species of this age originated by allopatric speciation.

Derek Ager, the world leader in studying later Mesozoic brachiopods, summed up his lifelong effort in several papers towards the end of his career. He wrote (1973, p. 20): "In twenty years work on the Mesozoic Brachiopods, I have found plenty of relationships, but few if any evolving lineages . . . What it seems to mean is that evolution did not normally proceed by a process of gradual change of one species into another over long periods of time." Ten years later (1983, p. 563), Ager reiterated: "The general picture seems to fit in with the Gouldian doctrine of 'hardly ever' [that is, documentation of gradualism only very rarely]. Certainly there is no evidence in the group as a whole of phyletic gradualism happening throughout a species at any one moment in time. Species A never changes into species B everywhere simultaneously and gradually."

When we consider trilobites, the exemplars of Paleozoic invertebrate "complexity," Robison (1975, p. 220) concluded from extensive study of Middle Cambrian agnostid trilobites in Western North America: "I have found a conspicuous lack of intergradation in species-specific characters, and I have also found little or no change in these characters throughout the observed stratigraphic ranges of most species."

Fortey (1985) spent many years studying a particularly favorable sequence for fine-scale temporal resolution from the early Ordovician of Spitzbergen. He examined 111 trilobite and 56 graptoloid species, finding a predominance of punctuated equilibrium in both groups—with gradualism in "less than 10 percent of the total" for trilobites, and, for graptoloid species, with punctuational origins "at least four times as important as gradualistic ones" (1985, p. 27). Fortey's case becomes particularly convincing because he could calibrate punctuational sequences against rarer cases of gradualism in the same strata—and therefore be confident that punctuations do not merely represent the missing strata of conventional gradualistic rates. In a later paper, Fortey, who is, by the way, no partisan of punctuated equilibrium, reaches the following general conclusion, and also affirms our point about respect for the age-old knowledge of biostratigraphic practitioners: "Many invertebrate paleontologists would agree that the fossil record of species of their groups is dominated by lack of change—by stasis—and that where phylogenies have been worked out then the evidence direct from the rocks shows punctuated lineages in a majority of cases. For reasons I have explained, it is likely that stratigraphic paleontologists would *always* have maintained such a view, but the difference is that now this would be accepted by paleobiologists as well" (1988, p. 13).

Moving to a different arthropod group from another time, Coope's famous studies of Late Cenozoic fossil beetles (summarized in Coope, 1979) provide one of our best cases for dominance of the punctuational mode. Unusually good preservation greatly increases the power of this example. Coope discusses his best case (for beetles extracted from the carcasses of woolly rhinos in the western Ukraine), but then extends his argument to most examples:

> Here the complete beetles were preserved down to the tarsal and antennal joints; when the elytra were raised, the wings could be unfolded and mounted; and parasitic mites, both larvae and adults, were found underneath the wings. Although this was quite exceptional preservation, it is common to find intact abdomens from which the genitalia can be dissected; the fre-

> quently transparent integument often reveals detailed structures of the internal sclerites. Preservation is frequently adequate to enable details of the microstructure of the surface of the hairs and scales to be examined with scanning electron microscopy (1979, p. 248).

Coope concluded that most species showed extensive stasis, even with such detail available for observation: "The early Pleistocene fossils, probably dating from over a million years ago, are referable to living species and some existing species extend well back into the late Tertiary" (1979, p. 250).

In what I regard as the most fascinating and revealing comment of all, George Gaylord Simpson, the greatest and most biologically astute paleontologist of the 20th century (and a strong opponent of punctuated equilibrium in his later years), acknowledged the literal appearance of stasis and geologically abrupt origin as *the* outstanding general fact of the fossil record, and as a pattern that would "pose one of the most important theoretical problems in the whole history of life" if Darwin's argument for artifactual status failed. Simpson stated at the 1959 Chicago centennial celebration for the *Origin of Species* (in Tax, 1960, p. 149):

> It is a feature of the known fossil record that most taxa appear abruptly. They are not, as a rule, led up to by a sequence of almost imperceptibly changing forerunners such as Darwin believed should be usual in evolution. A great many sequences of two or a few temporally intergrading species are known, but even at this level most species appear without known intermediate ancestors, and really, perfectly complete sequences of numerous species are exceedingly rare. . . . These peculiarities of the record pose one of the most important theoretical problems in the whole history of life: is the sudden appearance . . . a phenomenon of evolution or of the record only, due to sampling bias and other inadequacies?

Such a discordance between theoretical expectation and actual observation surely falls within the category of troubling "anomalies" that, in Kuhn's celebrated view of scientific change (1962), often spur a major reformulation.

Darwinian Solutions and Paradoxes

Only one chapter of the *Origin of Species* bears an apologetic title—ironically, for the subject that should have provided the crown of direct evidence for evolution in the large: the archive of life's actual history as displayed in the fossil record. Yet Darwin entitled Chapter 9 "On the Imperfection of the Geological Record."

In Chapter 2 of *SET* (pp. 146–155), I discussed Darwin's convictions about gradualism, and the crucial link between his defense of natural selection and one of the three major and disparate claims subsumed within this complex concept: the insensibility of intermediacy. The theory of punctuated equilibrium does not engage this important meaning for two reasons: first, our theory does not question the operation of natural selection at its conventional organismic level; second, as a theory about the deployment of speciation events in macroevolutionary time, punctuated equilibrium explains how the insensible intermediacy of human timescales can yield a punctuational pattern in geological perspective—thus requiring the treatment of species as evolutionary individuals, and precluding the explanation of trends and other macroevolutionary patterns as extrapolations of anagenesis within populations.

Rather, punctuated equilibrium refutes the third and most general meaning of Darwinian gradualism, designated in Chapter 2 of *SET* (see pp. 152–155) as "slowness and smoothness (but not constancy) of rate." Natural selection does not require or imply this degree of geological sloth and smoothness, though Darwin frequently, and falsely, linked the two concepts—as Huxley tried so forcefully to advise him, though in vain, with his famous warning: "you have loaded yourself with an unnecessary difficulty in adopting *Natura non facit saltum* so unreservedly." The crucial error of Dawkins (1986) and several other critics lies in their failure to recognize the theoretical importance of this third meaning, the domain that punctuated equilibrium does challenge. Dawkins correctly notes that we do not question the second meaning of insensible intermediacy. But since his extrapolationist view leads him to regard only this second meaning as vital to the rule of natural selection, he dismisses the third meaning—which we do confute—as trivial. Since Dawkins rejects the hierarchical model of selection, he does not grant himself the conceptual space for weighing the claim that punctuated equilibrium's critique of the third meaning undermines the crucial Darwinian strategy for rendering all scales of evolution by smooth extrapolation from the organismic level. For this refutation of extrapolation by punctuated equilibrium validates the treatment of species as evolutionary individuals, and establishes the level of species selection as a potentially important contributor to macroevolutionary pattern.

This broadest third meaning of gradualism may not be required for natural selection at the organismic level, but gradualism as slowness and smoothness of rate (not just as insensible intermediacy between endpoints of a transition) forms the centerpiece of Darwin's larger worldview, indeed of his entire ontology—as illustrated (again, see Chapter 2 of *SET*) in the crucial role played by this style of gradualism throughout the corpus of his works—from his first book on the origin of coral atolls (1842) to his last on the formation of topsoil by the action of worms (1881).

Lest anyone doubt that Darwin strongly advocated this most inclusive form of gradualism as slowness and smoothness (in addition to the less comprehensive claim for insensible intermediacy of transitions), I shall cite a few examples from the full documentation of Chapter 2 of *SET*—cases where Darwin clearly meant "slow and steady over geological scales," not just "insensibly intermediate at whatever rate."

For example Darwin argues that species may arise so slowly that the process generally takes longer than the *entire* duration of a geological formation (usually several million years)—thus explaining apparent stasis *within* a formation as gradual evolution over insufficient time to record visible change! Darwin writes (1859, p. 293): "Although each formation may mark a very long lapse of years, each perhaps is short compared with the period requisite to change one species into another." Darwin even argued that the pace of evolutionary change might be sufficiently steady to serve as a rough geological clock: "The amount of organic change in the fossils of consecutive formations probably serves as a fair measure of the lapse of actual time" (1859, p. 488). I also show in Chapter 2 of *SET* that Darwin's conviction about extreme slowness and steadiness of change can be grasped, perhaps best of all, as the common source of his major errors—particularly his fivefold overestimate for the denudation of the Weald, and his conjecture that complex metazoan life of modern form must have undergone an unrecorded Precambrian history as long, or longer, than its known Phanerozoic duration.

Despite this strong belief in geological gradualism, Darwin knew perfectly well—as all paleontologists always have—that stasis and abrupt appearance represent a norm for the *observed* history of most species. I needn't rehearse Darwin's solution to this dilemma, for his familiar argument represents more than a twice-told tale. Following the lead of his mentor, Charles Lyell, Darwin attributed this striking discordance between theoretical expectation and actual observation to the extreme imperfection of the fossil record.

(As discussed more fully in *SET*, pp. 479–484, this argument served as the centerpiece for Lyell's system, and for the entire uniformitarian school. But then, what alternative could they embrace? The literal appearance of the geological record so often suggested catastrophe, or at least "moments" of substantial change, especially in faunal turnover. To assert a gradualism of geological rate against this sensory evidence, one had to declare the evidence illusory by advancing the general claim—quite legitimate as a philosophical proposition—that science must often work by probing "behind appearance" to impose the expectations of a valid theory upon an empirical record that, for one reason or another, cannot directly express the actual mechanisms of nature. Moreover, the "argument from imperfection" holds substantial merit and cannot be dismissed as "special pleading." Like most chronicles of history, and far more so than many others, the geological record is extremely spotty. To cite Lyell's famous metaphor once again, if Vesuvius erupted again and buried a modern Italian city atop Pompeii, later stratigraphers might find a sequence of Roman ruins capped by layers of volcanic ash and followed by the debris of modern Italy. Taken literally, this sequence would suggest a catastrophic end to Rome followed by a saltation, linguistically and technologically, to the industrial age—an artifact of nearly 2000 years of missing data that would have recorded the evolution of Italian from Latin and a gradual passage from walled cities to traffic jams.)

To quote the two most famous statements on this subject from the *Origin of Species,* Darwin summarizes his entire argument by closing Chapter 9 with Lyell's metaphor of the book (1859, pp. 310–311):

> For my part, following out Lyell's metaphor, I look at the natural geological record, as a history of the world imperfectly kept, and written in a changing dialect; of this history we possess the last volume alone, relating only to two or three countries. Of this volume, only here and there a short chapter has been preserved; and of each page, only here and there a few lines. Each word of the slowly-changing language, in which the history is supposed to be written, being more or less different in the interrupted succession of chapters, may represent the apparently abruptly changed forms of life, entombed in our consecutive, but widely separated, formations.

In epitomizing both geological chapters, Darwin begins with a long list of reasons for such an imperfect record, and then concludes with his characteristic honesty (1859, p. 342): "All these causes taken conjointly, must have tended to make the geological record extremely imperfect, and will to a large extent explain why we do not find interminable varieties, connecting together all the extinct and existing forms of life by the finest graduated steps.

He who rejects these views on the nature of the geological record, will rightly reject my whole theory." (Huxley must have been thinking of this line when he issued his warning that Darwin's unswerving support of *natura non facit saltum* represented "an unnecessary difficulty." Darwin's "whole theory"—the mechanism of natural selection—does not require, as Huxley pointed out, this geological style of gradualism in rate.)

The paradoxes set by Darwin's solution for the current practice of paleontology and macroevolutionary theory receive their clearest expression in another remarkable statement from the *Origin of Species* (1859, p. 302), a testimony to Darwin's sophisticated understanding that nature's "facts" do not stand before us in pristine objectivity, but must be embedded within theories to make any sense, or even to be "seen" at all. Darwin acknowledges that he only understood the extreme imperfection of the geological record when paleontological evidence of stasis and abrupt appearance threatened to confute the gradualism that he "knew" to be true: "But I do not pretend that I should ever have suspected how poor a record of the mutations of life, the best preserved geological section presented, had not the difficulty of our not discovering innumerable transitional links between the species which appeared at the commencement and close of each formation, pressed so hardly on my theory."

THE PARADOX OF INSULATION FROM DISPROOF

The "argument from imperfection" (with its preposition purposefully chosen by analogy to the "argument from design") works adequately as a device to save gradualism in the face of an empirical signal of quite stunning contrariness when read at face value. But if we adopt openness to empirical falsification as a criterion for strong and active theories in science, consider the empty protection awarded to gradualism by Darwin's strategy. For the data that should, *prima facie,* rank as the most basic empirical counterweight to gradualism—namely the catalog of cases, and the resulting relative frequency, for observed stasis and geologically abrupt appearances of fossil morphospecies—receive *a priori* interpretation as signs of an inadequate empirical record. How then could gradualism be refuted from within?

The situation became even more insidious in subtle practice than a bald statement of the dilemma might suggest. Abrupt appearance (the punctuations of punctuated equilibrium) might well be attributed to the admittedly gross imperfection of our geological archives. The argument makes logical sense, must certainly be true in many instances, and can be tested in a variety of ways on a case by case basis (particularly when we can obtain independent evidence about rates of sedimentation).

But how can imperfection possibly explain away stasis (the equilibrium of punctuated equilibrium)? Abrupt appearance may record an absence of information, but *stasis is data.* Eldredge and I became so frustrated by the failure of many colleagues to grasp this evident point—though a quarter century of subsequent debate has finally propelled our claim to general acceptance (while much else about punctuated equilibrium remains controversial)—that we urged the incorporation of this little phrase as a mantra or motto. Say it ten times before breakfast every day for a week, and the argument will surely seep in by osmosis: "stasis is data; stasis is data . . ."

The fossil record may, after all, be 99 percent imperfect, but if you can, nonetheless, sample a species at a large number of horizons well spread over several million years, and if these samples record no net change, with beginning and end points substantially the same, and with only mild and errant fluctuation among the numerous collections in between, then a conclusion of stasis rests on the *presence* of data, not on absence! In such cases, we must limit our lament about imperfection to a wry observation that nature, rather than human design, has established a sampling scheme by providing only occasional snapshots over a full interval. We might have preferred a more even temporal spacing of these snapshots, but so long as our samples span the temporal range of a species, with reasonable representation throughout, why grouse at nature's failure to match optimal experimental design—when she has, in fact, been very kind to us in supplying abundant information. *Stasis is data.*

So if stasis could not be explained away as missing information, how could gradualism face this most prominent signal from the fossil record? The most negative of all strategies—a quite unconscious conspiracy of silence—dictated the canonical response of paleontologists to their observations of stasis. Again, a "culprit" may be identified in the ineluctable embedding of observation within theory. Facts have no independent existence in science, or in any human endeavor; theories grant differing weights, values, and descriptions, even to the most empirical and undeniable of observations. Darwin's expectations defined evolution as gradual change. Generations of paleontologists learned to equate the potential documentation of evolution with the discovery of insensible intermediacy in a sequence of fossils. In this context, stasis can only record sorrow and disappointment.

Paleontologists therefore came to view stasis as just another failure to document evolution. Stasis existed in overwhelming abundance, as every paleontologist always knew. But this primary signal of the fossil record, defined as an absence of data for evolution, only highlighted our frustration—and certainly did not represent anything worth publishing. Paleontology therefore fell into a literally absurd vicious circle. No one ventured to document or quantify—indeed, hardly anyone even bothered to mention or publish at all—the most common pattern in the fossil record: the stasis of most morphospecies throughout their geological duration.

All paleontologists recognized the phenomenon, but few scientists write papers about failure to document a desired result. As a consequence, most nonpaleontologists never learned about the predominance of stasis, and simply assumed that gradualism must prevail, as illustrated by the exceedingly few cases that became textbook "classics": the coiling of *Gryphaea*, the increasing body size of horses, etc. (Interestingly, nearly all these "classics" have since been disproved, thus providing another testimony for the temporary triumph of hope and expectation over evidence—see Gould, 1972.) Thus, when punctuated equilibrium finally granted theoretical space and importance to stasis, and this fundamental phenomenon finally emerged from the closet, nonpaleontologists were often astounded and incredulous. Mayr (1992, p. 32) wrote, for example: "Of all the claims made in the punctuationalist theory of Eldredge and Gould, the one that encountered the greatest opposition was the observation of 'pronounced stasis as the usual fate of most species,' after having completed the phase of origination . . . I agree with Gould that the frequency of stasis in fossil species revealed by the recent analysis was unexpected by most evolutionary biologists."

(To cite a personal incident that engraved this paradox upon my consciousness early in my career, John Imbrie served as one of my Ph.D. advisors at Columbia University. This distinguished paleoclimatologist began his career as an evolutionary paleontologist. He accepted the canonical equation of evolution with gradualism, but conjectured that our documentary failures had arisen from the subtlety of gradual change, and the consequent need for statistical analysis in a field still dominated by an "old-fashioned" style of verbal description. He schooled himself in quantitative methods and applied this apparatus, then so exciting and novel, to the classic sequence of Devonian brachiopods from the Michigan Basin—where rates of sedimentation had been sufficiently slow and continuous to record any hypothetical gradualism. He studied more than 30 species in this novel and rigorous way—and found that all but one had remained stable throughout the interval, while the single exception exhibited an ambiguous pattern. But Imbrie did not publish a triumphant paper documenting the important phenomenon of stasis. Instead, he just become disappointed at such "negative" results after so much effort. He buried his data in a technical taxonomic monograph that no working biologist would ever encounter (and that made no evolutionary claims at all)—and eventually left the profession for something more "productive.")

Paradoxes of this sort can only be resolved by input from outside—for gradualism, having defined contrary data either as marks of imperfection or documents of disappointment, could not be refuted from within. Reassessment required a different theory that respected stasis as a potentially fascinating phenomenon worthy of rigorous documentation, not merely as a failure to find "evolution." Eldredge and I proposed punctuated equilibrium in this explicit context—as a framework and different theory that, if true, could validate the primary signal of the fossil record as valuable information rather than frustrating failure. We therefore began our original article (Eldredge and Gould, 1972) with a philosophical discussion, based on work of Kuhn (1962) and Hanson (1961), on the necessary interbedding of fact and theory. We ended this introductory section by writing (1972, p. 86):

> The inductivist view forces us into a vicious circle. A theory often compels us to see the world in its light and support. Yet we think we see objectively and therefore interpret each new datum as an independent confirmation of our theory. Although our theory may be wrong, we cannot confute it. To extract ourselves from this dilemma, we must bring in a more adequate theory; it will not arise from facts collected in the old way . . . Science progresses more by the introduction of new world-views or "pictures" than by the steady accumulation of information . . . We believe that an inadequate picture has been guiding our thoughts on speciation for 100 years. We hold that its influence has been all the more tenacious because paleontologists, in claiming that they see objectively, have not recognized its guiding sway. We contend that a notion developed elsewhere, the theory of allopatric speciation, supplies a more satisfactory picture for the ordering of paleontological data.

THE PARADOX OF STYMIED PRACTICE

This second paradox cascades from the first. If a theory—geologically insensible gradualism as the anticipated expression of evolution in the fossil record, in this case—can insulate itself against disproof from within by defining contrary data as artifactual, then proper assessments of relative frequencies can never be achieved—for how many scientists will devote a large chunk of a limited career to documenting a phenomenon that they view as a cardinal restriction recording a poverty of available information?

Paleontological studies of evolution therefore became warped in a lamentable way that precluded any proper use of the fossil record but seemed entirely honorable at the time. We practitioners of historical sciences, as emphasized throughout this book, work in fields that decide key issues by assessment of relative frequencies among numerous possible outcomes, and only rarely by the more "classical" technique of "crucial experiments" to validate universal phenomena. Therefore, any method that grossly distorts a relative frequency by excluding a common and genuine pattern from consideration must seriously stymie our work. When traditional paleontologists eliminated examples of abrupt appearance and stasis from the documentation of evolution, they only followed a conventional precept—for they believed that both patterns recorded an artifact of imperfect data, thus debarring such cases from consideration. The relative distributions of evolutionary rates would therefore emerge only from cases of gradualism—the sole examples judged as sufficiently data-rich to record the process of evolution in adequate empirical detail.

But this project could not even succeed in its own terms, for gradualism occurs too rarely to generate enough cases for calculating a distribution of rates. Instead, paleontologists worked by the false method of exemplification: validation by a "textbook case" or two, provided that the chosen instances be sufficiently persuasive. And even here, at this utterly minimal level of documentation, the method failed. A few examples did enter the literature (see Fig. 1-2 for comparison of an original claim with a secondary textbook version)—where they replicated by endless republication in the time-honored fashion of textbook copying (see Gould, 1988a). But, in a final irony, almost all these famous exemplars turned out to be false on rigorous restudy—see Hallam, 1968, and Gould, 1972, for stasis rather than gradual increase in coiling in the Liassic oyster *Gryphaea;* Prothero and Shubin, 1989, on stasis within all documented species of fossil horses, and with frequent overlap between ancestors and descendants, indicating branching by punctuational speciation rather than anagenetic gradualism; and Gould, 1974, on complete absence of data for the common impression that the enormous antlers of *Megaloceros* (the "Irish Elk") increased gradually in phylogeny, with positive allometry as body size enlarged.

Traditional paleontology therefore placed itself into a straightjacket that made the practice of science effectively impossible: only a tiny percentage of cases passed muster for study at all, while the stories generated for this minuscule minority rested so precariously upon hope for finding a rare phenomenon—and received such limited definition by the primitive statistical methods then available (or, more commonly, remained unidentified by any statistical practice at all)—that even these textbook exemplars collapsed upon restudy with proper quantitative procedures. But consider what might have occurred, if only paleontologists had recognized that stasis is data (I will grant some validity to the standard rationale for regarding the second

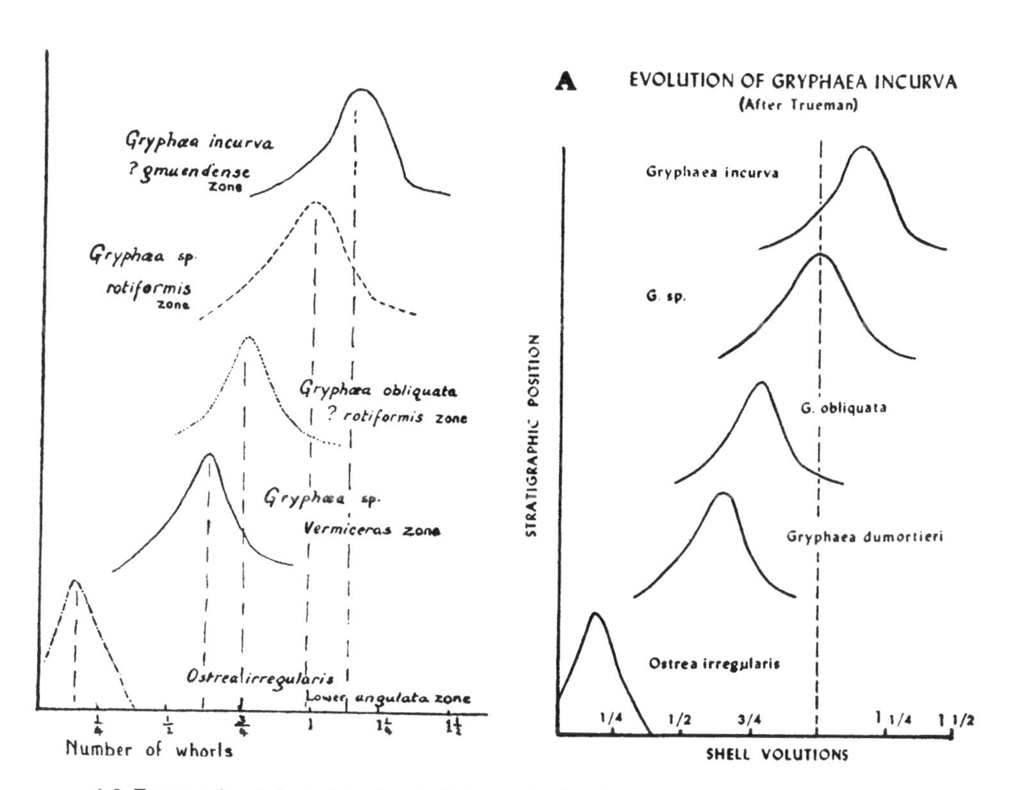

1-2. Trueman's original claim for phyletic gradualism in the increased coiling of *Gryphaea* in Lower Jurassic rocks of England (left). To the right, a textbook smoothing and simplification of the same figure. Trueman's claim has been invalidated for two reasons: first, *Gryphaea* did not evolve from *Ostrea;* and, second, subsequent studies have not validated any increase of coiling within *Gryphaea,* despite Trueman's graphs. Nonetheless, once such figures become ensconced in textbooks, they tend to persist even when their empirical justification has long been refuted in professional literature.

phenomenon of punctuation as an artifact of an imperfect record). As Hallam said to me many years ago, after he had disproved the classical story of gradualism in *Gryphaea:* more than 100 other species of mollusks, many with records as rich as *Gryphaea*'s, occur in the same Liassic rocks, yet no one ever documented the stratigraphic history of even a single one in any study of evolution, for all demonstrate stasis. Scientists picked out the only species that seemed to illustrate gradualism, and even this case failed.

Despite the widespread use of proper quantitative methods today, and despite increasing attention to the validity of stasis as an evolutionary phenomenon, this bias still persists. I do not doubt that several species of Cenozoic planktonic Foraminifera display gradual transitions (see pp. 84–96), but I know that these examples have been extracted for study from a much larger potential sample of species never documented in detail because their apparent stasis seems "boring" to students of evolution. An eager young statistician goes to a lifelong expert and says: I want to devote my doctoral thesis to a statistical study of evolution in a species of foram (the most promising of major taxa, thanks to a hyperabundance of specimens and excellent stratigraphic data in oceanic cores); which species shall I choose? And the expert advises: why not study *Graduloconoides gradualississima;* I know that this species shows interesting changes during the upper Miocene in cores A through Z. Meanwhile, poor old boring *Stasigerina punctiphora,* just as abundant in the same cores, and just as worthy of study, gets bypassed in silence.

I find this situation particularly frustrating as paleontology's primary example of an insidious phenomenon in science that simply has not been recognized for the serious and distorting results perpetrated under its aegis. Most scientists do not even recognize the problem—though some do, particularly in the medical and social sciences, where the error has been named "publication bias," and has inspired a small but important literature (Begg and Berlin, 1988). In publication bias, prejudices arising from hope, cultural expectation, or the definitions of a particular scientific theory dictate that only certain kinds of data will be viewed as worthy of publication, or even of documentation at all. Publication bias bears no relationship whatever with the simply immoral practice of fraud; but, paradoxically, publication bias may exert a far more serious effect (largely because the phenomenon must be so much more common than fraud)—for scientists affected by publication bias do not recognize their errors (and their bias may be widely shared among colleagues), while a perpetrator of fraud operates with conscious intent, and the wrath of a colleague will be tremendous upon any discovery.

Begg and Berlin (1988) cite several documented cases of publication bias. We can hardly doubt, for example, that a correlation exists between socioeconomic status and academic achievement, but the strength and nature of this association can provide important information, for both political practice and social theory. White (1982, cited in Begg and Berlin) found a progressively increasing intensity of correlation with prestige and permanence of published source. Studies published in books reported an average correlation coefficient of 0.51 between academic achievement and socioeconomic status; articles in journals gave an average of 0.34, while unpublished studies yielded a value of 0.24. Similarly, Coursol and Wagner (1986, cited in Begg and Berlin) found publication bias both in the decision to submit an article at all, and in the probability for acceptance. In a survey of outcomes in psychotherapy, they noted that 82 percent of studies with positive outcomes led to submission of papers to a journal, while only 43 percent of negative outcomes provoked an attempt at publication. Of papers submitted, 80 percent that report positive outcomes were accepted for publication, but the figure fell to 50 percent for papers claiming negative results.

In my favorite study of publication bias, Fausto-Sterling (1985) tabulated claims in the literature for consistent differences in cognitive and emotional styles between men and women. She does not deny that genuine differences often exist, and in the direction conventionally reported. But she then, so to speak, surveys her colleagues' file drawers for studies not published, or for negative results published and then ignored, and often finds that a great majority report either a smaller and insignificant disparity between sexes, or no differences at all. When she collated all studies, rather than only those published, the much-vaunted differences often dissolved into statistical insignificance or triviality.

For example, a recent favorite theme of pop psychology attributed different cognitive styles in men and women to the less lateralized brains of women. Some studies have indeed reported a small effect of greater male lateralization; none has found more lateralized brains in women. But most experiments, as Fausto-Sterling shows, detected no measurable differences in lateralization at all and this dominant relative frequency (even in published literature) should be prominently reported in the press and in popular books, but tends to be ignored as "no story."

Paleontology's primary example of publication bias—the nonreporting of stasis under the false belief that such stability represents "no data" for evolution—illustrates a particularly potent form of the general phenomenon, a category that I have called "Cordelia's dilemma" (Gould, 1995) to memorialize the plight of King Lear's honest but rejected daughter. When asked by Lear for a fulsome protestation of love in order to secure her inheritance, Cordelia, disgusted by the false and exaggerated speeches of her sisters Goneril and Regan, chose to say nothing, for she knew that "my love's more ponderous than my tongue." But Lear mistook her silence for hatred or indifference, and cut her off entirely (with tragic consequences later manifested in his own madness, blindness, and death), in proclaiming that "nothing will come of nothing."

Cordelia's dilemma arises in science when an important (and often predominant) signal from nature isn't seen or reported at all because scientists read the pattern as "no data," literally as nothing at all. This odd status of "hidden in plain sight" had been the fate of stasis in fossil morphospecies until punctuated equilibrium gave this primary signal some theoretical space for existence. Apparent silence—the overt nothing that actually records the strongest something—can embody the deepest and most vital meaning of all. What, in western history, has been more eloquent than the silence of Jesus before Pilate, or Saint Thomas More's date with the headsman because he acknowledged that fealty forbade criticism of Henry VIII's marriage to Anne Boleyn, but maintained, literally to the death, his right to remain silent, and not to approve?

In summary, the potentially reformative role of punctuated equilibrium resides in an unusual property among scientific innovations. Most new theories in science arise from fresh information that cannot be accommodated under an old explanatory rubric. But punctuated equilibrium merely honored the firmest and oldest of all paleontological observations—the documentable stasis of most fossil morphospecies—by promoting this pattern to central recognition as an expected result of evolution's proper expression at the scale of geological time. This reformulation cast a bright light upon stasis, a preeminent fact that had formerly been mired in Cordelia's dilemma as a grand disappointment, and therefore as "no data" at all, a pattern fit only for silence in a profession that accepted Darwin's argument for gradualism as the canonical expression of evolution in the fossil record.

2 THE PRIMARY CLAIMS OF PUNCTUATED EQUILIBRIUM

Data and Definitions

First of all, the theory of punctuated equilibrium treats a particular level of structural analysis tied to a particular temporal frame. G. K. Chesterton (1874–1936), the famous English author and essayist, wrote that all art is limitation, for the essence of any painting lies in its frame. The same principle operates in science, where claiming too much, or too broad a scope of application, often condemns a good idea to mushy indefiniteness and consequent vacuity.

Punctuated equilibrium is not a theory about all forms of rapidity, at any scale or level, in biology. Punctuated equilibrium addresses the origin and deployment of species in geological time. Punctuational styles of change characterize other phenomena at other scales as well (see Chapter 5)—catastrophic mass extinction triggered by bolide impacts, for example—and proponents of punctuated equilibrium would become dull specialists if they did not take an interest in the different mechanisms responsible for similarities in the general features of stability and change across nature's varied domains, for science has always sought unity in this form of abstraction. But punctuated equilibrium—a particular punctuational theory of change and stability for one central phenomenon of evolution—does not directly address the potentially coordinated history of faunas, or the limits of viable mutational change between a parental organism and its offspring in the next generation.

The theory of punctuated equilibrium attempts to explain the macroevolutionary role of species and speciation as expressed in geological time. Its statements about rapidity and stability describe the history of individual species; and its claims about rates and styles of change treat the mapping of these individual histories into the unfamiliar domain of "deep" or geological time—where the span of a human life passes beneath all possible notice, and the entire history of human civilization stands to the duration of primate phylogeny as an eyeblink to a human lifetime. The claims of punctuated equilibrium presuppose the proper scaling of microevolutionary processes into this geological immensity—the central point that Darwin missed when he falsely assumed that "slowness" of modification in domesticated animals or crop plants, as measured in ordinary human time (where all of our history, and so many human generations, have witnessed substantial change within populations, but no origin of new species), would translate into geological time as the continuity and slowness of phyletic gradualism.

Once we recognize that definitions for the two key concepts of stasis and punctuation describe the *history of individual species scaled into geological time,* we can establish sensible and operational criteria. As a central proposition, punctuated equilibrium holds that the great majority of species, as evidenced by their anatomical and geographical histories in the fossil record, originate in geological moments (punctuations) and then persist in stasis throughout their long durations (Sepkoski, 1997, gives a low estimate of 4 million years for the average duration of fossil species; mean values vary widely across groups and times, with terrestrial vertebrates at lesser durations and most marine invertebrates in the higher ranges; in any case, geological longevity achieves its primary measure in millions of years, not thousands). As the primary macroevolutionary implication of this pattern, species meet all definitional criteria for operating as Darwinian individuals (see *SET,* pp. 602–613) in the domain of macroevolution.

This central proposition embodies three concepts requiring definite operational meanings: stasis, punctuation, and dominant relative frequency. (I am not forgetting the thorny problems associated with the definition of species from fossil data, where anatomy prevails as a major criterion and reproductive isolation can almost never be assessed directly—and also with the putative correspondence of morphological "packages" that paleontologists designate as species with the concept as understood and practiced by students of modern populations of sexually reproducing organisms. I shall treat these issues on pages 62–76.)

Stasis does not mean "rock stability" or utter invariance of average values for all traits through time. In the macroevolutionary context of punctuated equilibrium, we need to know, above all, whether or not morphological change tends to accumulate through the geological lifetime of a species and, if so, what part of the average difference between an ancestral and descendant species can be attributed to incremental change of the ancestor during its anagenetic history. Punctuated equilibrium makes the strong claim that, in most cases, effectively no change accumulates at all. A species, at its last appearance before extinction, does not differ systematically from the anatomy of its initial entry into the fossil record, usually several million years before.

Of course we recognize that mean values will fluctuate through time. After all, measured means would vary even if true population values remained utterly constant—which they do not. And, with enough samples in a vertical sequence, some must include mean values (for some characters) outside conventional bounds of statistically insignificant difference from means for the oldest sample. Such fluctuation also implies that the final population will not be identical with the initial sample.

In operational terms, therefore, we need to set criteria for permissible fluctuation in average values through time. Two issues must be resolved: the amount of allowable difference between beginning and ending samples of a species, and the range of permissible fluctuation through time. Since we wish to test a hypothesis that little or no change accumulates by anagenesis during the history of most species, and since we have no statistical right to expect that (under this hypothesis) the last samples will be identical with the first, we should predict either that (i) the final samples shall not differ statistically, by some conventionally chosen criterion, from the initial forms; and at very least (ii) that the final samples shall not generally lie outside the range of fluctuation observed during the history of the species. (If final samples tend to lie outside the envelope of fluctuation for most of the species's history, then anagenesis has occurred.)

For the permissible range of fluctuation, we should, ideally, look to the extent of geographic variation among contemporary populations within the species, or its closest living relative. If the temporal range of variation stays within the spatial range for any one time, then the species has remained in stasis. Obviously, we cannot apply this optimal criterion for groups long extinct, but a variety of proxies should be available, including comparison of a full temporal range with the known geographic variation of a well-documented and widespread nearest living relative. Studies of stasis in Neogene species can often use the optimal criterion because the actual species, or at least some very close relatives, are often still extant. In the most elegant documentation of stasis for an entire fauna of molluscan species, Stanley and Yang (1987) used this best criterion to find that temporal fluctuation remained within the range of modern geographic variation for the same species. They could therefore affirm stasis in the most biologically convincing manner.

Since stasis is data, but punctuation generally records an unresolvable transition when assessed by the usual expression of fossil data in geological time, we need to formulate an appropriate definition of rapidity. (Punctuated equilibrium makes no claim about the possibility of substantial change at rates that would be called rapid by measuring rods of a human lifetime. Therefore, and especially, punctuated equilibrium provides no insight into the old and contentious issue of saltational or macromutational speciation.) As a first approach, the duration of a bedding plane represents the practical limit of geological resolution. Any event of speciation that occurs within the span of time recorded by most bedding planes will rarely be resolvable because evidence for the entire transition will be compressed onto a single stratigraphic layer, or "geological moment."

However, the limits of stratigraphic resolution vary widely, with bedding planes representing years or seasons in rare and optimal cases of varved sediments, but several thousand years in most circumstances. We therefore cannot formulate a definition equating punctuation with "bedding plane simultaneity." (After all, such a definition would, almost perversely, preclude the "dissection" of a punctuation in admittedly rare, but precious, cases of sedimentation so complete and so rapid that an event of speciation will not be compressed, as usual, onto a single bedding plane, but will "spread out" over a sufficient stratigraphic interval to permit the documentation of its rapid history.)

Punctuations must, instead, be defined relative to the subsequent duration of the derived species in stasis—for punctuated equilibrium, as a theory of relative timing, holds that species develop their distinctive features effectively "at birth," and then retain them in stasis for geologically long lifetimes. (These timings play an important role in the recognition of species as Darwinian individuals—see discussion on "vernacular" criteria of definable birth, death, and sufficient stability for individuation—Chapter 8 of *SET*, pp. 602–608).

I know no rigorous way to transcend the arbitrary in trying to define the permissible interval for punctuational origin. Since definitions must be theory-bound, and since the possibility of recognizing species as Darwinian individuals in macroevolution marks the major theoretical interest of punctuated equilibrium, an analogy between speciation and gestation of an organism may not be ill conceived. As the gestation time of a human being represents 1–2 percent of an ordinary lifetime, perhaps we should permit the same general range for punctuational speciation relative to later duration in stasis. At an average species lifetime of 4 million years, a 1-percent criterion allows 40,000 years for speciation. When we recognize that such a span of time would be viewed as gradualistic—and extremely slow paced at that—by any conventional microevolutionary scaling in human time; and when we also acknowledge that the same span represents the resolvable moment of a single bedding plane in a great majority of geological circumstances; then we can understand why the punctuations of punctuated equilibrium do not represent de Vriesian saltations, but rather denote the proper scaling of ordinary speciation into geological time.

Punctuation does suffer the disadvantage of frequently compressed recording on a single bedding plane (so that the temporal pattern of the full event cannot be dissected); moreover, an observed punctuation often represents the even less desirable circumstance of missing record (Darwin's classic argument from imperfection), or only partial pattern (as when a punctuation in a single geological section marks the first influx by migration of a species that originated earlier and elsewhere). Since stasis, on the other hand, provides an active (and often excellent) record of stability, empirical defenses of punctuated equilibrium have understandably focused on the more easily documentable claims for equilibrium, and less frequently on more elusive predictions about punctuation. But we must not conclude, as some authors have suggested, that punctuation therefore becomes untestable or even impervious to documentation—and that the thesis of punctuated equilibrium must therefore depend for its empirical support only upon the partial data of stasis. The documentation of punctuation may be both more difficult and less frequently possible, but many good cases have been affirmed and several methods of rigorous testing have been developed.

In the first of two general methods, one may document the reality of a punctuation (as opposed to interpretation as a Darwinian artifact based on gaps in sedimentation) by finding cases of gradualism within a stratigraphic sequence (which must then be sufficiently complete to record such an anagenetic transition), and then documenting punctuational origins for other species in the same strata. Using this technique for Ordovician trilobites from Spitzbergen, Fortey (1985) found a ratio of about 10:1 for cases of punctuation compared with gradualism.

In a second, and more frequently employed, method, one searches explicitly for rare stratigraphic situations, where sedimentation has been sufficiently rapid and continuous to spread the usual results of a single bedding plane into a vertical sequence of strata. Williamson (1981), for example, published a famous series of studies on speciation of freshwater mollusks in African Pleistocene lakes. (These articles provoked considerable debate [Fryer, Greenwood and Peake, 1983], and Williamson died young before he could complete his work. However, in my admittedly partisan judgment, Williamson more than adequately rebutted his critics [1985, 1987].)

These African lakes form in rift valleys, where sedimentation rates are unusually high because the rift-block foundations of the lake sink continuously, and sediments can therefore accumulate above, without interruption. Thus, the thousand-year duration of a speciation event may span several layers of foundering sediment. With this unusual degree of resolution, Williamson was even able to demonstrate a remarkable phenomenon in change of variability within a speciating population—a pattern that appeared over and over again in several events of speciation, and may therefore be viewed as potentially general (see Fig. 2-1): Williamson found limited variation around parental mean values in the oldest samples; intermediacy of mean values within speciating samples, but accompanied by a greatly expanded range of variation (though still normal in distribution); and subsequent "settling down" of variation to the reduced level of the ancestral population, but now distributed around the altered mean value of the derived species.

If this kind of unusual circumstance spreads a punctuational event of speciation through a sufficient stratigraphic interval for resolution, another strategy of research will sometimes permit the dissection of a punctuation in conventional cases of full representation on a single bedding plane. Goodfriend and Gould (1996) documented such a case because they could establish absolute dates for the individual shells on a single bedding plane. (Admittedly, this technique cannot be generally applied—especially to sediments of appreciable age, where errors of measurement for any method of dating must greatly exceed the full span of the bedding plane. But this method can be used for late Pleistocene and Holocene samples.)

On a single mud flat (a modern "bedding plane," if you will) on the island of Great Inagua, we found a complete morphological transition between the extinct fossil pulmonate species *Cerion excelsior* and the modern species

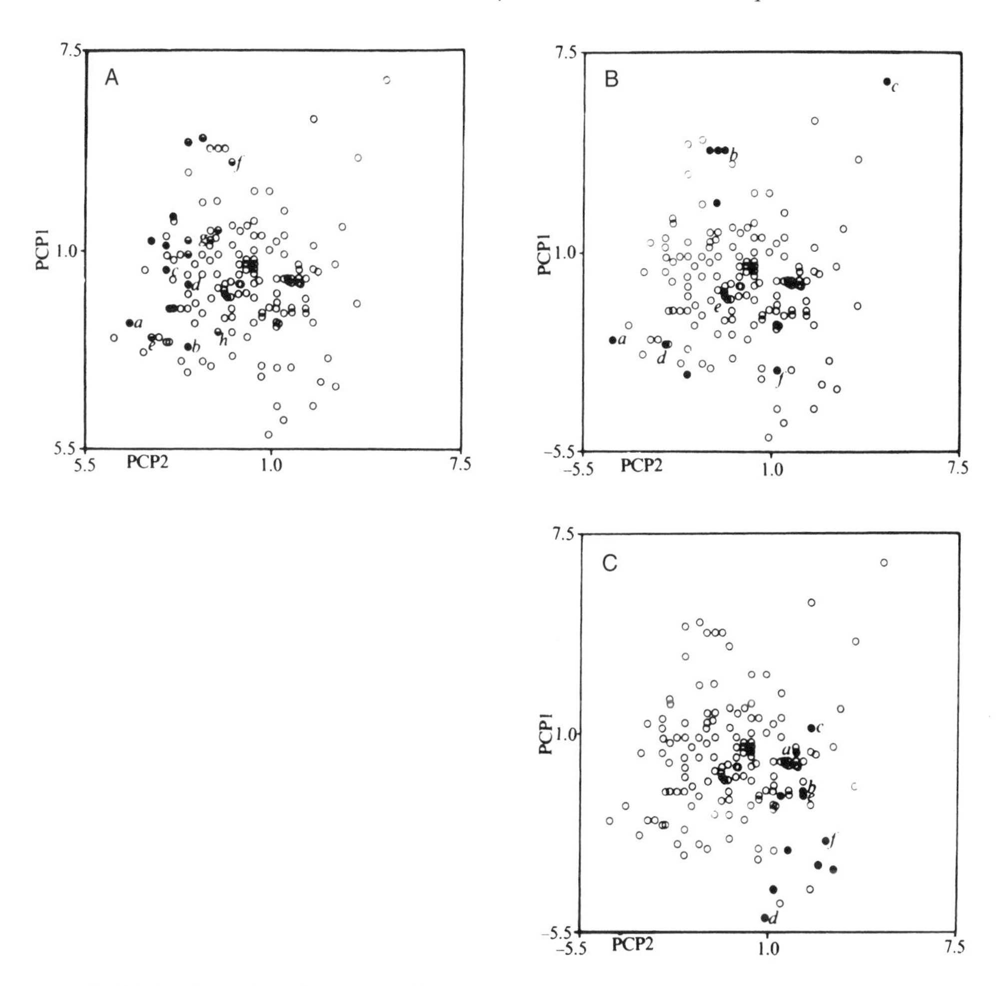

2-1A. The dissection of a punctuation made possible by unusually high sedimentation rates. Williamson's analysis of variation and central tendency during a punctuation in the *B. unicolor* lineage of Pleistocene fresh water pulmonate snails from the African rift valley. Each diagram shows all the specimens from the entire sequence, with only those specimens for the relevant interval depicted in black. A. Parental form before the punctuation with multivariate modal morphology concentrated to the left of the range. B. Expanded variation throughout the range during the time of the punctuation itself. C. Restricted variation again, but settling down upon the morphology of a new taxon following the punctuation, as seen in the reduction of variation with change in modal position towards the right side of the array. From Williamson, 1981.

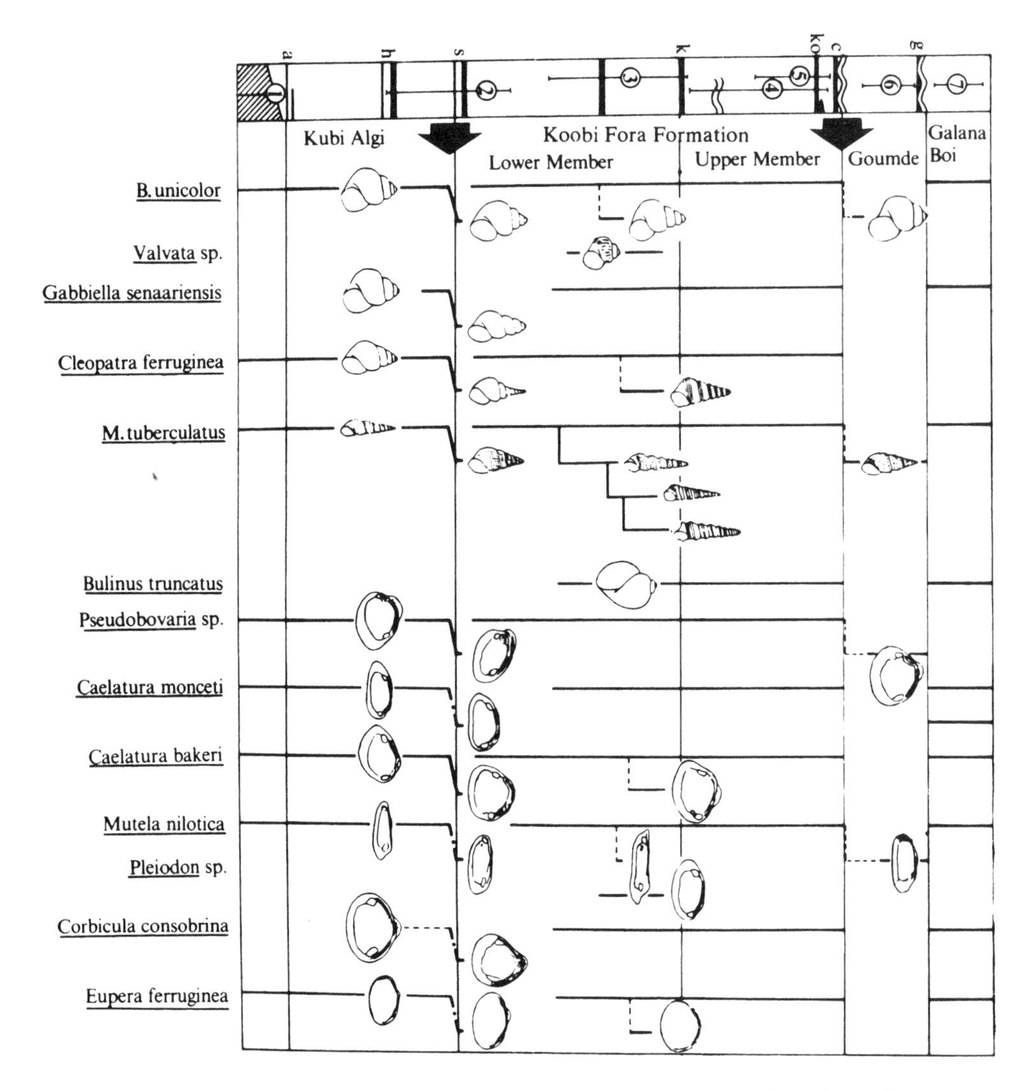

2-1B. Relative timings of punctuational events throughout Williamson's entire series. From Williamson, 1981.

Cerion rubicundum. Many lines of evidence indicate that this transition occurred by hybridization, as *C. rubicundum* migrated to an island previously inhabited only by *C. excelsior* among large species of *Cerion.* Ordinarily, we would find such a complete morphological transition on a single bedding plane, but be unable to perform any fine-scale analysis in the absence of methods for dating individual shells. That is, we would be unable to discover whether the unusual morphological range represented a temporal transition or a standing population with enhanced variation. But Goodfriend and I could date the individual shells by amino acid racemization for all specimens, keyed to radiocarbon dates for a smaller set of marker shells. We found an excellent correlation between measured age and multivariate morphometric position on the continuum between ancestral *C. excelsior* and descendant *C. rubicundum* (see Fig. 2-2). The transition lasted between 15,000 and 20,000 years—a good average value for a punctuational event, and a fact that we could ascertain only because the individual specimens of a single bedding plane could be chemically dated independently of their morphology.

We can therefore define stasis and punctuation in operational terms, with stasis available for testing in almost any species with a good fossil record, but punctuation requiring an unusual density of information, and therefore not routinely testable, but requiring a search for appropriate cases (not an unusual situation in sciences of natural history, where nature sets the experiments, and scientists must therefore seek cases with adequate data). The third key issue of relative frequency may be easier to operationalize—as one

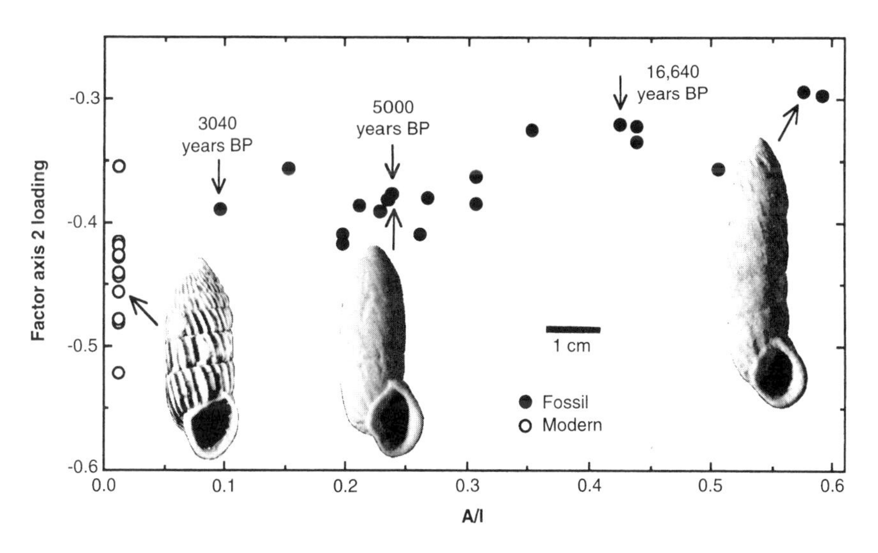

2-2. Another way to dissect a punctuation by obtaining absolute age dates for all specimens on a bedding plane, and thus obtaining temporal distinctions within the compression. The ancestral and high spired *Cerion excelsior,* over no more than 15,000 to 20,000 years (well within the range of punctuational dynamics), hybridizes with invading *Cerion rubicundum,* with gradual fading out of all morphometric influence from the unusually shaped ancestor.

need only tabulate cases pro and con within well-documented faunas—but remains harder to define.

As the most important ground rule, the theory of punctuated equilibrium makes a claim about dominating pattern, or relative frequency, not just an assertion for the existence of a phenomenon. Such issues cannot be resolved by anecdote, or the documentation, however elegant, of individual cases. If anyone ever doubted that punctuated equilibrium exists as a phenomenon, then this issue, at least, has been put to rest by two decades of study following the presentation of our theory, and by clear and copious documentation of many cases (see Chapter 4). Nonetheless, as pleased as Eldredge and I have been both by the extent of this research and the frequency of its success, the "ideal case study" method cannot validate our theory.

Punctuated equilibrium does not merely assert the existence of a phenomenon, but ventures a stronger claim for a dominant role as a macroevolutionary pattern in geological time. But how can this vernacular notion of "dominant" be translated into a quantitative prediction for testing? At this point in the argument, we encounter the difficult (and pervasive) methodological issue of assessing relative frequency in sciences of natural history. If species were like identical beans in the beanbag of classical thought experiments in probability, then we could devise a sampling scheme based on enumerative induction. Enough randomly selected cases could establish a pattern at a desired level of statistical resolution. But species are irreducibly unique, and the set of all species does not exhibit a distribution consistent with requirement of standard statistical procedures. It matters crucially whether we study a clam or a mammal, a Cambrian or a Tertiary taxon, a species in the stable tropics, or at volatile high latitudes. Moreover—and especially—the "ideal case study" method has often failed, and led to parochialisms and false generalities, precisely because we tend to select unusual cases and ignore, often quite unconsciously, a dominant pattern. Indeed, proclamations for the supposed "truth" of gradualism—asserted against every working paleontologist's knowledge of its rarity—emerged largely from such a restriction of attention to exceedingly rare cases under the false belief that they alone provided a record of evolution at all! The falsification of most "textbook classics" upon restudy only accentuates the fallacy of the "case study" method, and its root in prior expectation rather than objective reading of the fossil record.

Punctuated equilibrium must therefore be tested by relative frequencies among all taxa (or in a truly randomized subset) in a particular fauna, a particular clade, a particular place and time, etc. If we can say, as Ager did (see

p. 24) that all but one Mesozoic brachiopod species displays stasis, or as Imbrie did (see pp. 32–33) that all but one Devonian species from the Michigan Basin shows no change, then we have specified a dominant pattern, at least within a particular, well-defined and evolutionarily meaningful package. I cannot give a firm percentage for what constitutes a "dominant" relative frequency—for, again, we encounter a theory-bound claim, where "dominant" specifies a weight, beyond which the morphological history of a clade must be explicated primarily by the differential success of species treated as stable entities, or Darwinian individuals in macroevolution—and not by anagenetic change within species. More research must be done, largely in the testing of mathematical models under realistic circumstances, to learn the relative frequencies and rates required to impart such dominance to species-individuals in the course of macroevolution. For now, and for empirically minded paleontologists, the study of relative frequencies in entire faunas, rather than the extraction of apparently idealized cases, should be pursued as a primary strategy of research.

Critics have sometimes stated that punctuated equilibrium rests upon declaration rather than documentation. (Maynard Smith once compared the theory to "Aunt Jobisca's" maxim about ancient verities "known" by folk wisdom *a priori.*) We do indeed assert that working paleontologists know the fact of dominant stasis in their bones—but this claim represents a fair consensus about the history of a field, and does underscore a paradox of nonconcordance between deep practical knowledge and imposed theoretical expectation. We have never tried to argue that such a "professional feeling" constitutes documentation for punctuated equilibrium. As with all scientific theories, punctuated equilibrium will live or die by concrete and quantifiable evidence. As with any good hypothesis, punctuated equilibrium becomes operational when workable definitions can be provided for key claims and expectations—in this case, for *stasis, punctuation,* and *relative frequency.* Contrary to the impression of some critics who have not followed the primary literature of paleobiology during the last 25 years, punctuated equilibrium has proven its fruitfulness and operational worth by being tested—and usually confirmed, but sometimes confuted—in a voluminous literature of richly documented cases (see Chapter 4).

MICROEVOLUTIONARY LINKS

Eldredge and I coined the term punctuated equilibrium in a paper first presented (Gould and Eldredge, 1971) at a symposium entitled "Models in Paleobiology" at the 1971 Annual Meeting of the Geological Society of

America. T. J. M. Schopf, the organizer of the symposium, conceived the enterprise as a tutorial in modern evolutionary theory for professional invertebrate paleontologists. By accidents of history, invertebrate paleontologists generally receive their advanced academic degrees from geology departments, not from biology. Fossils became primary tools for stratigraphic correlation long before the development of evolutionary theory, and even before all scientists had accepted them as remains of ancient organisms! Given traditions of narrowness in postgraduate education—particularly in Europe, where students often attend no formal courses at all, and certainly no courses for credit, outside the department that will grant their degree—most paleontologists, before the present generation, did not receive any explicit training in evolutionary biology, and could not articulate the basic concepts of population genetics or theories of speciation. In paleontological usage, "evolution" designated little more than the inferred pathway of phylogeny. This "little learning" often became the "dangerous thing" of Alexander Pope's classic couplet, as paleontologists derived their understanding of evolution from memories of old textbooks, or from shared impressions amounting to little more than the blind leading the blind. This situation has now changed dramatically—and Eldredge and I do take pride in the role played by punctuated equilibrium in encouraging this shift of interest—as a profession of paleobiology, supported by several new journals dedicated to the subject (*Paleobiology, Historical Biology, Lethaea, Palaios,* and *Palaeogeography Palaeoclimatology Palaeoecology* (or *P-cubed* to aficionados), for example), has arisen to accommodate burgeoning research in the application of evolutionary theory to the fossil record, and in enlarging and revising the theory in the light of novel macroevolutionary data.

In any case, Schopf's symposium featured a series of presentations, each suggesting how one aspect of paleontological work might be enlightened by modern microevolutionary theory, particularly as expressed in the application of models, preferably quantitative in nature. Eldredge and I drew the topic of species and speciation—and our original article on punctuated equilibrium (Eldredge and Gould, 1972) emerged as a result. (As I have often stated, the basic idea had been presented in Eldredge, 1971. We had been graduate students together at the American Museum of Natural History, under the tutelage of Norman D. Newell. We had discussed these issues often and intensely throughout our graduate years. We had been particularly frustrated—for we had both struggled to master statistical and other quantitative methods—with the difficulty of locating gradualistic sequences for applying these techniques, and therefore for documenting "evolution" as paleontological tradition then defined the term and activity. When I received Schopf's invitation to talk on models of speciation, I felt that Eldredge's 1971 publication had presented the only new and interesting ideas on paleontological implications of the subject—so I asked Schopf if we could present the paper jointly. I wrote most of our 1972 paper, and I did coin the term punctuated equilibrium—but the basic structure of the theory belongs to Eldredge, with priority established in his 1971 paper.)

I mention this background to clarify the original context and continuing focus of the theory of punctuated equilibrium—a notion rooted in the explicit goal that Eldredge and I set for ourselves: to apply microevolutionary ideas about *speciation* to the data of the fossil record and the scale of geological time. Before we proposed the theory of punctuated equilibrium, most paleontologists assumed that the bulk of evolutionary change proceeded in the anagenetic mode—that is, by continuous transformation of a unitary population through time (see Fig. 2-3). In this context, most paleontological discussion about species centered itself upon a contentious issue that constantly circulated throughout our literature (see Imbrie, 1957; Weller, 1961; McAlester, 1962; Shaw, 1969) and even generated entire symposia dedicated

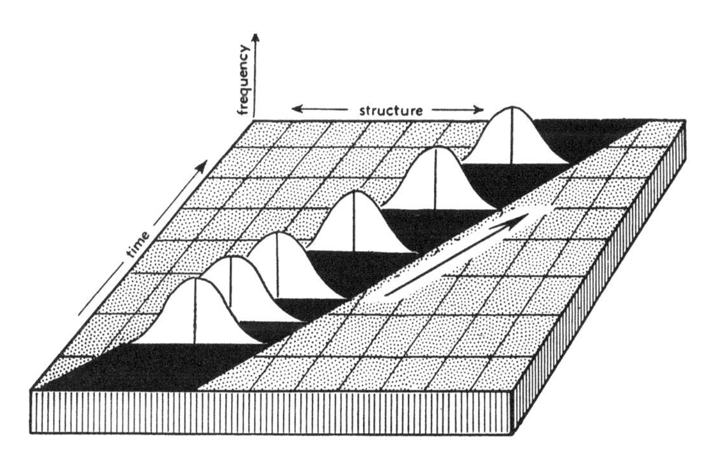

2-3. Typical textbook illustration of evolution by continuous anagenetic transformation of an unbranched population through time. This textbook labels the figure explicitly and exclusively as its icon of "evolution" itself, not of gradualism or any other subcategory of evolutionary change. From the standard paleontological textbook of my student generation, Moore, Lalicker, and Fischer, 1953.

to potential solutions (see Sylvester-Bradley, 1956): the so-called species problem in paleontology.

This supposed problem—more philosophical and definitional than empirical (once one accepts the underlying assumptions about anagenesis as a dominant factual reality)—arises because a true continuum cannot be unambiguously divided into segments with discrete names. If population A changes so extensively by anagenesis that we feel impelled to provide the resulting population with a new Linnaean name (as species B), then where should we place the breakpoint between A and B? Any boundary must be arbitrary—if only by the illogic of the unavoidable implication that the last parental generation of species A could not, in principle, breed with its own immediate offspring in species B. (We may abhor human incest for social reasons, but we can scarcely deny the biological possibility—hence the perceived societal need for a taboo.) This problem generated a large, tedious, and fruitless literature, primarily because the issue always remained available, unresolved and therefore ripe for yet another go-round whenever a paleontologist needed to deliver a general address and couldn't think of anything else to say.

Punctuated equilibrium took a radically different approach by admitting unresolvability under the stated assumptions, but then denying the focal empirical premise that new species usually (or even often) arise by gradualistic anagenesis. Instead, Eldredge and I argued that the vast majority of species originate by splitting, and that the standard tempo of speciation, when expressed in geological time, features origin in a geological moment followed by long persistence in stasis. Thus, the classic and endlessly fretted "species problem in paleontology" disappears because species act as well-defined Darwinian individuals, not as arbitrary subdivisions of a continuum. Species then gain definability because they almost always arise by speciation (that is, by splitting, or geographic isolation of a daughter population followed by genetic differentiation from the parental population), not by anagenesis (or transformation of the entire mass of an ancestral species). To be sure, a new species must pass through a short period of ambiguity during its initial differentiation from an ancestral population, but, in the proper scaling of macroevolutionary time, this period passes so quickly (almost always in the unresolvable geological moment of a single bedding plane) that operational definability encounters no threat.

Of course, gradualists did not deny that speciation often occurs by branching. They just didn't grant this process of splitting any formative role in the accumulation of macroevolutionary change for three reasons. First, they conceived speciation only as an engine for generating diversity, not as an agent for changing average form within a clade (that is, for the key macroevolutionary phenomenon of trends—see quotes of Huxley and Ayala, and Mayr's response, in *SET,* p. 563). Trends arose by anagenesis (see Fig. 2-4), and speciation only served the subsidiary (if essential) function of iterating a favorable feature, initially evolved by anagenesis, into more than one taxon—thus providing a hedge against extinction.

Second, they granted little quantitative weight to the role of speciation (splitting as opposed to anagenesis) in the totality of evolutionary change. In a famous estimate that became canonical, Simpson (1944) stated that about 10 percent of evolutionary change occurred by speciation, and 90 percent by anagenesis.

Third, when gradualists portrayed speciation at all (see Fig. 2-5), they depicted the process as two events of anagenesis proceeding at characteristically slow rates. Thus, they identified nothing distinctively different about change by speciation. Some contingency of history, they argued, splits a population into two separate units, and each proceeds along its ordinary

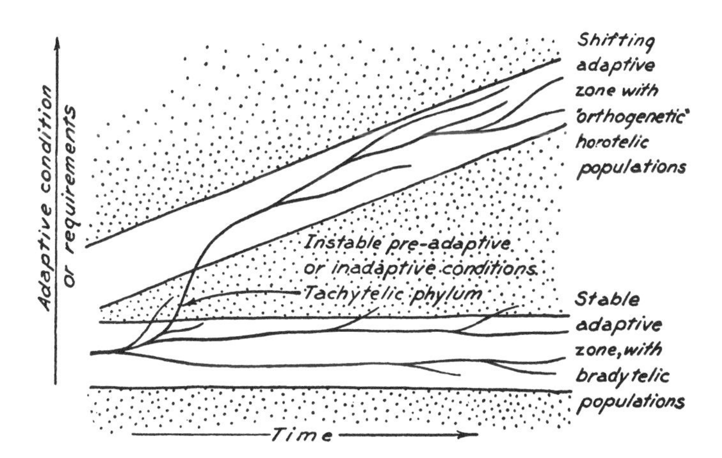

2-4. A standard illustration from Simpson (1944), showing that all trends, and all stability for that matter, originate primarily in the anagenetic mode—that is, by change during the lifetime of individual species, with branching serving primarily to diversify and iterate the favorable designs originated by anagenesis, and thus to prevent extinction of the lineage.

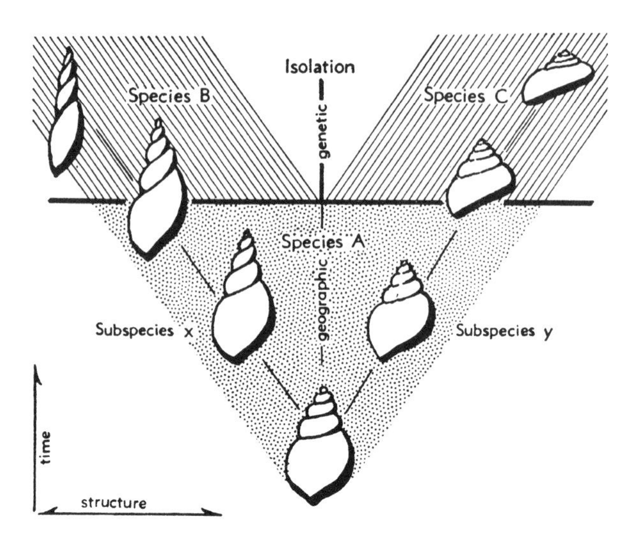

2-5. Another illustration from the standard student paleontological textbook of the 1950s, with speciation depicted merely as two events of gradualistic change, following a separation of lineages. From Moore, Lalicker, and Fischer, 1953.

anagenetic way. Punctuated equilibrium, on the other hand, proposes that the geological tempo of speciation differs radically from gradualistic anagenesis. (We also argue, of course, that such anagenesis rarely occurs at all!)

The theory of punctuated equilibrium therefore began as a faithful response to Schopf's original charge to Eldredge and me: to show how standard microevolutionary views about speciation, then unfamiliar to the great majority of working paleontologists, might help our profession to interpret the history of life more adequately. (As a best testimony to this unfamiliarity, I note that most paleontologists didn't even recognize the conceptual and terminological distinction between "speciation" defined as a process of splitting, and the accumulation of enough change by anagenesis to provoke the coining of a new Linnaean name for an unbranched single population.)

In this crucial sense, the theory of punctuated equilibrium adopts a very conservative position. The theory asserts no novel claim about modes or mechanisms of speciation; punctuated equilibrium merely takes a standard microevolutionary model and elucidates its expected expression when properly scaled into geological time. This scaling, however, did provoke a radical reinterpretation of paleontological data—for we argued that the literal appearance of the fossil record, though conventionally dismissed as an artifact of imperfect evidence, may actually be recording the workings of evolution as understood by neontologists.[1] This empowering switch enabled paleontologists to cherish their basic data as adequate and revealing, rather than pitifully fragmentary and inevitably obfuscating. Paleontology could emerge from the intellectual sloth of debarment from theoretical insight imposed by poor data—a self-generated torpor that had confined the field to a descriptive role in documenting the actual pathways of life's history. Paleontology could now take a deserved and active place among the evolutionary sciences.

The major and persisting misunderstanding of punctuated equilibrium among neontologists—a great frustration for us, and one that we have tried to explicate and resolve again and again (Gould and Eldredge, 1977, 1993; Gould, 1982b, 1989b), though without conspicuous success—involves the false assumption that if we are really saying something radical, we must be staking a claim for a novel mechanism of speciation, or for a different (read non-Darwinian) style of genetic change. When our critics then join this false assumption to our terminology of "unresolvable geological moments" or "punctuations," they begin to fear that the dreaded specter of saltationism must be lurking just around the corner, trying yet again to raise its ugly head after such a well-deserved burial. Vituperation then trumps logic, angry assumption precludes careful reading, and punctuated equilibrium becomes a loathed doctrine of ignorant and grandstanding paleontologists who ought to stay in their own limited bailiwick, and get on with the job of documenting large-scale patterns generated by mechanisms that can be recognized and comprehended only by neontologists.

But punctuated equilibrium makes no iconoclastic claim about speciation at all. The radicalism of punctuated equilibrium lies in the extensive consequences of its key implication that conventional mechanisms of speciation scale into geological time as the observed punctuations and stasis of most species, and not as the elusive gradualism that a century of largely fruitless paleontological effort had sought as the only true expression of evolution in the fossil record. The central intellectual strategy of our original 1972 paper rests upon this premise. We took Mayr's allopatric theory (as expressed in his classic treatise of 1963, deemed "magisterial" by Huxley), and tried to elucidate its implied expression when scaled into geological time. We did not select this theory to fit a paleontological pattern that we wished to validate. We choose Mayr's formulation because his allopatric theory represented the most orthodox and conventional view of speciation then available in neontological literature—and we had been given the task of applying *standard* evolutionary views to the fossil record. I recognize, with 30 years of hindsight, that our original assessment both of Mayr's theory and of professional consensus may have been both naive and overly dichotomous, but we could not have stated our intent more clearly—the reform of paleontological practice by the paradoxical route of applying a *fully conventional* apparatus of neontological theory. We wrote (1972, p. 94): "During the past thirty years, the allopatric theory has grown in popularity to become, for the vast majority of biologists, *the* theory of speciation. Its only serious challenger is the sympatric theory. Here we discuss only the implications of the allopatric theory for interpreting the fossil record of sexually-reproducing metazoans. We do this simply because it is the allopatric, rather than the sympatric, theory that is preferred by biologists."

Mayr's version of allopatry fit the paleontological pattern of punctuation and stasis particularly well. If most new species arise from small populations peripherally isolated at the edges of a parental range, then we cannot expect to document a gradual transition by analyzing the stratigraphic sequence of samples for a common species. For we will usually be collecting from the population's central range during its period of stability. Daughter species originate in three circumstances that virtually guarantee a punctuational expression in the fossil record: (1) they arise rapidly (usually instantaneously) in geological time, and they originate both (2) in a small geographic region (the peripheral isolate), and (3) elsewhere (beyond the borders of the parental range that provides the exclusive source for standard paleontological collections). The "sudden" entrance of a daughter species into strata previously occupied by parents usually represents the inward migration of a peripheral isolate, now "promoted" by reproductive isolation to full separation, not the origin of a new species *in situ.*

Eldredge and I have often been asked what we think of sympatric speciation, or of various models, like polyploidy, for rapid origin even in human time. We do not mean to be evasive or obscure in our assertions of agnosticism. (I am intensely interested in the literature on speciation, and I would love to know the relative frequencies of these other models vs. classical Mayrian peripatry. But this important issue does not strongly impact punctuated equilibrium, and surely cannot be resolved by paleontological data.) Punctuated equilibrium simply requires that any asserted mechanism of speciation, whatever its mode or style, be sufficiently rapid and localized to appear as a punctuation *when scaled into geological time.* If I understand them correctly, most alternative models to peripatry generally operate even more rapidly than the conventional Mayrian mode that we invoked to anchor our theory—as obviously true for polyploidy, and also for most versions of sympatric speciation (if only because the constant threat of dilution by gene flow from surrounding parentals can best be overcome by rapid achievement of reproductive isolation in ecological time). Therefore, punctuated equilibrium can only gain strength if these alternative mechanisms become validated at meaningful relative frequencies. (The faster the better, one might say.) But punctuated equilibrium does not require this boost—and we therefore remain agnostic—because the most conventional form of Mayrian peripatry already yields the full set of phenomena predicted by punctuated equilibrium when properly scaled into the immensity of geological time. (Punctuated equilibrium, on the other hand, does not maintain a similar agnosticism towards any putative mechanism of speciation that conceives the process of splitting as no more rapid than imagined rates for the gradual anagenesis of large central populations. Some models of so-called "dumbbell allopatry"—or the splitting of a parental population into two effectively equal moieties, with subsequent anagenesis in each—do construe speciation as consequently slow in geological expression, and therefore do threaten punctuated equilibrium. But I do not think that such models enjoy much support among biologists, especially for operation at a high relative frequency.)

Geological time can be both a wonder and a snare because we grasp the idea in our heads (all scientists know how many zeroes follow the one in expressing millions or billions), but we face a primal, and fundamentally psychological, difficulty in trying to incorporate this central concept into the guts of our intuition. We can lose information in upward scaling when glacial slowness in human history becomes a passing and unresolvable geological moment. But we can also gain when operational invisibility at our scale (inability to distinguish a small effect from measurement error) becomes palpable and prominent in the large, or when the almost inconceivable rarity of an event that averages one expression in ten thousand years achieves guaranteed repetition across millions.

Macroevolutionary Implications

If punctuated equilibrium has broader utility beyond the reform of paleontological practice, then we must look to potential implications for macroevolutionary theory, and for consequent enrichment in our general understanding of mechanisms that regulate the history of life. I have linked my treatments of punctuated equilibrium and the hierarchical theory of natural selection (presented as two chapters in *SET*, 8 and 9; this book reprints Chapter 9 of that volume) because I believe that punctuated equilibrium supplies the central argument for viewing species as effective Darwinian individuals at a relative frequency high enough to be regarded as general—thereby validating the level of species as a domain of evolutionary causality, and establishing the effectiveness and independence of macroevolution by two of the three criteria featured throughout *SET* and this book as indispensable foundations of Darwinism.

First, punctuated equilibrium secures the hierarchical expansion of selectionist theory to the level of species, thus moving beyond Darwin's preference for restricting causality effectively to the organismic realm alone (leg one on the essential tripod). *Second,* by defining species as the basic units or atoms of macroevolution—as stable "things" (Darwinian individuals) rather than as arbitrary segments of continua—punctuated equilibrium precludes the explanation of all evolutionary patterns by extrapolation from mechanisms operating on local populations, at human timescales, and at organismic and lower levels (leg three on the tripod of Darwinian essentials). Thus, as emphasized in the last section, punctuated equilibrium presents no radical proposal in the domain of microevolutionary mechanics—in particular (and as so often misunderstood), the theory advances no defenses for saltational models of speciation, and no claims for novel genetic processes. Moreover, punctuated equilibrium does not attempt to specify or criticize the conventional mechanisms of microevolution at all (for punctuated equilibrium emerges as the anticipated expression, by proper scaling, of microevolutionary theories about speciation into the radically different domain of "deep," or geological time). But punctuated equilibrium does maintain, as the kernel of its potential novelty for biological theory, that these unrevised microevolutionary mechanisms do not hold exclusive sway in evolutionary explanation, and that their domain of action must be restricted (or at least shared) at the level of macroevolutionary pattern over geological scales—for punctuated equilibrium ratifies an effective realm of macroevolutionary mechanics based on recognizing species as Darwinian individuals. In other words, *punctuated equilibrium makes its major contribution to evolutionary theory, not by revising microevolutionary mechanics, but by individuating species (and thereby establishing the basis for an independent theoretical domain of macroevolution).*

As discussed in Chapter 8 of *SET* (see pp. 648–652), punctuated equilibrium wins this role by refuting Fisher's otherwise decisive argument for the impotence (despite the undeniable existence) of species selection. So long as most new species arise by branching (speciation) rather than by transformation (anagenesis), species can be individuated by their uniquely personal duration, bounded by birth in branching and death by extinction. But if anagenesis, fueled by Darwinian organismic selection, operates to substantial effect during the lifetimes of most species, then, by Fisher's argument, such microevolutionary transformation must overwhelm species selection in building the overall pattern of macroevolutionary change—for the number of organism-births must exceed species-births by several orders of magnitude, and if every event of birthing, at each level, supplies effective variation for evolutionary transformation, then the level of species can contribute virtually nothing to the totality of change. But if stasis rules and anagenesis rarely occurs, then speciation becomes the more effective level of evolutionary variation. And if speciation unfolds in geological moments, then species in geological time match organisms on our ordinary yearly scales in both distinctness and discreteness. Thus, the pattern of punctuated equilibrium establishes species as effective individuals and potential Darwinian agents in the mechanisms of macroevolution.

In summary, G. G. Simpson gave a singularly appropriate title to his epochal 1944 book that defined the potential of paleontology to devise insights about evolutionary mechanisms: *Tempo and Mode in Evolution.* If we accept Simpson's focus on tempo and mode as primary subjects, then punctuated equilibrium has provoked substantial revisions of macroevolutionary theory and practice in both domains.

TEMPO AND THE SIGNIFICANCE OF STASIS

For tempo, punctuated equilibrium reverses our basic perspective. We must abandon our concept of constant change operating within a sensible, stately range of rates as the normal condition of an evolving entity. We must then reformulate evolutionary change as a set of rare episodes, short in duration relative to periods of stasis between. Stability becomes the normal state of a lineage, with change recast as an infrequent and concentrated event that, nonetheless, renders phylogeny as a set of summed episodes through time. The implications of this fundamental shift resonate afar by impacting a set of issues ranging from the most immediately practical to the most broadly philosophical (including, in the latter category, an interesting consonance with the atomism and quantization invoked to define the general intellectual movement known as "modernism"—as expressed in disparate disciplines from Seurat's pointillism in art, to Schönberg's serial style in music; and as opposed to the smooth continuationism favored by earlier mechanistic views of causality). In a theme more immediately relevant to biology, the same shift ineluctably places much greater emphasis upon chance and contingency, rather than predictability by extrapolation—for the ordinary condition of stasis provides little insight into when and how the next punctuation will occur, whereas the fractal character of gradualism suggests that causes of change at any moment will, by extrapolation, predict and explain the larger effects accumulated through longer times.

On the practical side, punctuated equilibrium's formulation of tempo has validated the study of stasis—paleontology's prevalent pattern within species—as a source of insight about evolution, rather than a cause of chagrin best bypassed and ignored as a testimony to an embarrassing poverty of evidence. Punctuated equilibrium has broken "Cordelia's Dilemma" of silence about the supposed "nothing" of stasis, and has established a burgeoning subfield of research in the documentation of stability at several levels. In pursuing and valuing this documentation, scientists then feel compelled to postulate explanations for the puzzling frequency of this previously "invisible" phenomenon—and theoretical inquiry about the "why" of stasis has also flourished following the prod from punctuated equilibrium (see pp. 175–185 for fuller discussion).

MODE AND THE SPECIATIONAL FOUNDATION OF MACROEVOLUTION

For mode, as discussed throughout this chapter, punctuated equilibrium has established a speciational basis for macroevolution. By supplying crucial data and arguments for defining species as effective Darwinian individuals—that is, as basic units for describing macroevolution in Darwinian terms as an outcome of patterns in differential birth and death of species treated as stable individuals, just as microevolution works by the same process applied to births and deaths of organisms—punctuated equilibrium validates the hierarchical theory of selection. This hierarchical theory (explicated in Chapter 8 of *SET*) establishes the independence of macroevolution as a theoretical subject (not just as a domain of description for accumulated microevolutionary mechanics), thereby precluding the full explanation of evolution by extrapolation of microevolutionary processes to all scales and times.

In practical terms, the implications of punctuated equilibrium for evolutionary mode have strongly impacted two prominent subjects, heretofore almost always rendered by extrapolation as consequences of adaptation within populations writ large: evolutionary trends within clades, and relative waxing and waning of diversity within supposedly competing clades through time. Punctuated equilibrium suggests novel, and irreducibly macroevolutionary, explanations for both phenomena (see pp. 185–222).

Finally, the role of punctuated equilibrium in establishing an independent field of macroevolution includes both a weak and a strong version. The first, undoubtedly valid as a generality, "uncouples" macro from microevolution as a descriptive necessity, while not establishing independent causal principles of macroevolution. The second clearly regulates many cases, but has not yet been validated as commanding a high relative frequency; this second, or strong, version establishes irreducible causal principles of macroevolution.

The weak version, based on "species sorting" rather than "species selection," holds that evolution must be described as differential success in birth and death of stable species, but allows that the causality behind reasons for differential success might emerge from the conventional Darwinian level of struggling organisms within successful populations—the effect hypothesis of Vrba (see *SET,* p. 658). In this version, we need a descriptive, but not a causal, account of macroevolution based on species as individuals.

However, in the strong version, based on true species selection, the differential success of species arises from irreducible fitness defined by the interaction of species-individuals with their environments. Chapter 8 of *SET* presents an extensive argument for the efficacy of true species selection at high relative frequency. Validation of this argument would establish a genuinely causal and irreducible theory of macroevolution. This difficult issue stands far from resolution, but represents the most exciting potential for punctuated equilibrium as an impetus in formulating a revised structure for evolutionary theory.

3 THE SCIENTIFIC DEBATE ON PUNCTUATED EQUILIBRIUM: CRITIQUES AND RESPONSES

Critiques Based on the Definability of Paleontological Species

EMPIRICAL AFFIRMATION

The issue of whether true biospecies (or entities operationally close enough to biospecies) can be recognized in fossils has prompted long and intense debate in paleontology (see Sylvester-Bradley, 1956, and other references previously cited), and does not represent a new or special difficulty raised by punctuated equilibrium. But given the reliance of punctuated equilibrium on speciation as the mechanism behind the pattern, this old problem does legitimately assume a central place in debates about our theory (as emphasized in all negative commentary, particularly clearly by Turner, 1986, and in the book-length critiques of Levinton, 1988, and Hoffman, 1989).

At least we may begin by exposing the canonical issue of the older literature as a *Scheinproblem* (literally an "appearance problem" with no real content): the logical impossibility of defining a species boundary within a gradualistic continuum (see my previous discussion on p. 52). I think we may now accept that the punctuational pattern exists at high relative frequency, and that few gradualistic and anagenetic continua have been documented between fossil species. Turner's (1986) sharp critique, for example (and I do accept his formulation, though not his resolution), depicts the chief claims of punctuated equilibrium as a three-pronged fork. He accepts the first tine—the existence of the punctuational pattern itself—as sufficiently demonstrated by enough empirical cases in the fossil record. He regards the third tine—macroevolutionary invocation of the theory to explain trends by species sorting—as "an important extension of evolutionary theory into a hitherto little explored territory" (1986, p. 206). But he then rejects the second tine as both unlikely and too difficult to test in any case—explanation of the punctuational pattern as a consequence of speciation scaled into geological time.

If we accept that temporal sequences of fossils generally don't appear in the geological record as unbreakable continua, but usually as morphological "packages" with reasonably defined boundaries and sufficient stability within an extended duration, how can we assert that these packages represent biospecies, or at least that they approximate these neontologically defined units with sufficient closeness to bear comparison? After all, we cannot apply conventional tests of observed ecological interaction or interbreeding to fossils—and, whereas biospecies may be *recognized* by morphological differentia in everyday practice, they are not supposed to be so *defined.* Can the temporally extended "morphospecies" of paleontology really be equated with the "nondimensional species concept" (Mayr's words) of neontology?

I certainly accept the centrality and difficulty of these issues, but I do not regard them as insuperable, and I do not view the species concept as untestable with fossils. After all, the overwhelming majority of modern species in our literature and museum drawers have also been phenotypically, not ecologically, defined. Once we accept that no special paleontological riddles arise from the *Scheinproblem* of temporal continua, then most paleospecies have been no worse characterized than the majority of neospecies. Still, I will not advance this excuse as exculpatory for the fossil record, for a neontologist could reply, with impeccable logic, that neospecies so defined should also be regarded as uncertain, if not vacuous, and that no paleontological defense can be mounted by arguing that ordinary practice with fossils follows the worst habits (majoritarian though they may be) of neontological taxonomy.

But a best defense of phenotypically defined neospecies would follow from demonstrations that taxa so established usually do match true biospecies upon proper behavioral and ecological study—a line of research often pursued with success (see references in Jackson and Cheetham, 1994, and in Jablonski, 1999). Similarly, my main source for confidence about paleospecies arises from proven correspondences with true biospecies in favorable cases providing sufficient information for such a test (particularly for extant species with lengthy fossil records). I do not, of course, argue that all named paleospecies are true biospecies, or that I can even estimate the percentage properly so defined (any more than we know the relative frequency of modern taxa that represent true biospecies). But I do not see why the probability that well-defined paleospecies, based on good collections from many times and places, might represent proper biospecies should be any lower than the corresponding figure for equally well documented, but entirely morphologically defined, modern taxa. (In fact, one might argue that well-documented paleospecies probably maintain a higher probability for representing biospecies, because we know their phenotypes, and have measured their stability, across long periods of time and wide ranges of environment—whereas modern "morphospecies" may arise as ecophenotypic expressions of a single time and place, therefore ranking only as local populations, rather than true species.)

When well-defined paleospecies have been tested for their correspondence with modern biospecies, such status has often been persuasively affirmed. Two recent studies seem particularly convincing. Michaux (1989) studied four living species of the marine gastropod genus *Amalda* from New Zealand. Fossils of this genus date to the upper Eocene of this region, while all four species extend at least to the Miocene-Pliocene boundary. The four taxa represent good biospecies, based on absence of hybrids in sympatry, and on extensive electrophoretic study (Michaux, 1987) showing distinct separation among species and "no detectable cryptic groupings" (Michaux, 1989, p. 241) within any species. Michaux then used canonical discriminant analysis to achieve clear morphometric distinction among the species based on 10 shell measurements for each of 671 live specimens.

He then made the same measurements on 662 fossil specimens from three of the species (the fourth did not yield enough shells for adequate characterization). Mean values, in multivariate expression based on all 10 variables, fluctuated mildly through time (see Fig. 3-1), but never departed from the range of variation within extant populations—an excellent demonstration of stasis as dynamic maintenance within well-defined biospecies through several million years. Michaux concluded (1989, pp. 246–248): "Fossil members of three biologically distinct species fall within the range of variation that is exhibited by extant members of these species. The phenotypic trajectory of each species is shown to oscillate around the modern mean through the time period under consideration. This pattern demonstrates oscillatory change in phenotype within prescribed limits, that is, phenotypic stasis."

Jackson and Cheetham's (1990, 1994) extensive studies of cheilostome bryozoan species provide even more gratifying affirmation, especially since these "simple" sessile and colonial forms potentially express all the attributes of extensive ecophenotypic variation (especially in molding of colonies to substrates, and in effects of crowding) and morphological simplicity (lack of enough complex skeletal characters for good definition of taxa) gen-

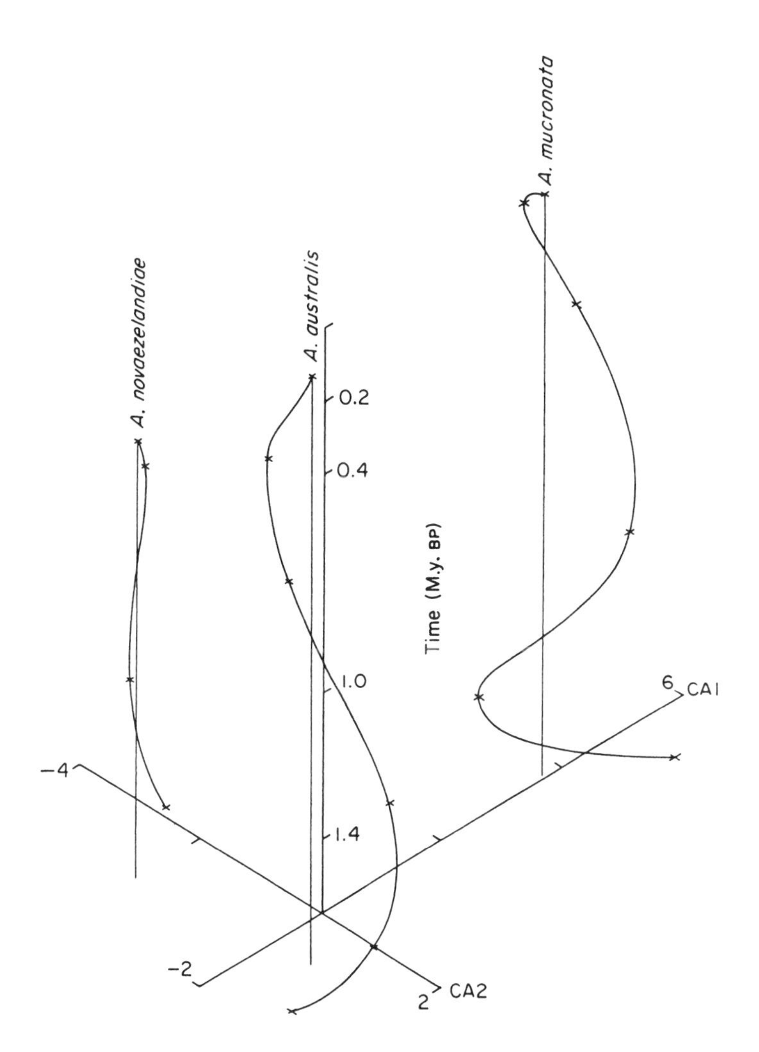

3-1. Stasis in three genetically well defined extant species of the gastropod *Amalda* from New Zealand, based on 662 fossil specimens. Mean values in multivariate expression based on all ten variables fluctuate mildly through time, but never depart from the range of variation within extant populations. From Michaux, 1989.

erally regarded as rendering the identification of biospecies hazardous, if not effectively impossible, in fossils. Moreover, Cheetham had begun his paleontological studies (see discussion on pp. 161–167) under the assumption that careful work would reveal predominant gradualism and refute the "new" hypothesis of punctuated equilibrium—so the conclusions eventually reached were not favored by any *a priori* preference!

In a first study—devoted to determining whether biospecies could be recognized from skeletal characters (of the sort used to define fossil taxa) in several species within three genera of extant Caribbean cheilostomes—Jackson and Cheetham (1990) examined heritability for skeletal characters in seven species. In a "common garden" experiment (under effectively identical conditions at a single experimental site), they grew F_1 and F_2 generations from embryos derived from known maternal colonies collected in disparate environments and places. Multivariate discriminant analysis assigned all but 9 of 507 offspring into the same morphospecies as their maternal parent. The authors then used electrophoretic methods to study enzyme variation in 402 colonies representing 8 species in the three genera. They found clear and complete correspondence between genetic and morphometric clusterings, and also determined (p. 581) that "genetic distances between morphospecies are consistently much higher than between populations of the same morphospecies"; moreover, they found no evidence for any cryptic division (potential "sibling species") within skeletally defined morphospecies.

In a concluding and gratifying observation—indicating that paleontologists need not always humble themselves before the power of neontological genetic analysis of biospecies—Jackson and Cheetham (1990, p. 582) make an empirical observation about the capacity of morphometric data (of the sort generated from fossils):

> The identity of quantitatively defined morphospecies of cheilostome bryozoans is both heritable and unambiguously distinct genetically. The importance of rigorous quantitative analysis was underlined by our discovery of three species of *Stylopoma* previously classified as one, a separation subsequently confirmed genetically. The widely supposed lack of correspondence between morphospecies and biospecies may result as much out of uncritical acceptance of outdated, subjectively defined taxa as from any fundamental biologic differences between the two kinds of species.

Jackson and Cheetham (1994) then followed this study with a more extensive documentaiton, this time using large numbers of fossil species as well as living forms, of phylogenetic patterns in two Caribbean cheilostome genera, *Stylopoma* (included in the first study as well), and *Metrarabdotos* (the subject of Cheetham's earlier and elegant affirmations of punctuated equilibrium from morphometric data alone—Cheetham, 1986 and 1987, and extensively discussed on pp. 161–167). Again, and for both genera, they found strict correspondence between genetically defined clusters and taxa established by skeletal characters accessible from fossils.

With increased confidence that the taxa of his classical studies on punctuated equilibrium in *Metrarabdotos* represent true biospecies, Cheetham (now writing with Jackson) could affirm his earlier work (Jackson and Cheetham, 1994, p. 420): "Morphological stasis over millions of years punctuated by relatively sudden appearances of new morphospecies was demonstrated previously for *Metrarabdotos.* Our updated results strengthen confidence in that pattern, with 11 morphospecies persisting unchanged for 2–6 m.y., all at $p > 0.99$, and no evidence that intraspecific rates of morphological change can account for differences between species."

For *Stylopoma,* where fossil evidence had not previously been analyzed morphometrically, results also affirmed punctuated equilibrium throughout (1994, p. 420): "The excellent agreement between morphologically and genetically defined species used in this taxonomy suggests that morphological stasis reflects genuine species survival over millions of years, rather than a series of morphologically cryptic species. Morover, eleven of the 19 species originate fully formed at $p > 0.9$, with no evidence of morphologically intermediate forms, and all ancestral species but one survived unchanged all with their descendants."

In a concluding paragraph about both genera, Jackson and Cheetham wrote (p. 407): "Stratigraphically rooted trees suggest that most well-sampled *Metrarabdotos* and *Stylopoma* species originated fully differentiated morphologically and persisted unchanged for > 1 to > 16 m.y., typically alongside their putative ancestors. Moreover, the tight correlation between phenetic, cladistic, and genetic distances among living *Stylopoma* species suggests that changes in all three variables occurred together during speciation. All of these observations support the punctuated equilibrium model of speciation."

Despite the encouragement provided by these and other cases, problems continue to surround the definition of paleontological species—a subject of central importance to punctuated equilibrium, given our invocation of speciation as the quantum of change for life's macroevolutionary history, and the source of raw material for higher-level selection and sorting. These problems center upon three main issues (in both the inherent logic of the case and by recorded debate in the literature): the first untroubling, the second potentially serious, and the third largely resolved in empirical terms. All three issues raise the possibility that paleospecies systematically misrepresent the nature and number of actual biospecies. (If paleospecies don't correspond with biospecies in all cases—an undeniable proposition of course—but if these discrepancies show no pattern and produce no systematic bias, then we need not be troubled unless the relative frequency of noncorrespondence becomes overwhelmingly high, an unlikely situation given the excellent alignments found in the few studies explicitly done to investigate this problem, as discussed just above.) The following three subsections treat these three remaining issues *seriatim*.

REASONS FOR A POTENTIAL SYSTEMATIC UNDERESTIMATION OF BIOSPECIES BY PALEOSPECIES

Might we be missing a high percentage of actual speciation events because paleontologists can only recognize a cladogenetic branch with clear phenotypic consequences (for characters preserved as fossils), whereas many new species arise without substantial morphological divergence from their ancestors? In the clearest case, paleontologists (obviously) cannot detect sibling species, a common phenomenon in evolution (see Mayr, 1963, for the classic statement). Moreover, we may also miss subtle changes in phenotype, or substantial alterations (of color, for example) in features that are often important in recognizing species, but do not achieve expression in the fossil record.

Our harshest critics have urged this point as particularly telling against punctuated equilibrium. Levinton, for example (1988, p. 182), holds that "the vast majority of speciation events probably beget no significant change." He then views the consequences as effectively fatal for punctuated equilibrium (1988, p. 211): "The punctuated equilibrium model argues that morphological change is associated with speciation and that species are static during their history due to some internal stabilizing mechanism. There is no evidence coming from living species to support this. If anything, recent research has demonstrated that speciation occurs typically with little or no morphological change; hence the large-scale occurrence of sibling species."

Hoffman (1989, p. 115) invokes this argument to assert the untestability, hence the nonscientific status, of punctuated equilibrium:

> Long-term evolutionary stasis of species, however, simply cannot be tested in the fossil record. Paleontological data consist solely of a small sample of phenotypic traits—little more than morphology of the skeletal parts—which does not allow us to make any inference about changes in a species's genetic pool or even about changes of the frequency distribution of phenotypes in a phyletic lineage. The non-preserved portion of the phenotype of each fossil species is so extensive that it may always undergo considerable evolutionary changes that remain undetectable by the paleontologist. What appears then to the paleontologist as a species in complete evolutionary stasis may in fact represent a succession of fossil species or perhaps a whole cluster of species, a phylogenetic tree with a sizable number of branching points, or speciation events.

While I freely admit all these arguments for underrepresentation of true species in fossil data, I do not comprehend how punctuated equilibrium could be thus rendered untestable, or even seriously compromised (see further arguments in Gould, 1982b and 1989b; and Gould and Eldredge, 1993). I base my argument on two logical and methodological principles, not on the probable empirical record (where I largely agree with our critics).

THE PROPER STUDY OF MACROEVOLUTION. By consensus, and accepting a criterion of testability, science does not include, within its compass of inquiry, fascinating questions that cannot be answered (even if they address potentially empirical subjects). For example, and for the moment at least, we know no way to ask a scientific question about what happened before the big bang, for compression of universal matter to a single point of origin wipes out all traces of any previous history. (Perhaps we will eventually devise a way to obtain such data, or perhaps the big bang theory will be discarded. The question might then become scientifically tractable.) Similarly, we know that many kinds of evolutionary events leave no empirical record—and that we therefore cannot formulate scientific questions about them. (For example, I doubt that we will be able to resolve the origins of human language, unless written expression occurred far earlier than current belief and evidence now indicate.)

The nature of the fossil record leads us to define macroevolution as the study of phenotypic change (and any inferable correlates or sequelae) in lineages and clades throughout geological time. Punctuated equilibrium proposes that such changes generally occur in discrete units or quanta in geological time, and that these quanta represent events of branching speciation.

Thus, we do identify speciation as the source of raw material for macroevolutionary change in lineages. But we do not, and cannot, argue (or attempt to adjudicate at all) the quite different proposition that all speciation events produce measurable quanta of macroevolutionary change. The statement—our proposition—that nearly all macroevolutionary change occurs in increments of speciation carries no implications for the unrelated claim, often imputed to punctuated equilibrium by our critics (but largely irrelevant to our theory), that nearly all events of speciation produce an increment of macroevolutionary change. This conclusion flows from elementary logic, not from empirical science. The argument that all B comes from A does not imply that all A leads to B. All human births (at least before modern interventions of medical technology) derived from acts of sexual intercourse, but all acts of intercourse don't lead to births.

To draw a more relevant analogy: in the strict version of Mayr's peripatric theory of speciation, nearly all new species arise from small populations isolated at the periphery of the parental range. But the vast majority of peripheral isolates never form new species; for they either die out or reamalgamate with the parental population. Similarly, most new species may never be recorded in the fossil record; but, if the theory of punctuated equilibrium holds, when changes do appear in lineages of fossils, speciation provides the source of input in a great majority of cases. Thus, most speciation could be cryptic (and unknowable from fossil evidence), while effectively all macroevolutionary change still arises from the minority of speciation events with phenotypic consequences. Just as peripheral isolates might represent "the only game in town" for forming new species (though few isolates ever speciate), cladogenetic speciation may be "the only game in town" for inputting phenotypic change into macroevolution (though few new species exhibit such change).

THE TREATMENT OF INELUCTABLE NATURAL BIAS IN SCIENCE. In an ideal world—the one we try to construct in controlled laboratory experiments—no systematic bias distorts the relative frequency of potential results. But the real world of nature meets us on her own terms, and we must accept any distortions of actual frequencies that directional biases of recording or preservation inflict upon the archives of our evidence. At best, we may be able to correct such biases if we can make a quantitative estimate of their strength. (This general procedure, for example, has been widely followed to correct the systematic undermeasurement of geological ranges imposed by the evident fact that observed first and last occurrences of a fossil species can only provide a minimal estimate for actual origins and extinctions, for the observed geological range of a species must be shorter (and at least cannot be longer) than the actual duration. Studies of "waiting times" between sequential samples within the observed range, combined with mathematical models for constructing error bars around first and last occurrences, have been widely used to treat this important problem—see Sadler, 1981; Schindel, 1982; Marshall, 1994.)

Often, however, we can specify the direction of a bias, but do not know how to make a quantitative correction. In such cases, the sciences of natural history must follow a cardinal rule: if the direction of bias coincides with the predicted effect of the theory under test, then researchers face a serious, perhaps insurmountable, problem; but if a systematic bias works against a theory, then researchers encounter an acceptable impediment—for if the theory can still be affirmed in the face of unmeasurable biases working *against* a favored explanation, then the case for the theory gains strength.

For proponents of punctuated equilibrium, speciation represents the primary source for morphological changes that, by summation of increments, build trends in the history of lineages. If a systematic bias in the nature of paleontological evidence leads us to underestimate the number of speciation events, and if we can still explain trends by this observed number (necessarily less than the actual frequency), then the case for punctuated equilibrium becomes stronger by affirmation in the face of a bias working against full expression of the theory's effect. Thus, although we regret the existence of any bias that we cannot correct, a systematic underrepresentation of speciation events does not subvert punctuated equilibrium because such a natural skewing of evidence makes the hypothesis even more difficult to affirm—and support for punctuated equilibrium therefore emerges in a context even more challenging than the unbiased world of controlled experimentation.

Moreover, one might even stress the bright side and recognize that such biases may exist for interesting reasons in themselves—reasons that might even enhance the importance of punctuated equilibrium and its implications. I doubt that Levinton (1988, p. 379) intended the following passage in such a positive light, but I would suggest such a reading: "One cannot rule out the possibility that speciation is rampant, but morphological evolution only occurs occasionally when a population is forced into a marginal environment and subjected to rapid directional selection. What then becomes interesting is why the character complexes evolved in the daughter species remain constant. This is, again, the issue of stasis, which I believe to be the legitimate problem spawned by the punctuated equilibrium model."

Finally, I am not sure that fossil species do strongly and generally underestimate the frequency of true biospeciation—although I do accept that a bias, if present at all, probably operates in this direction. The most rigorous empirical studies on correspondence between well-defined paleospecies and true biospecies—the works of Michaux and of Jackson and Cheetham discussed above—affirm a one-to-one link between paleontological morphospecies and extant, genetically defined biospecies.

REASONS FOR A POTENTIAL SYSTEMATIC OVERESTIMATION OF BIOSPECIES BY PALEOSPECIES

If a bias did exist in this opposite direction, the consequences for punctuated equilibrium would be troubling (as implied in the previous section on acceptable and unacceptable forms of unavoidable natural biasing). For if we systematically name too many species by paleontological criteria, then we might be affirming punctuated equilibrium by skewing data in the direction of our favored theory, rather than by genuine evidence from the fossil record. However, I doubt that such a problem exists for punctuated equilibrium, especially since all experts—both strong advocates and fierce critics alike (as the preceding discussion documented)—seem to agree that if any systematic bias exists, the probable direction lies in the acceptable opposite claim for underestimation of biospecies by paleospecies.

I don't doubt, of course, that past taxonomic practice, often favoring the erection of a species name for every recognizable morphological variant (even for odd individuals rather than populations), has greatly inflated the roster of legitimate names in many cases, particularly for fossil groups last monographed several generations ago. (Our literature even recognizes the half-facetious term "monographic burst" for peaks of diversity thus artificially created. But this problem of past oversplitting cannot be construed as either uniquely or even especially paleontological, for neontological systematics then followed the same practices as well.) The grossly uneven, and often greatly oversplit, construction of species-level taxonomy in paleontology has acted as a strong impediment for the entire research program of the prominent school of "taxon-counting" (Raup, 1975, 1985). For this reason, the genus has traditionally been regarded as the lowest unit of rough comparability in paleontological data (see Newell, 1949). Sepkoski (1982) therefore compiled his two great compendia—the basis for so much research in the history of life's fluctuating diversity—at the family, and then at the genus, level (but explicitly not at the species level in recognition of frequent oversplitting and extreme imbalance in practice of research among specialists on various groups).

Although this problem has proved far more serious for taxon-counters than for proponents of punctuated equilibrium, a potential bias towards overrepresentation also poses a threat for our theory, as Levinton (1988, p. 364) rightly recognizes: "The problem is not very new. Meyer (1878) claimed that the ability to recognize gradual evolutionary change in *Micraster* [a famous sequence of Cretaceous echinoids] was obscured by the rampant naming of separate species by previous taxonomists."

This issue would cause me serious concern—for the claim of overestimation does, after all, fall into the worrisome category of biases favoring a preferred hypothesis under test—if two arguments and realities did not obviate the danger. First, if supporters of punctuated equilibrium did try to affirm their hypothesis by using names recorded in the literature as primary data for judging the strength and effect of speciation upon evolutionary trends, then we would face a serious difficulty. But I cannot think of any study that utilized this invalid approach—for paleontologists recognize and generally avoid the dangers of this well-known directional bias. Punctuated equilibrium, to my knowledge, has never been defended by taxon counting at the species level. All confirmatory studies employ measured morphometric patterns, not the geological ranges of names recorded in literature.

Second, as stated above, all students of this subject seem to agree that if a systematic bias exists in relative numbers of paleospecies and biospecies, fossil data should be skewed in the opposite direction of recognizing *fewer* paleospecies than biospecies—an acceptable bias operating against the confirmation of punctuated equilibrium.

REASONS WHY AN OBSERVED PUNCTUATIONAL PATTERN MIGHT NOT REPRESENT SPECIATION

Suppose that we have empirical evidence for a punctuational event separating two distinct morphological packages regarded as both different enough to be designated as separate paleospecies by any standard criterion, and also genealogically close enough to support a hypothesis of direct ancestry and descent. What more do we need? Does this situation not affirm punctuated equilibrium *ipso facto?*

But critics charge (and I must agree) that such evidence cannot be persuasive by itself, because punctuated equilibrium explicitly links punctuational patterns to events of branching speciation. Therefore, recorded punctuations produced for other reasons do not affirm punctuated equilibrium—and may even challenge the theory if their frequency be high and, especially, if they cannot be distinguished in principle (or frequently enough in practice) from events of cladogenetic branching.

Punctuational patterns often originate (at all scales in evolutionary hierarchies of levels and times) for reasons other than geologically instantaneous speciation—and I welcome such evidence as an affirmation of pervasive importance (see p. 229 *et seq.*) for a general style of nongradualistic change, with punctuated equilibrium as its usual mode of expression at the speciational scale under consideration here. But testable, and generally applicable, criteria have been formulated for distinguishing punctuated equilibrium from other reasons for punctuational patterns—and available evidence amply confirms the importance and high relative frequency of punctuated equilibrium.

Of the two major reasons for punctuational patterns not due to speciation, Darwin's own classic argument of imperfection—geological gradualism that appears punctuational because most steps of a continuum have not been preserved in the fossil record—retains pride of place by venerable ancestry. I have already presented my reasons for regarding this argument as inconsequential (see pp. 38–49). I do not, of course, deny that many (or most) breaks in geological sequences only reflect missing evidence. But proponents of punctuated equilibrium do not base their claims on such inadequate examples that cannot be decided in either direction. The test cases of our best literature—whether their outcomes be punctuational or gradualistic—have been generated from stratigraphic situations where temporal resolution and density of sampling can make appropriate distinctions by recorded evidence, not conjectures about missing data.

The second reason has been highlighted by some critics, but unfairly I think, because punctuated equilibrium has always recognized the argument and has, moreover, enunciated and explicitly tested proper criteria for making the necessary distinctions. To state the supposed problem: what can we conclude when we document a truly punctuational sequence that cannot be attributed to imperfections of the fossil record? How do we know that such a pattern records an event of branching speciation, as the theory of punctuated equilibrium requires? When ancestral Species A abruptly yields to descendant Species B in a vertical sequence of strata, we may only be witnessing an anagenetic transformation through a population bottleneck, or perhaps an event of migration, where Species B, having evolved gradualistically from Species A in another region, invades the geographic range, and abruptly wipes out its ancestor.

But an appropriate and non-arbitrary criterion exists—and has been fully enunciated, featured as crucial, and subjected to frequent test, from the early days of punctuated equilibrium. We can distinguish the punctuations of rapid anagenesis from those of branching speciation by invoking the eminently testable criterion of ancestral survival following the origin of a descendant species. If the ancestor survives, then the new species has arisen by branching. If the ancestor does not survive, then we must count the case either as indecisive, or as good evidence for rapid anagenesis—but, in any instance, certainly not as evidence for punctuated equilibrium.

Moreover, by using this criterion, we obey the methodological requirement that existing biases must work against a theory under test. When ancestors do not survive following the first appearance of descendants, the pattern may still be recording an event of branching speciation—hence affirmation for punctuated equilibrium. But we cannot count such cases in our favor, for the plausible alternative of rapid anagenesis cannot be disproven. By restricting affirmations to cases where ancestors demonstrably survive, we accept only a subset of events actually caused by speciation. Thus, we underestimate the frequency of punctuated equilibrium—as we must do in the face of an unresolvable bias affecting a hypothesis under test.

In our first papers, we did not recognize or articulate the importance of tabulating cases of ancestral survival following punctuational origin of a descendant as a criterion for distinguishing punctuated equilibrium from other forms of punctuational change. (Both of our original examples in Eldredge and Gould, 1972, did feature—and prominently discuss—ancestral survival as an important aspect of the total pattern. We had a proper "gut feeling" about best cases, but we did not formalize the criterion.) But, beginning in 1982, and continuing thereafter, we have stressed the centrality of this criterion in claims for speciation as the mechanism of punctuated equilibrium. Contrasting the difference in paleontological expression between Wright's shifting balance and punctuated equilibrium by speciation, for example, I wrote (Gould, 1982b, p. 100): "Since punctuational events can occur in the phyletic mode under shifting balance, but by branching speciation under punctuated equilibrium, the persistence of ancestors following the abrupt appearance of a descendant is the surest sign of punctuated equilibrium."

This criterion has been actively applied, in an increasingly routine manner (as researchers recognize its importance), in the expanding literature on empirical study of evolutionary tempos and modes in well-documented fossil sequences. Cases of probable anagenetic transformation have been documented (no ancestral survival when good stratigraphic resolution should have recorded such persistence, had it occurred), especially in planktic marine Foraminifera, where long oceanic cores often provide unusually complete evidence (Banner and Lowry, 1985; Malmgren and Kennett, 1981, who coined the appropriate term "punctuated anagenesis" for this phenomenon).

However, abundant cases of ancestral survival, and consequent punctuational origin of descendant taxa by branching speciation, have also been affirmed as illustrations of punctuated equilibrium. These examples span the gamut of taxonomies and ecologies, ranging from marine microfossils (Cronin, 1985, on ostracodes); to "standard" macroscopic marine invertebrates (with Cheetham's famous studies of bryozoans, 1986 and 1987, as classic and multiply documented examples), to freshwater invertebrates (Williamson's 1981 work on multiple events of speciation in African lake mollusks, where ancestral species reinvade upon coalescence of lakes following periods of isolation that provided conditions for speciation); to terrestrial vertebrates (Flynn, 1986, on rodents; Prothero and Shubin, 1989, on horses). I shall discuss this important issue in more detail in Chapter 4, which presents evidence for punctuated equilibrium, but I have been particularly (if parochially) gratified by the increasing application of punctuated equilibrium to the resolution of hominid phylogeny. The criterion of ancestral survival has been prominently featured in this literature, as by McHenry (1994), who notes that "ancestral species overlap in time with descendants in most cases in hominid evolution, which is not what would be expected from gradual transformations by anagenesis."

In any case, punctuated equilibrium can be adequately and generally recognized by firm evidence linking observed punctuational patterns to branching speciation as a cause. The theory of punctuated equilibrium is eminently testable and has, indeed, passed such trials in cases now so numerous that a high relative frequency for this important evolutionary phenomenon can no longer be denied (see Gould and Eldredge, 1993).

Critiques Based on Denying Events of Speciation as the Primary Locus of Change

Once we overcome the problem of definability for species in the fossil record, punctuated equilibrium still faces a major issue rooted in the crucial subject of speciation. Punctuated equilibrium affirms, as a primary statement, that ordinary biological speciation, when properly scaled into geological time, produces the characteristic punctuational pattern of our fossil record. We must therefore be able to defend the central implication that morphological change should be preferentially associated with events of branching speciation. Our critics have strongly argued that such a proposition cannot be justified by our best understanding of evolutionary processes and mechanisms.

I believe that our critics have been correct in this argument, and that Eldredge and I made a major error by advocating, in the original formulation of our theory, a direct acceleration of evolutionary rate by the processes of speciation. This claim, I now think, represents one of the two most important errors that we committed in advocating punctuated equilibrium during the past 25 years. (The other error, as discussed and corrected in *SET,* pp. 670–673, lay first in our failure to recognize the phenomenon of species selection as distinct (by hierarchical reasoning) from classical Darwinian organismic selection, and then (see Gould and Eldredge, 1977) in our decision to advocate an overly broad and purely descriptive definition rather than a properly limited meaning based on emergent characters or fitnesses—see *SET,* pp. 656–670.)

We did not urge this correlation between speciation events and morphological change in a self-serving and circular manner—i.e., only because the pattern of punctuated equilibrium could be best defended thereby. We did, of course, recognize the logical link, as in the following statement from Gould, 1982b, p. 87 (see also Gould and Eldredge, 1977, p. 137): "Reproductive isolation and the morphological gaps that define species for paleontologists are not equivalent. Punctuated equilibrium requires either that most morphological change arise in coincidence with speciation itself, or that the morphological adaptations made possible by reproductive isolation arise rapidly thereafter." But we based our defense of this proposition upon a large, and then quite standard, literature advocating a strong negative correlation between capacity for rapid evolutionary change and population size. Small populations, under these models, maintained maximal prospects for rapid transformation based on several factors, including potentially rapid fixation of favorable variants, and enhancement of differences from ancestral populations by interaction of intense selection with stochastic reasons for change (particularly the founder effect) that can only occur with such effective speed in small populations. Large and stable populations, by the converse of these arguments, should be sluggish and resistant to change.

This literature culminated in Mayr's spirited defense for "genetic revolution" as a common component of speciation (first proposed in a famous 1954 article, and then defended *in extenso* in the 1963 book that served as the closest analog to a "bible" for graduate students of my generation). Since Mayr (who coined the name "founder effect" in this context) also linked his concept of "genetic revolution" to the small, peripherally isolated populations that served as "incipient species" in his influential theory of peripatric speciation—and since we had invoked this theory in our original formulation of punctuated equilibrium (Eldredge and Gould, 1972)—our defense of a link between speciation and concentrated episodes of genetic (and phenotypic) change flowed logically from the evolutionary views we had embraced. Thus, we correlated punctuations with the extensive changes that often occurred during events of speciation in small, peripherally isolated populations; and we linked stasis with the expected stability of large and successful populations following their more volatile and punctuational origins as small isolates.

I can claim no expertise in this aspect of neontological evolutionary theory, but I certainly acknowledge, and must therefore provisionally accept, the revised consensus of the past twenty years that has challenged this body of thought, and rejected any general rationale for equating the bulk of evolutionary change with events of speciation in small populations, or with small populations in any sense. As I read the current literature, most evolutionists now view large populations as equally prone to evolutionary transformation, and also find no reason to equate times of speciation—the attainment of reproductive isolation—with acceleration in general rates of genetic or phenotypic change (see, for example, Ridley, 1993; and Williams, 1992). (I do, however, continue to wonder whether the Mayrian viewpoint might still hold some validity, and might now be subject to overly curt and confident dismissal.)

This situation creates a paradox for our theory. The pattern of punctuated equilibrium has been well documented and shown to predominate in many situations (see Chapter 4), but its most obvious theoretical rationale has now fallen under strong skepticism. So either punctuated equilibrium is wrong—a proposition that this partisan views as unlikely (although obviously possible), especially in the face of such strong documentation—or we must identify another reason for the prominence of punctuated equilibrium as a pattern in the history of life. In our article on the "majority" (21st birthday!) of punctuated equilibrium, Eldredge and I expressed this dilemma in the following manner (Gould and Eldredge, 1993, p. 226): "The pattern of punctuated equilibrium exists (at predominant relative frequency, we would argue) and is robust. *Eppur non si muove;* but why then? For the association of morphological change with speciation remains as a major pattern in the fossil record." (Our Italian parody, missed by many readers of the original article, alters Galileo's famous, but almost surely legendary, rebuke to the Inquisition, delivered secretly and *sotto voce* after he had been forced to recant his Copernican views in public: *Eppur si muove*—nevertheless it does move. Our parody says "nevertheless it does *not* move"—a reference to the overwhelming evidence for predominant stasis in the history of species, even if our original evolutionary rationale, based on population size, must be reassessed.)

This paradox permits several approaches, including the following two that I would not favor. One might simply argue that the pattern of punctuated equilibrium demonstrably exists, so the task falls to evolutionary theorists to find a proper explanation. The current absence of a satisfactory account does not threaten the empirical record, but rather directs inquiry by posing a problem. Or one might doubt that any single explanation can render the phenomenon, and suspect that many rationales will yield the observed pattern (including Mayrian genetic revolutions, even if we now regard their relative frequency as low). Thus, we need to identify a set of enabling criteria from evolutionary theory, and then argue that their combination may render the observed phenomena of the fossil record.

Most researchers would regard a third approach as preferable in science: an alternate general explanation of different form from the previous, but now rejected, leading candidate. I believe that such a resolution has been provided by Douglas Futuyma (1986, 1988a and b, but especially 1987),[1] although his simple, yet profound, argument has not infused the consciousness of evolutionists because the implied and required hierarchical style of thinking remains so unfamiliar and elusive to most of us. (In fact, and with some shame, I am chagrined that I never recognized this evident and elegant resolution myself. After all, I am supposedly steeped in this alternative hierarchical mode of thinking—and I certainly have a strong stake in the problems of punctuated equilibrium.)

In short, Futuyma argues that we have been running on the wrong track, and thinking at the wrong level, in trying to locate the reason for a correlation between paleontological punctuations and events of speciation in a direct mechanism of accelerated change promoted by the process of speciation itself. Yet Futuyma does agree that a strong correlation exists (and has been demonstrated, in large part by research and literature generated by debate about punctuated equilibrium). Since we all understand (but do not always put into practice!) the important logical principle that correlation does not imply causality (the *post hoc* fallacy), an acknowledgement of the genuine link doesn't commit us to any particular causal scheme—especially, in this case, to the apparently false claim that mechanisms of speciation inherently enhance evolutionary rates.

Futuyma begins by arguing that morphological change may accumulate anywhere along the temporal trajectory of a species, and not exclusively (or even preferentially) during the geological moment of its origin. What then could produce such a strong correlation between events of branching speciation and morphological change from an ancestral phenotype to the subsequent stasis of an altered descendant? Futuyma—and I am somewhat rephrasing and extending his argument here—draws an insightful and original analogy between macroevolution and the conventional Darwinism of natural selection in populations.

The operation of natural selection requires that Darwinian individuals interact with environments in such a manner that distinct features of these individuals bias their reproductive success relative to others in the population. As a defining criterion of Darwinian individuality, entities that interact with the environment must show "sufficient stability" (see discussion in *SET,* pp. 611–613)—defined in terms of the theory and mechanism under discussion as enough coherence to perform as an interactor in the process of natural selection.

Darwin recognized that organisms operate as fundamental interactors for microevolution within populations. (Gene selectionists make a crucial error in arguing that sexual organisms are not stable enough to be regarded as units of selection because they must disaggregate in forming the next generation. But units of selection are interactors, and the "sufficient stability" required by the theory only demands persistence through one episode [generational at this level] of selective interaction to bias reproductive success—as organisms do in the classical Darwinian "struggle for existence," see full discussion in *SET*, pp. 619–625.) Organisms achieve this stability through ordinary mechanisms of bodily coherence (a protective skin, functional integration of parts, a regulated developmental program, etc.).

What, then, produces a corresponding stability for units of macroevolution? Species-individuals are constructed as complex units, composed of numerous local populations, each potentially separate (at any moment) due to limited gene flow, and each capable of adaptation to unique and immediate environments. Thus, in principle, substantial evolution can occur in any local population at any time during the geological trajectory of a species. A large and developing literature, much beloved by popular sources (media and textbooks) for illustrating the efficacy of evolution in the flesh of immediacy (that is, within a time frame viscerally understood by human beings), has documented these rapid and adaptive changes in isolated local populations—substantial evolution of body size in guppies (Reznick et al., 1997), or of leg length in anolid lizards (Losos et al., 1997), for example (see Gould, 1997c).

But these changes in local populations cannot gain any sustained macroevolutionary expression unless they become "locked up" in a Darwinian individual with sufficient stability to act as a unit of selection in geological time. Local populations—as a primary feature of their definition—do not maintain such coherence. They can in principle—and do, in the fullness of geological time, almost invariably in practice—interbreed with other local populations of their species. The distinctively evolved adaptations of local populations must therefore be ephemeral in geological terms, unless these features can be stabilized by individuation—that is, by protection against amalgamation with other Darwinian individuals. Speciation—as the core of its macroevolutionary meaning—provides such individuation by "locking up" evolved changes in reproductively isolated populations that can, thereafter, no longer amalgamate with others. The Darwinian individuation of organisms occurs by bodily coherence for structural and functional reasons. The Darwinian individuation of species occurs by reproductive coherence among parts (organisms), and by prevention of intermingling between these parts and the parts of other macroevolutionary individuals (that is, organisms of other species).

Rapid evolution in local population of guppies and anoles illustrates a fascinating phenomenon that teaches us many important lessons about the general process of evolution. But such changes can only be ephemeral unless they then become stabilized in coherent higher-level Darwinian individuals with sufficient stability to participate in macroevolutionary selection. These local populations usually strut and fret their short hour on the geological stage, and then disappear by death or amalgamation. They produce the ubiquitous and geologically momentary fluctuations that characterize and embellish the long-term stasis of species. They are, to use Mandelbrot's famous metaphor for fractals, the squiggles and jiggles on the coastline of Maine depicted at a scale that measures the distance around every boulder on every beach along the shore, and not at the resolution properly enjoined when the entire state appears on a single page in an atlas. Macroevolution represents the page of the atlas. The distance around each boulder (marking substantial but ephemeral changes in local populations of guppies and lizards)—however important in the immediacy of an ecological moment—becomes invisible and irrelevant (as the transient fluctuations of stasis) in the domain of sustained macroevolutionary change (Fig. 3-2).

In other words, morphological change correlates so strongly with speciation not because cladogenesis accelerates evolutionary rates, but rather

3-2. Stasis does not imply absolute stability, but rather directionless fluctuation that generally does not stray beyond the boundaries of geographic variation within similar species and, particularly, does not trend in any given direction, especially towards the modal morphology of descendant forms. This figure shows that, when a small segment in geological stasis becomes magnified so that change may be visualized on a generational scale, the natural fluctuations within local populations become more visible—but still do not, at the proper geological focus, exceed the bounds of stasis within the species.

because such changes, which can occur at any time in the life of a local population, cannot be retained (and sufficiently stabilized to participate in selection) without the protection provided by individuation—and speciation, via reproductive isolation, represents nature's preeminent mechanism for generating macroevolutionary individuals. Speciation does not necessarily promote evolutionary change; rather, speciation "gathers in" and guards evolutionary change by locking and stabilization for sufficient geological time within a Darwinian individual of the appropriate scale. If a change in a local population does not gain such protection, it becomes—to borrow Dawkins's metaphor at a macroevolutionary scale—a transient duststorm in the desert of time, a passing cloud without borders, integrity, or even the capacity to act as a unit of selection, in the panorama of life's phylogeny.

To cite Futuyma's summary of his powerful idea (1987, p. 465): "I propose that because the spatial locations of habitats shift in time, extinction of and interbreeding among local populations makes much of the geographic differentiation of populations ephemeral, whereas reproductive isolation confers sufficient permanence on morphological changes for them to be discerned in the fossil record." Futuyma directly follows this statement with the key implication of punctuated equilibrium for the explanation of evolutionary trends: "Long-term anagenetic change in some characters is then the consequence of a succession of speciation events."

Later in his article, Futuyma (p. 467) explicitly links speciation with sufficient stability (individuation) for macroevolutionary expression: "In the absence of reproductive isolation, differentiation is broken down by recombination. Given reproductive isolation, however, a species can retain its distinctive complex of characters as its spatial distribution changes along with that of its habitat or niche . . . Although speciation does not accelerate evolution within populations, it provides morphological changes with enough permanence to be registered in the fossil record. Thus, it is plausible to expect many evolutionary changes in the fossil record to be associated with speciation." And, at the end of his article, Futuyma (p. 470) notes the crucial link between punctuated equilibrium and the possibility of sustained evolutionary trends: "Each step has had a more than ephemeral existence only because reproductive isolation prevented the slippage consequent on interbreeding with other populations . . . Speciation may facilitate anagenesis by retaining, stepwise, the advances made in any one direction . . . Successive speciation events are the pitons affixed to the slopes of an adaptive peak."

I hope that Futuyma's simple yet profound insight may help to heal the remaining rifts, thereby promoting the integration of punctuated equilibrium into an evolutionary theory hierarchically enriched in its light.

Critiques Based on Supposed Failures of Empirical Results to Affirm Predictions of Punctuated Equilibrium

I shall treat the specifics of this topic primarily in the next section on "the data of punctuated equilibrium." But the logic of this chapter's development also requires that I state the major arguments and my responses in this account of principal critiques directed at the theory—for the totality of attempted rebuttals has not only posited theoretical objections in an effort to undermine the theory's logic or testability (as discussed in the first two parts of this section), but has also proceeded by accepting the theory's program of research as valid, and then arguing that the bulk of data thus accumulated refutes punctuated equilibrium empirically. I shall summarize discussion on the two major strategies pursued under this rubric: refutation by accumulation of important cases, and rejection by failure of actual data to fit models for predicted phylogenetic patterns.

CLAIMS FOR EMPIRICAL REFUTATION BY CASES

Phenotypes

Despite some early misunderstandings, long since resolved by all parties to the discussion, we recognize that no individual case for or against punctuated equilibrium, however elegantly documented, can serve as a "crucial experiment" for questions in natural history that must be decided by relative frequencies. No exquisite case of punctuated equilibrium—and many have been documented—can "prove" our theory; while no beautiful example of gradualism—and such have been discovered as well—can refute us. The key question has never been "whether," but rather "how often," "with what range of variation in what circumstances of time, taxon, and environment," and especially, "to what degree of control over patterns in phylogeny?" A single good case can only validate the reality of the phenomenon—and the simple claim for existence has not, surely, been an issue for more than 20 years. Similarly, an opposite case of gradualism can only prove that punctuated equilibrium lacks universal validity, and neither we nor anyone else ever made such a foolish and vainglorious claim in the first place. The empirical debate about punctuated equilibrium has always, and properly, focused upon issues of relative frequency.

I shall present the empirical arguments for asserting dominant relative frequency, rather than mere occurrence, for punctuated equilibrium on pages 147–171. If we ask, by contrast, whether strong evidence for predominant gradualism has been asserted for any major taxon, time or environment, one case stands out as a potentially general refutation of punctuated equilibrium in one important domain at least: the claim for anagenetic gradualism as a primary phylogenetic pattern in the evolution of Cenozoic planktonic Foraminifera.

This case gains potential power and generality from the unusually favorable stratigraphic context, and the consequent nature of sampling, in such studies. The data come from deep oceanic cores, with stratigraphic records presumably unmatched in general completeness, for these environments receive a continuous supply of sediment (including foraminiferal tests) from the water column above. Moreover, these microscopic organisms can usually be extracted in large and closely spaced samples (sieved from disaggregated sediments), even from the restricted volume of a single oceanic core. Thus, forams in oceanic cores should provide our most consistently satisfactory information—in terms of large samples with good stratigraphic resolution—for the study of phylogenetic pattern. If gradualistic anagenesis prevails in such situations of maximal information—even if punctuated equilibrium predominates in the conventional fossil record of marine invertebrates from shallow water sediments—shouldn't we then conclude that Darwin's old argument must be valid after all; that punctuational patterns represent an artefact of missing data; and that more complete information will affirm genuine gradualism as the characteristic signal of phylogeny?

I acknowledge the highest relative frequency of recorded gradualism for foraminiferal data of this type, and I also admire the procedural rigor and informational richness in several of these studies. But I do not regard this case as a general argument against punctuated equilibrium—and neither, I think, do most of my paleontological colleagues, whatever their overall opinion about our theory, for the following reasons based upon well-known features of the fossil record in general, and the biology of forams in particular.

1. As emphasized in my previous discussion of publication bias (see p. 36), I remain unconvinced that a predominant relative frequency for gradualism—as opposed to genuine documentation of several convincing

cases—has been established, even for this maximally promising taxon. No one has ever compiled an adequately random, or even an adequately numerous, sample of planktonic species drawn from the entire clade. Gradualistic lineages have been highlighted for study as a consequence of their greater "interest" under conventional views, while putatively stable lineages have tended to remain in unexamined limbo as supposedly uninformative, or even dull. Thus, the fact that gradualism prevails in a high percentage of published studies tells us little about the relative frequency of gradualism in the clade as a whole.

A telling analogy may be drawn with a crucial episode in the history of genetics. With classical techniques based on the Mendelian analysis of pedigrees, only variable genes could be identified. (If every *Drosophila* individual had red eyes, earlier researchers could legitimately assume some genetic bases for the invariance, but no genes could be specified because traits could not be traced through pedigrees. But once a white-eyed mutant fly appeared in the population, geneticists gained a necessary tool for identifying relevant genes by crossbreeding the two forms and tracing the alternate phenotypes through successive generations. In other words, genes had to vary before they could be specified at all.)

Therefore, under these methodological constraints (which prevailed during most of the 20th century history of genetics), a dominant *measured* frequency for variable genes taught us nothing about the *actual* frequency of variable genes across an entire genome—for we knew no way to generate a random or unbiased sample by selecting genes for study *prior* to any knowledge about whether or not they varied. The fact of variation in all *known* genes only recorded a methodological limitation that precluded the identification of nonvariable genes.

I don't, of course, claim that methodological strictures on paleontological lineages have ever been so strong—that is, we could always have selected stable lineages for study, had we chosen to do so. But, in practice, I'm not sure that the actual procedural bias has operated with much less force in paleontology than in genetics, so long as researchers confined their attention to lineages that appeared (by initial qualitative impression) to evolve by gradual anagenesis. Just as all known genes might be variable (while variable genes actually represent only a few percent of the total complement, because the remaining 95 percent of invariant genes could not be recognized at all), most studied species might illustrate gradual trends (while gradualistic species represent a small minority of all lineages because no one chooses to study stable species).

Genetics resolved this problem by inventing techniques—with electrophoresis as the first and historically most important—for identifying genes prior to any knowledge about whether or not they varied. This methodological advance permitted the resolution of several old and troubling questions, most notably the calculation of average genetic differences among human races. This central problem of early Mendelian genetics could never be addressed—even to counter the worst abuses of biological determinism and social Darwinism—because biologists could not generate random samples of genes, and could therefore only overestimate average distances by ignoring the unknowable invariant genes among races, while studying the (potentially small) fraction capable of recording differences among groups. With electrophoretic techniques, and the attendant generation of a random sample with respect to potential variability, geneticists soon calculated the average genetic differences among races as remarkably small and insignificant—a conclusion of no mean practical importance in a xenophobic world. Similarly, a truly random sample (with respect to the distribution of anagenetic rates) might show a predominance for stasis, even if previous studies (with their strong bias for preselection of variable species) had generally affirmed gradualism.

I am encouraged to accept the probable validity of this argument by the important study of Wagner and Erwin (1995), who used the different and comprehensive technique of compiling full cladograms for two prominent Neogene families of planktonic forams: Globigerinidae and Globorotaliidae. In applying a set of methods for inferring probable evolutionary mode from cladistic topology (see full discussion and details on pp. 106–108), they found that, in both families, branching speciation in the mode favored by punctuated equilibrium (divergence of descendants with survival of ancestors in stasis) vastly predominated over the origin of new species by anagenetic transformation. Thus, the literature's apparent preference for anagenesis in tabulated studies of individual lineages may only record an artifact of biased selection in material for research.

2. Even if gradualism truly does prevail in planktonic forams, we could not infer that the observed predominance of punctuated equilibrium in marine Metazoa must therefore reflect the artifact of an imperfect geological record. The difference might record a characteristic disparity between the

taxa, not a general distinction in quality of geological evidence between deep oceanic cores and conventional continental sequences—a proposition defended in the third argument, just below. The deep oceanic record may *usually* be more complete, but the subset of best cases from conventional sequences surely matches the foram data in quality—and convincing studies of punctuated equilibrium and gradualism generally use these best records. Thus, the subset of most adequate metazoan examples should match, in quality of evidence, the usual records of forams from oceanic cores.

3. A third argument completes the trio of logical possibilities (all partially valid, I suspect, though I would grant most weight to this third point) for denying that a currently recorded maximal frequency of gradualism for planktonic foraminiferal lineages casts doubt on the general importance of punctuated equilibrium in evolution. The first argument attributes an apparent high frequency to biased sampling in the preselection, for rigorous study, of lineages already highlighted by taxonomic experts for suspected gradual change. The second and third arguments, on the other hand, hold that if high frequency truly characterizes this group, no general rebuttal of punctuated equilibrium follows thereby. The second argument denies the common assumption that high-frequency records uniquely complete geological evidence—and that gradualism will therefore prevail whenever the fossil record becomes good enough to preserve its true domination (with a high frequency for punctuated equilibrium then construed, by Darwin's original argument, as the artifact of a gappy record). This second argument maintains that, while foram data may be more complete on average, the best metazoan examples of punctuated equilibrium have been validated with excellent samples from admittedly rarer but equally complete geological sequences, thus precluding the explanation of punctuated equilibrium as artifactual.

The third argument also grants the reality of higher relative frequency for gradualism in forams, but argues against extrapolation to larger multicellular organisms on grounds of genuine difference in evolutionary mode, based on important biological distinctions between these single-celled creatures of the oceanic plankton and sexually reproducing metazoan species that, however parochially, have served as the basis for most of our evolutionary theory and, in any case, form the bulk of the known fossil record.

This third argument should not be viewed as special pleading by partisans, but as a positive opportunity for developing hypotheses about the importance (or insignificance) of punctuated equilibrium based on the correlation between differences in frequency and distinctive biological properties of various taxonomic groups—particularly in features related to speciation, the presumed evolutionary basis of punctuated equilibrium. Planktonic forams, with their asexuality, their small size and rapid turnover of generations, their unicellularity, their vast populations, and their geographic links to water masses, display maximal difference from most metazoans, and may therefore be especially suited for helping us to understand, by contrast, the prevailing mechanisms of evolution in multicellular and sexually reproducing organisms. The general nature of these differences does indeed point to a set of factors tied to the definition and division of populations, therefore granting plausibility to the claim that so-called "species" of planktonic forams should show more gradualism than metazoan taxa, while punctuated equilibrium may prevail in sexually reproducing multicellular species. The subject deserves much more attention and rigor, but to sketch a few suggested factors:

(1) Population characteristics. We conventionally name Linnaean species of asexual protistans, but even if adequately stable "packages" of form or genetic distinctness exist in sufficiently extended domains of space and time to merit a vernacular designation as "populations," what comparison do such entities bear with species of sexually reproducing multicellular organisms? (Needless to say, I raise no new issue here, but only recycle the perennial question of "the species problem" in asexual organisms.) Punctuated equilibrium posits a link of observed evolutionary rates to properties of branching speciation in populations. I don't even know how to think about such issues in planktonic forams, where vast populations may be coextensive with entire oceanic water masses, and where numbers must run into untold billions of organisms for every tiny subsection of a geographic range. How do new populations become isolated? How do favorable (or, for that matter, neutral) traits ever spread through populations so extensive in both space and number?

(2) Morphology and definition. If metazoan stasis can be attributed, at least in part, to developmental buffering, what (if any) corresponding phenomenon can keep the phenotypes of simple unicells stable? Perhaps foraminiferal phenotypes manifest substantial plasticity for shaping by forces of temperature, salinity, etc., in surrounding water masses (see Greiner, 1974)—as D'Arcy Thompson (1917, 1942) proposed for most of nature in his wonderfully iconoclastic classic, *On Growth and Form*—see pp. 1179–1208. (Thompson's claim that physical forces shape organisms directly holds

limited validity for complex and internally buffered multicellular forms, but his views may not be so implausible for several features of simpler unicells.) Could many examples of foraminiferal gradualism (compared with metazoan stasis in similar circumstances) reflect the plasticity of these protists in the face of gradual changes in the physical properties of enveloping oceanic water masses through time? If so, such gradual trends would not be recording evolutionary change in the usual genetic sense.

(3) Most interestingly (as a potential illustration of the main theoretical concern of this book), we must consider the potential for strongly allometric scaling of effects from a defined locus of change to other levels of an evolutionary hierarchy. To reiterate a claim that runs, almost like a mantra, throughout this text: punctuated equilibrium is a particular theory about a definite level of organization at a specified scale of time: the origin and deployment of species in geological perspective. The punctuational character of such change does not imply—and may even, in certain extrapolations to other scales, explicitly deny—a pervasive punctuational style for all change at any level or scale. In particular, punctuated equilibrium posits that tolerably gradual trends in the overall history of phenotypes within major lineages and clades (including such traditional tales as augmenting body size in hominids, increasing sutural complexity in ammonoids, or symmetry of the cup in crinoids) should reveal a punctuational fine texture when placed "under the microscope" of dissection to visualize the individual (speciational) "building blocks" of the totality—what we have long called the "climbing up a staircase" rather than the "rolling a ball up an inclined plane" model of fine structure for trends.

Similarly, in asking about evolutionary causality under selective models (see Chapter 8 of *SET*), we need to identify the primary locus of Darwinian individuality for the causal agents of any particular process—for only properly defined Darwinian individuals can operate as "interactors" in a selective process: that is, can interact with environments in such a way that their own genetic material becomes plurified in future generations because certain distinctive properties confer emergent fitness upon the individual in its "struggle for existence" (see *SET*, pp. 656–667). Punctuated equilibrium maintains that species, as well-defined Darwinian individuals, hold this causal status as irreducible components, or "atoms," of evolutionary trends in clades. The apparatus of punctuated equilibrium then explains why trends, when necessarily described as speciational, display a punctuated pattern at geological scales (as expressed in the theory's basic components of stasis and geologically abrupt appearance). In a larger sense, punctuational accounts of trends propose a similar allometric model for any relevant scale—that is, any microscope placed over higher-level smoothness may reveal an underlying "stair-step" pattern among constituent causal individuals acting as Darwinian agents of the trend.

In sexually reproducing metazoa, species clearly play this role as causal individuals (see Chapter 8 of *SET*). The theoretical validity of punctuated equilibrium depends upon such a claim and model. But when we turn to such asexually reproducing unicells as planktonic forams, designated "species" cannot be construed as proper Darwinian individuals, and therefore cannot be primary causal agents (or interactors) in evolutionary trends. To locate the proper agent, the legitimate analog of the metazoan species, we must move "down" a level to the clone—to what Janzen (1977), in a seminal paper, called the EI, or "evolutionary individual."

When we execute this conceptual downshift in levels to locate the focal evolutionary individual in asexual and unicellular lineages, we recognize that the foram "species" acts as an analog to the metazoan lineage or clade, not to the metazoan species. The foram "species" represents a temporal collectivity whose evolutionary pattern arises as a summed history of the Darwinian individuals—clones in this case—acting as primary causal agents.

We can now shift the entire causal apparatus one level down to posit a different locus of punctuational change in planktonic forams. Just as the punctuational history of species generates smooth trends in the collectivity of a lineage or clade in Metazoa, so too might the punctuational history of clones yield gradualism in the collectivities so dubiously designated as "species" in asexual unicells. In other words, foram "species" may exhibit gradualism because these supposed entities are really results or collectivities, not proper Darwinian individuals or causal agents. Eldredge and I first presented this argument in our initial commentary on the debate about punctuated equilibrium (Gould and Eldredge, 1977, p. 142, and Fig. 3-3):

> We predict more gradualism in asexual forms on biological grounds. Their history should be, in terms of their own unit, as punctuational as the history of sexual Metazoa. But their unit is a clone, not a species. Their evolutionary mode is probably intermediate between natural selection in populations, and species selection in clades: variability arises via new clones produced rapidly (in this case, truly suddenly) by mutation. The phenotypic distribution of these new clones may be random with respect to selection within an

asexual lineage (usually termed a "species," but not truly analogous with sexual species composed of interacting individuals). Evolution proceeds by selecting subsets within the group of competing clones. If we could enter the protists' world, we would view this process of "clone selection" as punctuational. But we study their evolution from our own biased perspective of species, and see their gradualism as truly phyletic—while it is really the clonal analog of a gradual evolutionary trend produced by punctuated equilibria and species selection.

Lenski's remarkable studies on controlled evolution of bacteria under laboratory conditions of replication provide striking evidence for this claim (see full discussion of this work on pp. 240–245). Lenski and colleagues (Lenski and Travisano, 1994; Papadopoulos et al., 1999) monitored average cell size for 10,000 generations in 12 lineages of *E. coli.* Cell size increased asymptotically in each lineage, steadily for the first 3000 generations or so, but remaining relatively stable thereafter. The fine structure of increase, how-

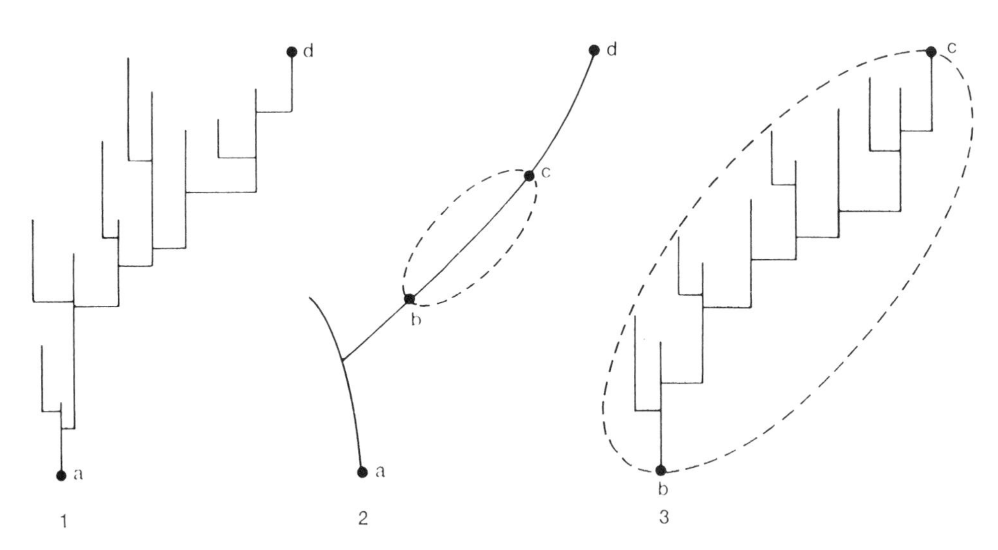

3-3. The supposed gradualism noted in many foram species may represent a view, from too high a level, of an overall trend within a phyletic sequence properly analyzed in terms of punctuational events at the level of clone selection—the appropriate mode in such asexual forms. 1 shows a conventional metazoan lineage in punctuated equilibrium. 2 shows the apparent gradualism in a foram lineage. 3 shows a gradualistic segment between B and C magnified so that the appropriate process of punctuational clone selection becomes visible. From Gould and Eldredge, 1977.

ever, proceeded in a punctuational manner in each lineage—a step-like pattern of stability in average cell size, followed by rapid ratcheting of the full population up to larger dimensions. This punctuational pattern presumably occurred because clones act as primary Darwinian individuals in this system. The full lineage must "wait" for sudden introduction of favorable variation in the form of occasional mutations, initiating novel clones that can then sweep through the entire lineage to yield a punctuational step in the overall phylogeny (at a scale of 10,000 generations in phenotypic history). Predictable, replicable size increase occurs by punctuational clone selection in each case (see Fig. 3-4). Lenski's powerful result does not illustrate a case of punctuated equilibrium, *sensu stricto,* but he does provide a challenging and instructive argument for considering the validity of punctuational change at all levels.

Just as the careful watchdog at any scientific meeting will unhesitatingly call out "what's the scale" when a colleague fails to include a measurement bar on a slide of any important object, we must always ask "what's the level" when we analyze the causal basis of any evolutionary pattern. Punctuational clone selection can yield gradualism within collectivities conventionally (if dubiously) called "species," just as punctuated equilibrium, acting on species as Darwinian individuals, can produce gradual trends in the overall history of lineages and clades.

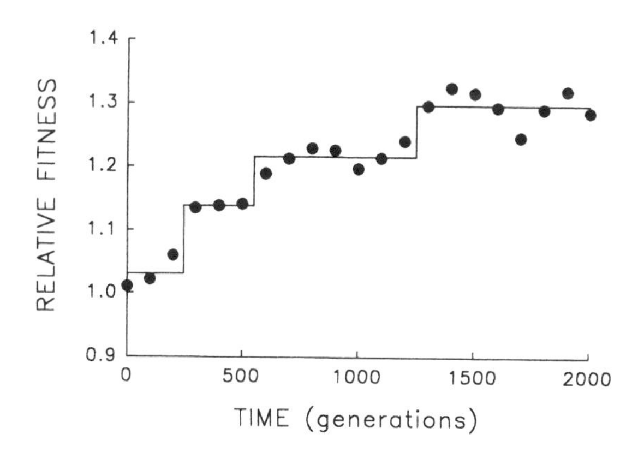

3-4. From Lenski and Travisano, 1994. At too broad a scale, the size increases within Lenski's bacterial lineages seem gradual. But a finer-scale analysis shows a stepwise punctuational pattern of clone selection with stasis recording the waiting time between favorable mutations, and punctuation caused by rapid sweep of these rarely favorable mutants through the population.

Genotypes

Punctuated equilibrium is a theory about the evolution of phenotypes (both in concept and in operational testability for paleontological hypotheses), and correlations with genotypic patterns provide neither a crucial test nor even any necessary prediction. For example, critics of punctuated equilibrium have often argued that the apparently cumulative character of overall genetic distances among members of an evolving clade, expressed as a high correlation between measured disparities and independently derived times since divergence from a common ancestor—the kind of information that, in idealized (but rarely encountered) situations, yields a rough "molecular clock"—should argue strongly against punctuational styles of evolution, while affirming anagenetic gradualism.

But, leaving aside the highly questionable empirical status of these claims, the hypothesis of punctuated equilibrium would not be affected by positive outcomes, even at much higher relative frequency than the known history of life apparently validates. In supposing that "molecular clocks" tick against the requirements of punctuated equilibrium, we fall into two bad habits of thinking that impede macroevolutionary theory in general, and therefore rank as important conceptual barriers against the theses of this book. First, reductionistic biases often lead us to seek an "underlying" genetic basis for any overt phenomenon at any scale, and then to view data at this level as a fundamental locus for proper evolutionary explanation. (But consider only two among many rebuttals of such a position: (1) a genetic pattern may be non-causally correlated with coincident evolutionary expressions at other scales; and (2) in principle, genetic expressions of a common causal structure do not rank as intrinsically more "deep," "real," "fundamental," or "basic" than other manifestations in different forms and at different levels; causal relevance depends upon the questions we ask and the processes that organisms undergo.)

Second, the "allometric" effects of scaling either render the same process in a very different manner at various scales, or (perhaps more frequently, and primarily in this case at least) generate the distinctive patterns of different scales by independent processes, acting simultaneously, but with each process primarily responsible for results at its own appropriate level.

If I could affirm, as may well often be the case, that punctuated equilibrium regulated the phenotypic pattern of evolution in a given clade, while genotypic distances conformed closely to a "molecular clock," I would not conclude that punctuated equilibrium had therefore been downgraded, or exposed as incorrect, superficial, or illusory—with genetic continuity as a physically underlying (and conceptually overarching) reality. Rather, I would regard each result as true and appropriate for its own scale and realm—with the full pattern of legitimate difference standing as an intriguing example of resolvable complexity in evolutionary scaling and causality. Moreover, this particular pattern might easily result from a highly plausible scenario of complex and multileveled causation—namely, that neutral substitutions at the nucleotide level impart a signal sufficiently like a genomic metronome to dominate the molecular results, while ordinary speciation both regulates the phenotypic history of populations and works by the expected pattern of punctuated equilibrium. The genomic results, *in principle,* need *not* extrapolate to encompass the pattern of speciational (macroevolutionary) change. After all, we do understand that gene trees do not entirely match organism trees in phylogeny!

In this way—as in the foregoing example of predictable differences between asexual unicells and sexually reproducing metazoa—punctuated equilibrium proves its value primarily by hypothesizing sensible distinctions: that is, by operating at scales and biological conditions where cladogenetic speciation plausibly sets evolutionary pattern. Punctuated equilibrium should not prevail where species cannot exist as Darwinian individuals, or where continuously occurring, and largely nonadaptive, substitution of nucleotides probably regulates the bulk of genomic change. In this crucial sense, punctuated equilibrium becomes a valuable hypothesis by delineating such testable distinctions, rather than allowing evolution to be conceptualized monistically as a single style of alteration, or a single kind of process either flowing from, or applicable to, all scales of change.

The question of consistency between observed genetic patterns in living species, and the relative frequency of punctuated equilibrium in their phylogeny, shall be treated in the next section on the correspondence of punctuated equilibrium with predictions of evolutionary modeling. But one genetic issue has been widely discussed in the literature, and should be included in this section on empirical results. Several researchers have noted that punctuated equilibrium implies a primary prediction about patterns of genetic differences among species: if most change accumulates at ruptures of stasis during events of speciation, and not continuously along the anagenetic history of a population, then overall genetic differences between pairs of species should correlate more closely with the estimated number of speciation events separating them, than with chronological time since divergence from common ancestry. (This prediction might be clouded by several factors, including the foregoing discussion on attributing the bulk of genomic change to continuity at a lower level, and a number of potential reasons for discordance between phenotypic effect and extent of responsible genetic change. But I certainly will not quibble, and I do allow that punctuated equilibrium suggests the broad generality of such a result.)

In the early days of debate about punctuated equilibrium, Avise (1977) performed an interesting and widely discussed test. In comparing genetic and morphological differences among species in two fish clades of apparently equal age but markedly different frequencies of speciation, Avise found a higher correlation of distances with age than with frequency of branching, and therefore favored gradualism over punctuated equilibrium as an explanation of his results. But Mayden (1986) then showed that Avise's test did not apply well to his chosen case (primarily because we cannot be sure of roughly equal antiquity for the two clades). He then argued, as several supporters of cladistic methodology had urged, that such tests should be applied only to well-confirmed cladistic sister groups—for, in such cases, even if paleontological data permit no certainty about the actual time of joint origin from common ancestry, at least we can be confident that the two clades are equally old! Mindel et al. (1989) then performed such a properly constituted test on the reptilian genus *Sceloporus*, and more loosely on allozymic data in general, and found a positive correlation between evolutionary distance and frequency of speciation—thus validating the primary prediction of punctuated equilibrium.

EMPIRICAL TESTS OF CONFORMITY WITH MODELS

Limitations of the fossil record restrict prospects for testing punctuated equilibrium by inductive enumeration of individual species and lineages. Cases with sufficient resolution may not be common enough to establish a robust relative frequency; or systematic biases based on imperfections in the fossil record may lead to artifactual preferences for punctuated equilibrium—thus making the data unusable as a fair test for a minimal frequency. (I do not regard these problems as particularly serious, and I will provide several examples of adequate resolution in the next section of this chapter. But we should, in the light of these difficulties, also be exploring other ways of testing punctuated equilibrium, as considered below.)

In another strategy that has been pursued by some researchers, but could (and should) be exploited to a much wider and more varied extent, we might characterize, in quantitative fashion, broader patterns in the deployment of diversity through time and space in major taxonomic groups—and then devise tests to distinguish among contrasting causes: anagenetic vs. cladogenetic; gradual vs. punctuational. If certain well-defined patterns can only be generated, say, by branching speciation rather than by anagenetic transformation (or vice versa, of course), then we can use the fit of broad results with distinctive models, rather than minute documentation on a case-by-base basis, to establish the relative frequency of punctuated equilibrium.

In an important study, for example, Lemen and Freeman (1989) investigated "the properties of cladistic data sets from small monophyletic groups (6–12 species) . . . using computer simulations of macroevolution" (p. 1538). They contrasted the differing outcomes of data generated under anagenetic gradualism vs. punctuated equilibrium, and then examined cladograms of extant monophyletic groups for consistency with these "abstract, end-member" alternatives. They claimed better support for gradualism, but several flaws in their logic and data render their conclusions moot, much as their pioneering approach may be applauded and recommended for further study.

Lemen and Freeman tested actual data against three modelled differences between cladograms generated by gradualism vs. punctuated equilibrium.

1. Punctuated equilibrium should produce a strongly positive correlation between number of branching points and apomorphies of species—because change occurs at speciation and does not further correlate with passage of time *per se.* Lemen and Freeman's models did affirm this expected result, and real data, somewhat ironically, revealed "higher correlations of apomorphies and branch points than could be explained by either mode of macroevolution" (p. 1549). But the authors rejected an interpretation of this information as favorable to punctuated equilibrium because anagenesis, under certain conditions, also yields positive correlations, and because high correlations can also arise artificially by errors in establishing cladograms: "a consistent error in polarity can profoundly affect the correlation of total apomorphies and branch points" (p. 1551). Fair enough, but Lemen and Freeman are not nearly so circumspect when equally flawed data seem to favor their preferred alternative of gradualism.

2. The modal (but not the mean or median) number of autapomorphies will always be zero under strict punctuated equilibrium. This odd-sounding situation arises because, in the cladograms, an event of branching produces a daughter species with some autapomorphies and a persisting parental species remaining in stasis with none. With no change except at branching

points, the value of zero autapomorphies must remain most common across all species on the cladogram. Under gradualism, autapomorphies simply accumulate through time, whatever the pattern of branching, so zero should not mark a preferred or particularly common value.

Lemen and Freeman never found a mode of zero autapomorphies in real data, and therefore rejected punctuated equilibrium as a predominant style of evolution. But had they pursued explanations based on artifacts (as they did so assiduously when the data seemed to favor punctuated equilibrium), they would have realized that taxonomic practice precludes the definition (or even the recognition) of species without autapomorphies. Such species arise frequently in the modelled system as a necessary consequence of the chosen rules of generation and the general logic of cladistic analysis. But, in neontological practices of naming, a species without autapomorphies represents an oxymoronic concept, and such taxa could never be designated at all. Lemen and Freeman recognize this point in writing about their various forms of gradualistic modeling (p. 1551): "When distinctness of species is demanded the lack of autapomorphies may not be the most expected condition."

3. Under punctuated equilibrium, "as the number of characters used in the analysis increases, the distribution of the number of autapomorphies per species becomes bimodal. Under gradualism, the distribution of autapomorphies remains unimodal under all conditions" (p. 1538). This situation, a spinoff from their second criterion, arises because each branch, in an event of punctuated equilibrium, produces one changed descendant and one persisting ancestor—and the more characters you measure, the more you pick up the differences between stasis on one branch and change on the other. Under gradualism, total change correlates only with elapsed time, so accumulating autapomorphies should form a unimodal distribution so long as species duration remains unimodal as well.

Lemen and Freeman found no bimodal distributions in real data, and therefore concluded again in favor of gradualism. But, once more, the differences between idealized modeling and data from real organisms scuttles this conclusion. In the models, we know for sure that long arms without branching are truly so constituted, for we have perfect information of all simulated events. These unbranched arms, under punctuated equilibrium, should accumulate no autapomorphies—and the low mode of the bimodal distribution arises thereby. But, in real data of cladograms based on living organisms, long unbranched arms usually (I would say, virtually always) record our ignorance of numerous and transient speciational branchings that quickly became extinct and left no fossil record. (Moreover, since Lemen and Freeman's cladograms only include living organisms, even if successful and well-represented fossil species existed, they would not be included.) When we note a long arm without branches on a modern cladogram, and then assume (as Lemen and Freeman did) that accumulated autapomorphies between node and terminus must have arisen gradually and anagenetically, we commit a major blunder. We have no idea how many unrecorded speciation events separate node and terminus, and we cannot assert that recorded autapomorphies did not occur at these (probably frequent) branchings. In other words, Lemen and Freeman's bimodality test assumes that unbranched arms of their cladograms truly feature no speciation events along their routes, whereas numerous transient and extinct species must populate effectively all of these pathways.

Other applications of this method—modeling of alternative outcomes and testing of contrasting predictions against patterns of real data—have yielded results favorable to punctuated equilibrium. In a pathbreaking paper, Stanley (1975—see elaboration in Stanley, 1979 and 1982) first proposed this style of testing and developed four putative criteria, all affirming punctuated equilibrium. (Stanley's tests may be reduced to three, as his second "test of the Pontian cockles" represents a particular instance of his first "test of adaptive radiations." Stanley argued:

TEST OF ADAPTIVE RADIATION. After calculating average species durations from the fossil record, one can affirm that pure anagenetic gradualism (or temporal stacking of species end-to-end) cannot account for the magnitude of recorded adaptive radiations in the time available—so rapid cladogenesis must be invoked.

TEST OF LIVING FOSSILS. Punctuated equilibrium associates realized amounts of change primarily with frequency of speciation, anagenetic gradualism primarily with elapsed time. If so-called "living fossils"—ancient groups with little recorded change—also show unvarying low diversity through time, then we can affirm the primarily prediction of punctuated equilibrium, and refute the corresponding expectation of gradualism (for these groups are ancient). Stanley then documented such a correlation between clades identified as "living fossils" and persistently low diversity in these clades.

TEST OF GENERATION TIME. Under gradualism, amounts of realized evolution should correlate strongly with generation time—for the time that should mark accumulated evolutionary change does not tick by an abstract Newtonian clock, but by number of elapsed generations, representing the number of opportunities for natural selection to operate. But, under punctuated equilibrium, amount of change correlates primarily with frequency of speciation—a property with no known relationship to generation time. Stanley then cited the well-documented lack of correlation between evolutionary rate and generation time as evidence for the prevalence of punctuated equilibrium (fast-evolving elephants vs. stable invertebrates with short generations).

Much as I regard Stanley's arguments as suggestive, I cannot accept them as conclusive for two basic reasons. First, other plausible explanations exist for the patterns noted. For example, many reasons other than the prevalence of punctuated equilibrium might explain a lack of correlation between realized evolution and generation time, even in a world of anagenetic gradualism. The correlation might simply be weak or too easily overwhelmed (and therefore rendered invisible) by such other systematic factors as variation in the intensity of selection. (Maybe elephants, on average, experience selection pressures higher by an order of magnitude than those affecting short-lived invertebrates; maybe population size overwhelms the factor of generation time.)

Second, most of Stanley's tests (particularly his key claim about adaptive radiation) don't really oppose punctuated equilibrium to gradualism, but rather contrast a more general claim about the speciational basis of change (whatever the mode of speciation) with anagenesis. Moreover, the tests employ a somewhat unfairly caricatured concept of gradualism. I doubt that the most committed gradualist ever tried to encompass the maximal change between ancestor and any descendant in an adaptive radiation by stacking species end to end, and then calculating whether the full effect could arise in the allotted time. A committed gradualist might fairly say of an adaptive radiation: "of course the magnitude of change in both form and diversity correlates with number of branching events (what else could a 'radiation' mean). But adaptive radiations only accelerate the frequency of branching in response to ecological opportunity ('open' environments just invaded or just cleared out by extinction); they do not affect the modality of change. I will allow that, in adaptive radiations, most new species arise in less time than usual, but still gradualistically. If full speciation takes half the average time (1 million rather than a modal 2 million years, for example), but still occurs imperceptibly and still occupies a large percentage of an average species's lifetime, then gradualism encounters no threat in adaptive radiation."

However, in another crucial sense, at least one of Stanley's tests does illustrate the most salutary potential role for punctuated equilibrium: its capacity to act as a prod for expansive thought and new hypotheses, whatever the outcome of the empirical debate about relative frequency. Paleontologists had been truly stymied in their thinking about the important and contentious topic of "living fossils." Neither of the two conventional explanations could claim any real plausibility. Every textbook that I ever consulted as a student dutifully repeated the old saw that living fossils had probably achieved optimal adaptation to their environment. Therefore, no alternative construction could selectively replace an ideal form achieved so long ago. But no one ever presented any even vaguely plausible evidence for such a confident assertion. Why should horseshoe crabs lie closer to optimality than any other arthropod? What works so well in the design of lingulid vs. other brachiopods? What superiority can a lungfish assert over a marlin or tuna? In fact, since living fossils also (by traditional depiction) present such a "primitive" or "archaic" look, the claim for optimality seemed specially puzzling.

The other obvious explanation, in a gradualistic and anagenetic world ruled by conventional selection, held that living fossils had stagnated because they lacked genetic variation, and therefore presented insufficient fuel for Darwinian change. This more plausible idea seemed sufficiently intriguing that Selander et al. (1970), in the early days of electrophoresis as a novel method for measuring overall genetic variation, immediately applied the technique to *Limulus,* the horseshoe crab—and found no lowering of genetic variability relative to known levels for other arthropods. This negative pattern has held, and no standard lineage of living fossils exhibits depauperate levels of genetic variability.

But punctuated equilibrium suggests another, remarkably simple, explanation once you begin to think in this alternative mode—an insight that ranks in the exhilarating, yet frustrating, category of obvious "scales falling from eyes" propositions, once one grasps the new phrasing of a basic question. If evolutionary rate correlates primarily with frequency of speciation—the cardinal prediction of punctuated equilibrium—then living fossils may simply represent those groups at the left tail of the distribution for numbers of speciation events through time. In other words, living fossils may be groups that have persisted through geological time at consistently and unvaryingly low species diversity. (Average species longevity need not be particularly high, for low species numbers, if consistently maintained without geological bursts of radiation, will yield the full effect.) Such groups cannot be common—for consistently low diversity makes a taxon maximally subject to extinction in our contingent world of unpredictable fortune, where spread and number represent the best hedges against disappearance, especially in episodes of mass extinction—but every bell curve has a left tail.

This explanation holds remarkably well, and probably provides a basic explanation of "living fossils." Such groups are neither mysteriously optimal, nor unfortunately devoid of variability. They simply represent the few higher taxa of life's history that have persisted for a long time at consistently low species number—and have therefore never experienced substantial opportunity for extensive change in modal morphology because species provide the raw material for change at this level, and these groups have never contained many species.

Westoll (1949), for example, published a classic study, summarized again and again in treatises and textbooks (Fig. 3-5), showing that lungfishes evolved very rapidly during their early history, but have stagnated ever since. The literature abounds in hypothesized explanations based on adaptation and ecological opportunity in an anagenetic world. The obvious alternative stares us in the face, but rises to consciousness only when theories like punctuated equilibrium encourage us to reconceptualize macroevolution in speciational terms: in their early period of rapid evolution, lungfishes maintained high species diversity, and could therefore change quickly in modal morphology. Their epoch of later stagnation correlates perfectly with a sharp reduction of diversity to very low levels (only three genera living today, for example) with little temporal fluctuation in numbers—thus depriving macroevolution of fuel for selection (at the species level), and relegating lungfishes to the category of living fossils.

A breakthrough in the application of quantitative modeling to cladistic patterns of evolution directly recorded in the fossil record has been achieved by Wagner (1995 and 1999) and Wagner and Erwin (1995). These authors show, first of all, the pitfalls of working only with cladistic information from living organisms, and they illustrate the benefits of incorporating stratophenetic data from the fossil record into any complete analysis (see Wagner and Erwin, 1995, pp. 96–98, in a section entitled "why cladistic topology is insufficient for discerning patterns of speciation"). They then build models based on three alternative modes of evolution, and characterize the differences in cladistic pattern expected from each: anagenetic gradualism, speciation by "bifurcation" (where, after branching, the two descendant species both accumulate differences from an ancestor then recorded as extinct), and speciation by "cladogenesis" (where one daughter species arises with autapomorphic differences, but the ancestral species persists in stasis). Cladogenesis is usually defined—both in this book and in the evolutionary literature in general—as any style of evolution by branching of lineages rather than by transformation of a single lineage (anagenesis). Wagner and Erwin restrict the term "cladogenesis" to the mode of speciation predicted by punctuated equilibrium—branching off of a descendant, leaving a persisting and unaltered ancestor. They contrast this mode with bifurcation—the style of speciation predicted by gradualism: splitting of an ancestral popula-

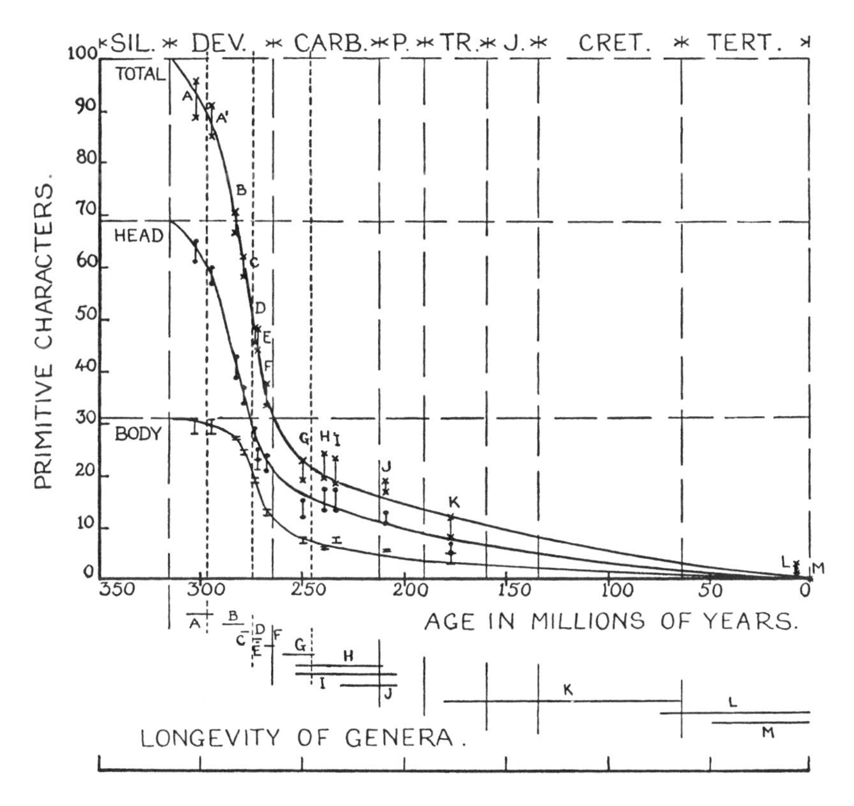

3-5A. The famous figure from Westoll (1949) showing rapid morphological change early in the history of lungfishes, followed by prolonged stagnation thereafter.

tion into two descendant species, both diverging steadily from the ancestor (which becomes extinct). I follow Wagner and Erwin's restricted use of "cladogenesis" only in discussing their work, and use the broader definition throughout the rest of this book.

The last two modes of bifurcation and cladogenesis both depict branching speciation in the definitional sense that two species emerge, where only one existed before. But note the crucial difference: bifurcation represents the

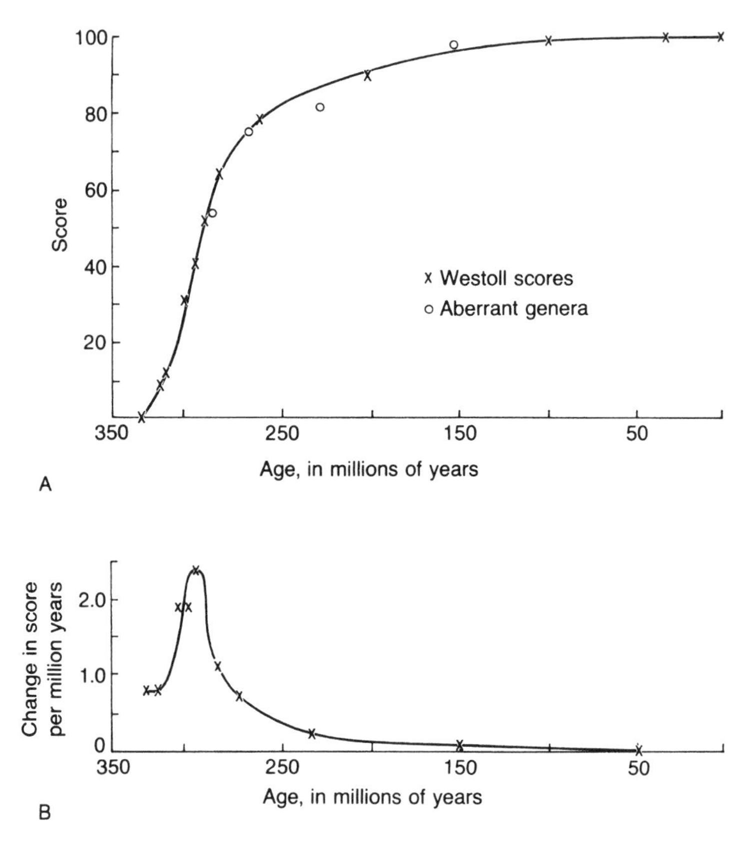

3-5B. Redrawing and simplification of these data in the excellent paleontological textbook of Raup and Stanley (1971). The bottom icon, showing an early mode and a right skew, has become canonical in textbooks. The data are firm and fascinating, but the interpretation has in general been faulty as a result of gradualistic and anagenetic assumptions. Lineages did not stagnate in any anagenetic sense; rather, species diversity became so dramatically lowered (and has always stayed so—only three genera of lungfishes remain extant today) that speciational processes have never again had enough fuel to power further extensive phyletic change.

operation of speciation in a gradualistic world, where an event of branching may be considered equivalent to two cases of gradualism following a separation of populations, and where the separation itself need not correlate with any acceleration in rate of evolutionary transformation. Cladogenesis, on the other hand, represents the predictions and expectations of punctuated equilibrium. Therefore, if we can model the differences between bifurcation and cladogenesis, and test these distinctive expectations against real patterns in nature, we may achieve our best and fairest potential evaluation for the relative frequency of punctuated equilibrium—for punctuated equilibrium cannot be affirmed merely by showing that realized evolutionary patterns must record speciation and cannot be rendered by anagenetic, end-to-end stacking. Even the most committed anagenetic gradualists never denied the importance and prevalence of speciation. They hold, rather, that speciation generally occurs in the gradualistic mode—as two cases of divergence at characteristic rates for unbranching lineages—and not, as supporters of punctuated equilibrium maintain, as geologically momentary bursts representing the budding of descendant populations from unchanged, and usually persisting, ancestral species in stasis. Thus, the best possible test for punctuated equilibrium must distinguish between the expectations of bifurcating vs. cladogenetic models of speciation.

I am embarrassed to say that neither I nor my colleagues working on the validation of punctuated equilibrium ever conceptualized the simple and obvious best test for distinguishing the bifurcating model of speciational gradualism from the cladogenetic model of punctuated equilibrium. In this case, the impediment may be clear, but I can offer no legitimate excuse for my opacity—and I congratulate Wagner and Erwin on their formulation.

The solution lies in the distribution and frequency of "hard" polytomies in cladogenetic topologies. I failed to appreciate the following point: under punctuated equilibrium, new species branch off from unchanged and persisting ancestors. The successful ancestor remains in stasis and may live for a long time. Therefore, these "stem" species may generate numerous descendants during their geological tenure, while remaining unchanged themselves. Now what cladistic pattern must emerge from such a situation? A group of species branching at different times from an unchanged ancestor must yield a cladistic polytomy. Cladograms cannot distinguish different times of origin from an unaltered ancestor, and can therefore only record the phenetic constancy of the common and unchanging ancestor as a polytomy, for all branches emerge from an invariant source. Bifurcation, on the other hand, can produce a range of cladistic topologies (Wagner and Erwin, 1995, p. 92), but not domination of the overall pattern by polytomies. Thus, gradualistic vs. punctuational models of speciation should be distinguishable by distributions of polytomies in the resulting cladogram.

I suspect that many of us never recognized this point because we have been trained to view polytomies negatively as an expression of insufficient data to resolve a true set of ordered dichotomies. (Shades of our profession's former failure to conceptualize punctuated equilibrium because we had been trained to view geologically rapid appearances as artifacts of an imperfect fossil record!) Thus, we never recognized that polytomies might also be denoting a positive and resolvable pattern—multiple branching through time of several species from an unaltered ancestral source. Of course—and, again, just as with punctuated equilibrium itself—polytomy can also result from imperfection, and we need criteria to separate "real" polytomies representing a signal from the history of life from polytomies that only record artifacts of an imperfect record. Wagner and Erwin (1995) develop such a criterion by distinguishing between "hard" polytomies that include the persisting ancestor and "soft" polytomies that arise from an inability to resolve true sets of ordered dichotomies.

Wagner and Erwin's modeling demonstrates the translation of punctuational speciation to a cladistic pattern of predominant polytomies. (Wagner and Erwin used my own model of punctuational phylogenies, done with D. M. Raup in the 1970s (Raup and Gould, 1974), to show this mapping of punctuational phylogeny to a polytomous cladogram—see Fig. 3-6—but I had never made the connection myself.)

Wagner and Erwin then applied their modeled differences to cladograms for two well-resolved, but maximally different (in taxon and time) species-level phylogenies in the fossil record: two Neogene clades of planktonic foraminifers (Globigerinidae and Globorotaliidae), and Ordovician representatives of the gastropod family Lophospiridae. In both cases, the cladograms indicated an overwhelming predominance of speciation by cladogenesis as a cause of phylogenetic patterning—thus affirming the predictions of punctuated equilibrium. For globigerinids, the cladistic topology revealed 40 speciation events, 5 probably anagenetic, 35 cladogenetic, and none bifurcating. Wagner and Erwin did not present full tabulations for the globorotaliids, but stated (p. 105): "The results are not presented here, but they were similar to those found for globigerinids: cladogenesis is significantly more common than anagenesis, a positive association exists between having long temporal ranges and leaving cladogenetic descendants, and no such association exists for anagenetic ancestors."

For the lophospirid gastropods, they write (p. 106): "Our preferred cladogram for lophospirids is rife with polytomies. Of the eleven polytomies, only two do not include plesiomorphic species. Thus, nine may represent hard polytomies." Of 42 implied speciation events, a maximum of six may have been anagenetic, while only one may represent a bifurcation. Again, clado-

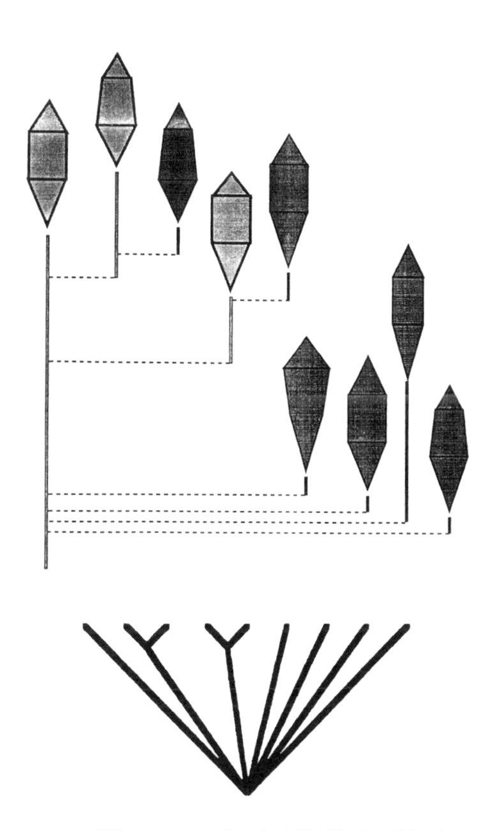

3-6. To my embarrassment, Wagner and Erwin (1995)—for I had not seen the obvious implication that would have enormously helped my argument—showed how phylogenies based upon iteration of several species from an unchanged parent stock (as Raup and Gould, 1974, had generated, and Wagner and Erwin reproduced, at the top of this figure) must yield, in cladistic representation, a polytomy. Thus, polytomies may provide evidence for punctuated equilibrium and do not necessarily represent the "signature" of missing data needed to resolve the system into dichotomies. If the ancestral form doesn't change throughout its geological range, all descendants must in principle arise at a polytomous junction of a cladogram.

genetic speciation, the expectation of punctuated equilibrium, dominates the phylogenetic pattern.

Wagner and Erwin's overall conclusion accords fully with patterns expected in phylogenies built primarily—one might say overwhelmingly—by punctuated equilibrium (Wagner and Erwin, p. 110):

- Cladogenesis is significantly more common than anagenesis.
- Species with longer temporal and geographic ranges are more likely to leave descendants via cladogenesis or the factors contributing to wider temporal and geographic ranges also contribute to the likelihood of cladogenetic evolution.
- If anagenesis occurs, it only applies to species with restricted temporal and geographic ranges.
- Bifurcation accounts for a negligible amount of speciation.

We cannot often obtain well-resolved species level phylogenies from paleontological data, and inferences from higher taxa will probably remain too murky and insecure to permit general use of such models for testing hypotheses explicitly based on the evolutionary behavior of species. Still, other data sets do exist in fair absolute abundance (while representing a low percentage of the total number of potential lineages in life's history). Studies like Wagner and Erwin's can be replicated and extended for many taxa—and such a strategy can provide powerful tests for the relative importance of punctuated equilibrium in the history of life and the generation of phylogenetic patterns. The first tests have been highly favorable, but we have scarcely any idea what an extended effort might teach us about the basic modalities of macroevolution.

4 SOURCES OF DATA FOR TESTING PUNCTUATED EQUILIBRIUM

Preamble

Punctuated equilibrium has generated a fruitful and far-ranging, if sometimes acrimonious, debate within evolutionary theory (see appendix). While we feel much pride (mixed with occasional frustration) for the role that punctuated equilibrium has played in instigating such extensive rethinking about the definitions and causes of macroevolution, we take even more pleasure in the volume of empirical study provoked by the theory of punctuated equilibrium, and pursued by paleontologists throughout the world. These carefully documented case studies (both pro and con) build a framework of proof for the value of punctuated equilibrium, as illustrated by the most important of all scientific criteria—operational utility. Such cases have been featured in numerous symposia and books dedicated to the empirical basis of punctuated equilibrium. This literature includes: the 1982 symposium in Dijon, France, entitled *Modalités, rythmes, méchanismes de l'évolution biologique: gradualisme phyletique ou équilibres ponctués* and published as Chaline, 1982; the 1983 Swansea symposium of the Palaeontological Association (United Kingdom), "Evolutionary case histories from the fossil record" and published as Cope and Skelton (1985); the book *The Dynamics of Evolution: The Punctuated Equilibrium Debate in the Natural and Social Sciences* (Somit and Peterson, 1992), which began as a symposium for the annual meeting of the American Association for the Advancement of Science and then appeared as a special issue (1989) of the *Journal of Biological and Social Structures;* the 1992 symposium of the Geological Society of America,

"Speciation in the Fossil Record," held to celebrate the 20th anniversary of punctuated equilibrium and published in book form as Erwin and Anstey (1995); and the 1994 Geological Society of America symposium on coordinated stasis, published in a special issue of the journal *Palaeogeography, Palaeoclimatology, Palaeoecology* in 1996 (volume 120, with Ivany and Schopf as editors). Several other unpublished symposia, including the notorious Chicago macroevolution meeting of 1980 (see pages 303–307), focused upon the topic of punctuated equilibrium. Finally, several books have treated punctuated equilibrium as an exclusive or major topic, including the favorable accounts of Stanley (1979), Eldredge (1985a, 1985b, 1995), and Vrba (1985a), and the strongly negative reactions of Dawkins (1986), Dennett (1995), Hoffman (1989), and Levinton (1988).

As emphasized throughout this book, most general hypotheses in natural history, with punctuated equilibrium as a typical example, cannot be tested with any single "crucial experiment" (that is, by saying "yea" or "nay" to a generality after resolving a case with impeccable documentation), but must stand or fall by an assessment of relative frequency. Moreover, we can't establish a decisive relative frequency by simple enumerative induction (as in classical "beans in a bag" tests of probability)—for individual species cannot be treated as random samples drawn from a totality with a normal (or any other kind of simple) distribution, but represent unique items built by long, complex and contingent histories. Time, taxon, environment and many other factors strongly "matter," and no global evaluation can be made by counting all cases equally. We may, however, be able to reach robust solutions for full populations within each factor—for planktonic forams, terrestrial mammals, Devonian brachiopods, or species of the Cambrian explosion, for example. The later sections of this chapter report several such studies, nearly all finding a predominant relative frequency for punctuated equilibrium.

Nonetheless, hundreds of individual cases have been documented since we proposed punctuated equilibrium in 1972. I do not think that most authors pursue such work under any illusion that they might thus resolve the general debate, but rather for the usual, and excellent reasons of ordinary scientific practice. Researchers pursue such studies in order to apply promising general concepts to cases of special interest that draw upon their unique skills and expertise. Such studies are pursued, in other words, to resolve patterns within *Australopithecus afarensis,* or among species in the genus *Miohippus,* not to adjudicate general issues in evolutionary theory.

Nonetheless, compendia of such studies do provide a "feel" for generalities of data in admittedly non-randomized samples, and they do establish archives of intriguing and well-documented cases both for pedagogical illustration, and simply for the general delight that all naturalists take in cases well treated and conclusively resolved. I shall therefore discuss this mode of documentation as practiced for two categories central to punctuated equilibrium: patterns of gradualism or stasis within unbranched taxa, and tempos and modes of branching events in the fossil record. I conclude with the more decisive theme of relative frequencies.

The Equilibrium in Punctuated Equilibrium: Quantitatively Documented Patterns of Stasis in Unbranched Segments of Lineages

As previously discussed (see pp. 30–33), the main contribution of punctuated equilibrium to this topic lies in constructing the theoretical space that made such research a valid and recognized subject at all. When paleontologists equated evolution with gradual change, the well-known stasis of most lineages only flaunted a supposed absence of desired information, and could not be conceptualized as a positive topic for test and study. By representing stasis as an active, interesting, and predictable feature of most lineages most of the time, punctuated equilibrium converted an unconceptualized negative to an intriguing and highly charged positive, thereby forging a field of study.

Nonetheless, we cannot argue that a proven predominance of stasis within lineages can establish the theory of punctuated equilibrium by itself. Punctuated equilibrium implies and requires such stasis, but remains, primarily, a theory about characteristic tempos and modes of branching events, and the primary patterning of phyletic change by differential birth and death of species.

Stasis has emerged from the closet of disappointment and consequent nonrecording. At the very least, paleontologists now write, and editors of journals now accept, papers dedicated to the rigorous documentation of stasis in particular cases—so skeptics, and scientists unfamiliar with the fossil record, need not accept on faith the assurances of experienced paleontologists about predominant stasis in fossil morphospecies (see pp. 22–26). Moreover, stasis has also become a subject of substantial theoretical interest (see pp. 172–185), if only as a formerly unexpected result now documented at far too high a frequency for resolution as an anticipated outcome within random systems (Paul, 1985); stasis must therefore be actively maintained. In any case, paleontologists are now free to publish papers with such titles as: "*Cosomys primus:* a case for stasis" (Lich, 1990), and "Apparent prolonged evolutionary stasis in the middle Eocene hoofed mammal *Hyopsodus*" (West, 1979).

The study of McKinney and Jones (1983) may be taken as a standard and symbol for hundreds of similar cases representing a characteristic mixture of satisfaction and frustration. These authors documented a sequence of three successional species of oligopygoid echinoids from the Upper Eocene Ocala Limestone of Florida. The two stratigraphic transitions are abrupt, and therefore literally punctuational. But available evidence cannot distinguish among the mutually contradictory explanations for such passages: gradualism, with transitions representing stratigraphic gaps; rapid anagenesis for a variety of plausible reasons including population bottlenecks or substantial environmental change; punctuated equilibrium based on allopatric speciation elsewhere (or unresolvably *in situ,* given coarse stratigraphic preservation), and migration of new species to the ancestral range. Hence, frustration. (Moreover, as this pattern represents the most frequent situation in most ordinary sequences of fossils, we can readily understand why the testing of punctuational claims within the theory of punctuated equilibrium requires selection of cases—fortunately numerous enough *in toto,* however modest in relative frequency—with unusual richness in both spatial and temporal resolution.)

At the same time, however, we gain satisfaction in eminent testability for the set of claims representing the second key concept of stasis. Any species, if well represented throughout a considerable vertical span marking the hundreds of thousands to millions of years for an average duration, can be reliably assessed for stasis vs. anagenetic gradualism by criteria outlined previously (pp. 39–49). McKinney and Jones (1983) compiled excellent evidence for stasis in each of their three species—the basis, after all, for using these taxa in establishing biozones for this section. (As argued on pp. 21–23, biostratigraphers have always used criteria of stasis and overlapping range zones in their practical work on the relative dating of strata.) McKinney and Jones conclude (1983, p. 21): "These observations suggest there is little chance of species misidentification due to ontogenetic or phylogenetic effects when using this lineage for biostratigraphic purposes."

Smith and Paul (1985) studied vertical variation of the irregular echinoid

Discoides subucula in a remarkably complete and well-resolved sequence of Upper Cretaceous sands. The species occurred throughout 8.6 m of section, apparently representing continuous sedimentation within one ammonite zone spanning less than 2 million years. The authors were able to sample meter by meter through a section with an interesting inferred environmental history: "The sediment that was then being deposited changed from clean, well-washed sand to a very muddy sand, and so one might expect to find evidence of phyletic gradualism in response to these changes" (1985, p. 36).

Smith and Paul did measure a steady change in shape towards a more conical form, a common response of irregular echinoids to muddy environments. But such an alteration can be ecophenotypically induced during ontogeny, and the authors see no reason to attribute this single modification to genetically based evolution (while not, of course, disproving the possibility of such genuine gradualism). Otherwise, stasis prevails throughout the section: "In other, more important characters, *D. subucula* remains morphologically static and shows no evidence of phyletic gradualism" (1985, p. 29).

This case becomes particularly interesting, and merits consideration here, as a demonstration of how far reliable inference can extend, even when the tempo and mode of origin for a descendant species cannot be directly resolved (the usual situation in paleontology). The potential descendant, *D.* favrina, enters the section near the top and overlaps in range with *D. subucula,* thus implying cladogenetic origin rather than anagenesis. The descendant's larger size and hypermorphic morphology suggest a simple heterochronic mechanism for the production of all major differences, hence increasing our confidence in (although clearly not proving) a hypothesis of direct evolutionary filiation. Finally, the fact that no morphological differentia of the species undergo any phyletic transformation within the lifetime of the putative ancestor further underscores the punctuational character of the transition, whatever the mode followed. The one character that does change during the tenure of *D. subucula* (perhaps only ecophenotypically, as discussed above) does not move towards the morphology of descendant *D. favrina.* The authors conclude (1985, pp. 36–37):

> Clearly the sedimentary record is complete enough and represents a sufficiently long period of time to be able to detect phyletic gradualism. Yet throughout this period *D. subucula* remains otherwise morphologically static. Characters that have been modified in closely related species show no

> evidence of undergoing gradual transformation within the duration of the species . . . The overlapping ranges of the two species and the total absence of phyletic gradualism in the characters that serve to distinguish the species suggests that punctuated equilibrium is a better model for speciation in this particular case.

In a later section (pp. 147–171), I shall discuss the generality of stasis within taxa or times under the more appropriate heading of empirical work on relative frequencies. But I shall also note this broader argument here, and in passing, if only to underscore the strong psychological bias that still pervades the field, thereby conveying a widespread impression that gradualism maintains a roughly equal relative frequency with punctuated equilibrium, whereas I would argue that, in most faunas, only a small minority of cases (surely a good deal less than 10 percent in my judgment) show evidence of gradualism. Under this largely unconscious bias, most researchers still single out rare cases of apparent gradualism for explicit study, while bypassing apparently static lineages as less interesting.

Johnson (1985), for example, studied 34 European Jurassic scallop species, and concluded (p. 91): "One case . . . was discovered where . . . the sudden appearance of a descendant form could fairly be ascribed to rapid evolution (within no more than 1 million years). Inconclusive evidence of gradual change over some 25 million years was discovered in one of the other lineages studied . . . but in the remaining 32 lineages morphology appears often to have been static." Yet Johnson virtually confines his biometrical study to the two cases of putative change, presenting only a single figure for just one of the 32 species in stasis. Johnson's title for his excellent article also records this bias in degrees of relative interest—for he sets the unmentioned but overwhelmingly predominant theme of stasis in opposition to his label for the entire work: "The rate of evolutionary change in European Jurassic scallops."

The most brilliantly persuasive, and most meticulously documented, example ever presented for predominant (in this case, exclusive) punctuated equilibrium in a full lineage—Cheetham's work on the bryozoan *Metrarabdotos,* more fully treated on pp. 134–136 and 164–167—began as an attempt to illustrate apparent gradualism. Cheetham wrote (*in litt.* to Ken McKinney, and quoted with permission from both colleagues): "The chronocline I thought was represented . . . is perhaps the most conspicuous casualty of the restudy, which shows that the supposed cline members largely overlap each other in time. Eldredge and Gould were certainly right about the danger of stringing a series of chronologically isolated populations together with a gradualist's expectations." Cheetham's biometry led him to the opposite conclusion of exclusive stasis: "In nine comparisons of ancestor-descendant species pairs, all show within-species rates of morphological change that do not vary significantly from zero, hence accounting for none of the across-species difference" (Cheetham, 1986, p. 190).

The establishment of stasis as an operational and quantifiable subject behooves us to develop methods and standards of depiction and characterization. Several studies have simply presented mean values for single characters in a vertical succession, but such minimalism scarcely seems adequate. At the very least, variances should be calculated (and included in published diagrams in the form of error bars, histograms, etc.)—if only to permit statistical assessment of significances for mean differences between levels, and for correlations of mean values with time.

Smith and Paul (1985), for example, presented both ontogenetic regressions and histograms for samples from each meter of sediment to illustrate stasis in relative size of the peristome in *Discoides subucula* (Fig. 4-1). Cronin (1985) also used both central tendency and variation to illustrate stasis throughout 200,000 years of intense climatic fluctuation (during Pleistocene ice cycles) for the ostracode *Puriana mesacostalis.* Cronin (Fig. 4-2) encircled all specimens of the species at three expanding levels of time in a multivariate plot of the first two canonical axes (encompassing 92 percent of total information): (1) variation in a single sample spanning 100 to 1000 years; (2) in one formation encompassing 20,000 to 50,000 years; and (3) across two formations, representing 100,000 to 200,000 years. Two features of this pattern provide insight into the anatomy of stasis: first, relatively small increase in the full range of variation over such marked extensions in lengths of time; second, the concentric nature of the enlarging ellipses, indicating no preferred direction in added variation, but merely the regular expansion anticipated in any random system with increasing sample size. As Cronin notes, this lack of directionality seems all the more surprising when we recognize that this lineage persisted in stasis through several ice-age cycles. Stasis must be construed as a genuine phenomenon, actively maintained—and not as an absence of anything. Cronin writes (1985, pp. 60–61): "Total within-sample variability representing 10^2 to 10^3 years is only slightly less than variability over 10^5 to 2×10^5 years. *Puriana mesacostalis* shows no secular trends in its morphology over this time interval that might be evident from a lack of concentricity of the ovals—stasis is directionless. Yet high-amplitude environmental fluctuations occurred during this time that could have catalyzed speciation or caused extinction."

Once we construe stasis as an interesting evolutionary phenomenon, actively promoted within species, we then become eager to know more about its fine-scale anatomy and potential causes. A remarkable series of studies by Michael A. Bell on the Miocene stickleback fish *Gasterosteus doryssus* (Bell and Haglund, 1982; Bell, Baumgartner and Olson, 1985; Bell and Legendre, 1987) provide evidence at a maximal level of paleontological resolution, for these fossils occur in abundance in varved sediments with yearly bands—

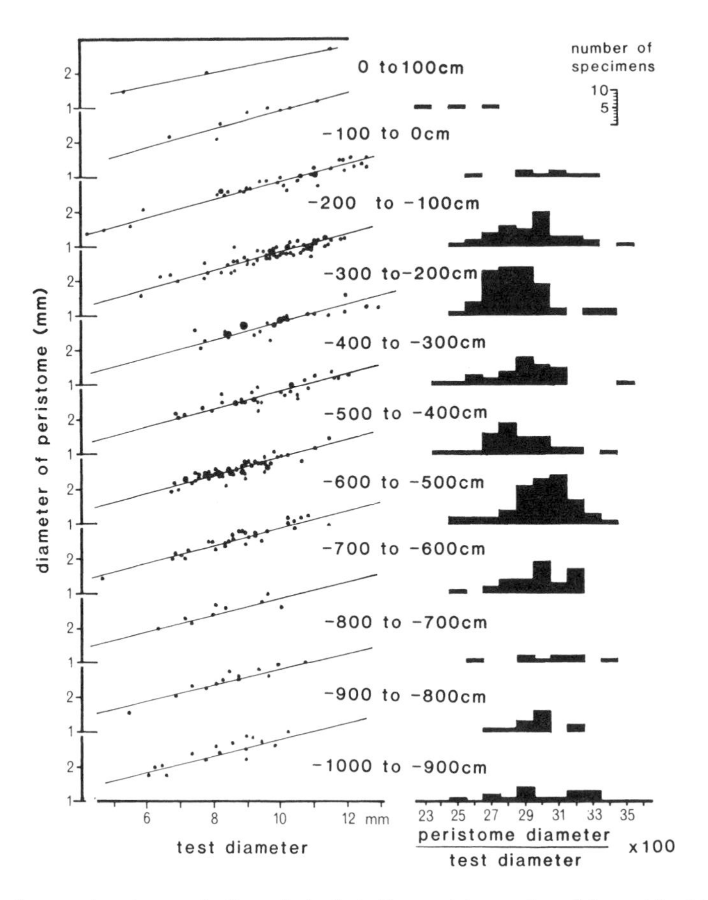

4-1. An impressive demonstration of stasis in the peristome size of the echinoid *Discoides subucula.* From Smith and Paul (1985). All specimens are shown for each narrow collecting interval, spaced one meter apart.

surely a *summum bonum* for attainable temporal precision! Bell and colleagues have documented extensive and complex temporal variability, both for single characters and correlated complexes, over tens of thousands of years (the 1985 study, for example, included several sampling pits covering about 1/3 of the total sedimentary record in a full sequence of approximately 110,000 years). They found some gradual trends in parts of the sequence, a great deal of fluctuation, and a few levels of abrupt alteration, often completely reversing a gradual change built through most of preceding

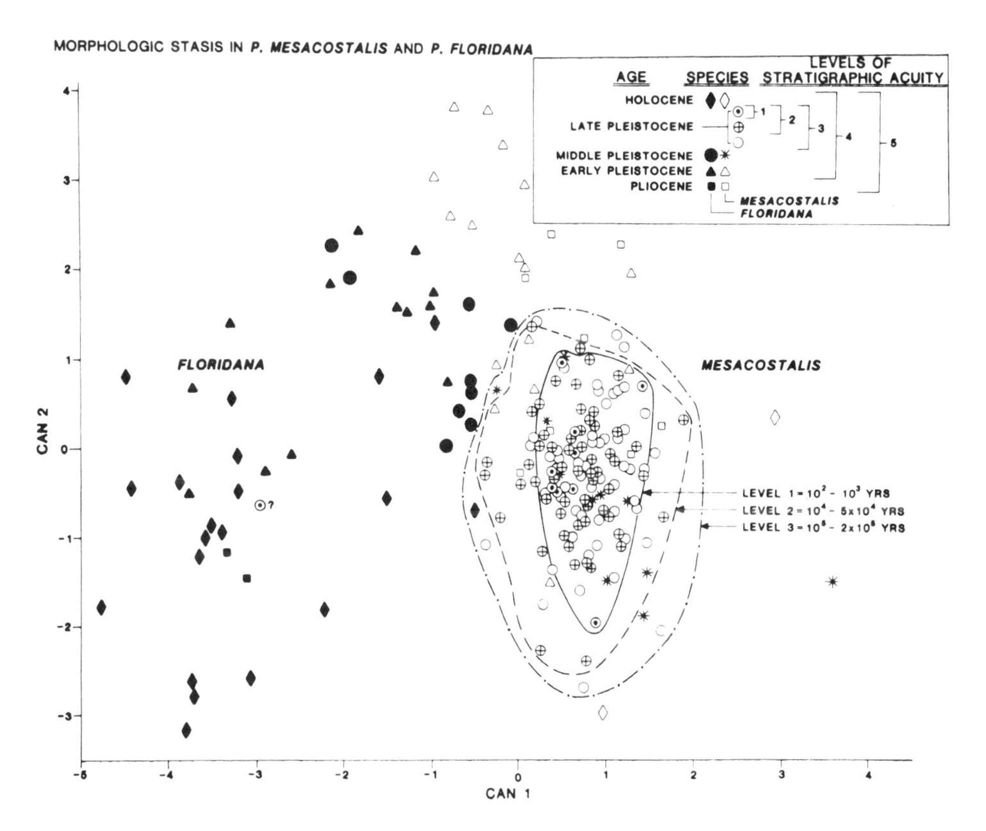

4-2. Two hundred thousand years of stasis, during intense climatic fluctuation of Pleistocene ice cycles, in the ostracod *Puriana mesacostalis.* From Cronin (1985). The three circles around the specimens of this species show increased variation with expanding amounts of time: a single sample representing 100 to 1000 years, one formation encompassing 20,000 to 50,000 years, and two formations representing 100,000 to 200,000 years. The range of variation expands but the modal values do not change at all—as a hypothesis of stasis would predict.

time. In sum, this extensive and multifarious variation includes no sustained or accomplished directionality, and means for most single characters end up about where they began, whatever the internal wanderings between end-points. Bell et al. (1985, p. 264) conclude: "Despite the temporal trends and heterogeneity of all characters through time, the end members [oldest and youngest sample] of only two of the time series (i.e., dorsal spine and dorsal fin ray numbers) are significantly different from each other; most characters return to their original states."

This complex anatomy of stasis again illustrates an active process of maintenance. Perhaps Futuyma's insight (see pp. 77–84) about linkage between speciation and achieved, stable morphological change can also help in this case. I suspect that much of the fluctuation, especially the occasional abrupt changes, represents a complex mosaic of shifting geographic borders for transient local populations. Vertical sequences probably record a mixture of temporary change within local populations and successive migrations of distinct local populations in and out over the same geographic spot (sample pit in this case). But both of these sources—short-term changes within a population and the mosaic of differences between demes—will be transient and fluctuating unless a set of differentia can be "locked up" by reproductive isolation within a newly-formed species. Since Bell's sequence includes no events of speciation, sustained changes do not accrue. The unusual extent of directionless fluctuation then records the especially high degree of temporal resolution.

Bell and colleagues (1985, p. 258) conclude correctly that "the irregular patterns and great magnitude of phenotypic changes that are observed indicate that conventional paleontological samples may miss important evolutionary phenomena and are not comparable to shorter-term evolution in extant populations." Fair enough; I asserted the same argument earlier (see p. 82) by metaphorical comparison to our current cliché for illustrating different scales of fractal self-similarity: one cannot measure around every headland of every sea-cove (transient changes over years in local populations) when calculating the coastline of Maine (macroevolutionary trends over millions of years) at the scale of a single page in an atlas. But one must firmly reject the tempting implication that either scale can be judged "better" or more complete. Yes, the paleontological scale misses "important evolutionary phenomena" of transient fluctuation in local populations. But measurement of details at this local scale cannot be extrapolated to encompass or explain a macroevolutionary trend either—and such local details therefore miss "important evolutionary phenomena" as well. Rather than accusing *any* level of insufficiency for its inevitable inability to resolve events at other unrecorded scales, we should simply acknowledge that any full understanding of evolution requires direct study and integration of the fascinating uniquenesses (as well as the common features) of all hierarchical levels in time and structure.

I have emphasized that one cannot achieve a proper "feel" for the relative frequency of punctuated equilibrium merely by tabulating published cases for individual lineages (whereas the study of relative frequency in a well-bounded taxon, time, or environment—provided that researchers do not preselect their circumstances based on well-documented subjective appearance in favor of one side or the other—has yielded valuable data, as discussed on pp. 147–171). Claims for gradualism attain their highest frequencies at "opposite" ends of the conventional chain of being—that is, for foraminifers and for mammals. In my partisan way, I suspect that the former case may be valid, but attributable to biological differences that predict gradualism within asexual protist "species" as the expected consequence of punctuational clone selection, and therefore a proper analog of gradual trends in metazoan lineages that arise from cumulated punctuational speciation (see pp. 90–93). For mammals, I suspect that published reports follow traditions of work and expectation more closely than they record actual relative frequencies of nature.

I certainly accept the numerous cases of well-documented gradualism in foraminiferal lineages, and I acknowledge MacLeod's argument (1991) that abrupt transitions without branching in some sequences (the "punctuated anagenesis" of Malmgren et al., 1983) may arise as artifacts of condensed intervals of sedimentation within truly gradualistic trends. But punctuated equilibrium has also been demonstrated with data of equal abundance and completeness (the elegant case of Wei and Kennett, 1988, stands out for thorough documentation of both the geography of origin *and* subsequent history of nontrending in *Globorotalia (Globoconella) pliozea*), and we have little idea, and no firm data, about overall relative frequencies for tempo and mode of evolution in this group.

Stasis has been demonstrated in other microfossil "species" with equally dense documentation—e.g., Nichols (1982) on lower Tertiary pollen, Wiggins (1986) on upper Miocene dinoflagellates, and Sorhannus (1990) on the Pliocene diatom *Rhizosolenia praebergonii.* Ross (1990) suggested that putative differences in relative frequencies might be tested by comparing foraminiferal lineages with microfossils of sexual metazoans preserved in comparable abundance in the same sediments—and that forams vs. ostracodes might provide a good test. Indeed, published cases for ostracodes seem to speak strongly for stasis and punctuation as a predominant pattern (in contrast with foraminiferal data). I have already discussed Cronin's (1985) work on Cenozoic ostracodes (see p. 115), and now cite his general conclusion (p. 60):

> Morphologic and paleozoogeographic analysis of Cenozoic marine Ostracoda from the Atlantic, Caribbean, and Pacific indicates that climatic change modulates evolution by disrupting long-term stasis and catalyzing speciation during sustained, unidirectional climatic transitions and, conversely, by maintaining morphologic stasis during rapid, high-frequency climatic oscillations. In the middle Pliocene, 4 to 3 million years ago, at least six new species of *Puriana* suddenly appeared as the Isthmus of Panama closed, changing oceanographic circulation and global climate. Since then morphologic stasis has characterized ancestral and descendant species during many glacial-interglacial cycles.

The origin of new species by branching in response to geographic opportunity (rise of the Panamanian isthmus), rather than by anagenetic gradualism as a selective consequence of changing environments, matches the predictions of punctuated equilibrium. Of the contribution made by stasis to this conclusion, Cronin writes (p. 61): "Morphologic stasis characterizes most shallow marine ostracodes from the western Atlantic that were subjected to these climatic changes, suggesting a pattern predicted by the model of punctuated equilibrium."

Whatley's (1985) study of the common and speciose ostracode genera *Poseidonamicus* and *Bradleya* in Tertiary and Quaternary sediments of the southwest Pacific also match the expectations of punctuated equilibrium throughout. Whatley found some gradualism in size changes (a common pattern), but only stasis for all defining features of shape and ornament. Whatley concludes for *Poseidonamicus* (p. 108): "Although over some 55 million years, the ornament of the genus underwent considerable change, several of its species remained morphologically very stable over long periods of time: 10 to 15 million years being not uncommon . . . This would seem to be evidence of virtual stasis between speciation events with respect to the evolution of the ornament of the various species." In an interesting comment, relating stasis to the major prediction of punctuated equilibrium for evolutionary trends—the stairstep rather than the ball-up-the-inclined-plane model—Whatley writes (p. 109):

> Although the morphological change from the ornate *P. rudis* to the smooth *P. nudus* [I do love the rhyme as well] took place over a time span of more than 50 million years and, therefore, from a generic standpoint represents a very gradual change, it must be emphasized that the individual species within this evolutionary series are effectively invariable with respect to their ornament. Morphological change was abrupt and coincided with speciation and further speciation was required to bring about yet further ornamental change. Ornamental change is clearly saltatory, very abrupt, and punctuated.

Gingerich's (1974, 1976) cases of gradualism in tooth size for several lower Eocene mammalian lineages in the Big Horn Basin of Wyoming instituted the empirical debate about punctuated equilibrium (see our response and critique in Gould and Eldredge, 1977). The tracing of gradualistic sequences for densely sampled series of small mammals (also based on dental evidence) then became an important research program for French paleontologists (see Godinot, 1985; and Chaline and Laurin, 1986, for sources more accessible to anglophonic readers). Large mammals have also furnished evidence for gradual anagenesis within species, as in Lister's study (1993a and b) of mammoths and moose—though he acknowledges that small sample sizes preclude a rigorous distinction of this interpretation from an alternative reading of several cladogenetic events, each perhaps punctuated, and all leading in the same direction of change (Lister, 1993a, p. 77).

But numerous examples of stasis in equally well sampled strata have also been documented for mammals (see pp. 147–171 for commentary on relative frequencies). The rodent sequences that form the empirical basis for most gradualistic studies of the French school have also yielded several examples of stasis (Lich, 1990; Flynn, 1986). Summarizing his work on rhizomyid rodents from the Miocene Siwalik deposits in Pakistan, Flynn (1986, p. 273) wrote: "Most early rhizomyid species survive on the order of millions of years, with at least two spanning about five million years, and display apparent stasis in most characters."

Several analogs of Gingerich's classic studies on gradualism have provided strong evidence for stasis, thus proving diversity of modes, even where gradualism had been most strongly asserted as an exclusive pattern. Gingerich had studied the small condylarth *Hyopsodus* in early Eocene rocks from the

Big Horn Basin of northwestern Wyoming. West (1979), however, found only stasis for the same genus from slightly younger Middle Eocene rocks from the Bridger Formation of southwestern Wyoming. West concluded (1979, p. 252): "Bridger Formation *Hyopsodus* data seems to show little size change through approximately one million years. This stasis or equilibrium condition . . . is the only well developed pattern in Bridger *Hyopsodus*." Schankler (1981) then analyzed another genus, the condylarth *Phenacodus*, from the Big Horn Basin strata used by Gingerich to document gradualism in different taxa, and found only stasis within species (with abrupt transitions between species—a pattern that Schankler interpreted, correctly in my view, as a probable result of migration into a local area, rather than punctuational speciation *in situ*). He concluded (1981, p. 137): "The long-term stasis in morphology and size shown by the four species of *Phenacodus* conforms to the pattern expected in a model of evolution by punctuated equilibria."

As for the mammal we all love best (see pp. 212–222 for a more complete analysis), gradualism had long reigned as an unquestioned (and often quite unconscious) assumption in hominid evolution. An extensive, historically sanctioned set of dogmata, from ideas about "missing links" to the "single species hypothesis," presupposed gradualism as a philosophical foundation. An early study by Cronin et al. (1980)—which would not be defended by several of its coauthors today—made the classic error of regarding a monotonically changing set of mean values as virtual proof for anagenetic gradualism. (Such data cannot distinguish the stair steps of punctuated equilibrium from the same empirical pattern produced by gradualism in highly incomplete sections.)

The spotty data of hominids offer little opportunity for adequate testing of such ideas (and we wouldn't even think of applying an apparatus of this kind to such a poor example if we didn't care so much about the particular case). Nonetheless, I am gratified by some strong hints of substantial stasis in several hominid species, especially for increasingly persuasive data on the importance of apparently punctuational speciation in this small clade during a crucial million year African interval (ca. 2–3 my B.P.) that featured the putative origin of at least half a dozen hominid species. Rightmire's early claims (1981, 1986) for stasis in *Homo erectus* have been strongly challenged (Wolpoff, 1984), though the jury has surely not yet come in (despite a tentative vote from this juror, despite his general biases in the other direction, for at least some fairly persuasive gradualism within this species).

But two apparently sound cases of stasis have attracted substantial attention while we should also not neglect, if only for its radical meaning in the light of previous assumptions, the short-term stasis of *Homo sapiens*, at least from the earliest Cro-Magnon records in Europe (about 40,000 years B.P. to our present circumstances). When we realize that the cave painters of Chauvet, Lascaux, and Altamira do not differ from us in any phenotypic features, their stunning achievement seems less mysterious. For the two more substantial cases, the 0.9 to 1.0 million years of stasis in the first well-documented hominid species, *Australopithecus afarensis* (aka "Lucy"), has been presented with much data and commentary (Kimbel, Johanson and Rak, 1994; see discussion of popular misapprehensions in Gould, 1995). Grine (1993) has also recorded 0.8 million years of stasis in *Australopithecus robustus* from Swartkrans cave in South Africa.

I am, in any case, gratified to note the changing presuppositions of this small, contentious and vital field of paleoanthropology. In early years of this debate, after refuting the Cronin et al. (1980) hypothesis, Jacobs and Godfrey (1982, p. 85) wrote: "The Hominidae can no longer be blissfully assumed to be safely above the punctuationist challenge to the gradualist orthodoxy." Just twelve years later, McHenry could assert in the closing line of his review (1994): "It is interesting, however, how little change occurs within most hominid species through time."

This elevation of stasis to visibility, respectability and even to expectation, has generated subtle and interesting repercussions for gradualism. When gradualism enjoyed high status as a virtually definitional consequence of evolution itself, few researchers thought to question such an anticipated result (but simply rejoiced in any rare instance of affirmation). However, once stasis emerges as an alternative norm, with gradualism designated as uncommon by the same analysis, then gradualism itself must fall under scrutiny for the first time.

With this shift of perspective, a paradox that should have been obvious from the start finally emerged into clear view: gradualism, *prima facie*, represents a "weird" result, not an anticipated and automatic macroevolutionary expression of natural selection—thus, perhaps, accounting for its rarity. Geological gradualism operates far too slowly to yield any workable effect at all when properly scaled down and translated to the immediacy of natural selection in local populations! (See Jablonski, 1999, for a forceful assertion of this paradox.)

Again, we encounter the major dilemma that I call (Gould, 1997c) "the paradox of the visibly irrelevant"—that is, phenomena prominent enough to be detectable and measurable at all in local populations during ordinary human time must cascade to instantaneous completion when scaled into geological time, whereas truly gradual effects in geological time must be effectively invisible at scales of human observation in ecological time. Consequently, what we see in our world can't be the direct stuff, by simple extrapolation, of sustained macroevolutionary change—while what we view as slow and steady in the geological record can't be visible at all (in the same form) by the measuring rod of our own life's duration.

Eldredge and I first raised this point explicitly in 1977 (Gould and Eldredge, 1977), for we had missed this implication in our original formulation of 1972. Here, on this issue, we finally caught the attention of many neontological colleagues who, before then, had been unmoved by punctuated equilibrium. How can geological gradualism be the extrapolated expression of natural selection within populations? Surely, if a doubling of tooth size (say) requires 2 million years to reach completion, then the process must be providing so small an increment of potential advantage in each generation that natural selection couldn't possibly "see" the effect in terms of reliably enhanced reproductive success on a generational basis. Can a tooth elongated by a tiny fraction of a single millimeter possibly confer any evolutionary advantage in a selective episode during one generation of a population's history? Conversely, if bigger teeth provide such sustained advantages, why stretch the process over millions of years? Neontological studies have amply confirmed that natural selection can be a powerful force—the lesson, after all, of our entire, and burgeoning, literature of measurable change in Darwin's finches, anolid lizards, peppered moths, etc. So why shouldn't such a doubling of tooth length be achieved over the palpable span of a few human generations? Of course we all recognize a host of standard arguments for reining in the speed of selective response: negative consequences through discoordination with other parts of the body, slowing by networks of correlated effects upon other anatomical features. But I doubt that even the summation of all such effects could generate sufficient restraining power to spread the blessings of a moment over 2 million years of plodding achievement. (See, however, p. 540 of *SET* for Mayr's confident assertion, *a priori* and without evidence, of this evolutionary style and rate as canonical.)

In other words, gradualism should be viewed as a *problem and a potential anomaly,* not as an expectation. In an important early recognition of this principle, Lande (1976), who (to say the least) is no friend of punctuated equilibrium, calculated that Gingerich's measured trends confer such a small effect upon the immediacy of ecological moments that, for one case, Lande calculated an advantage corresponding to elimination of individuals four or more standard deviations from the mean in regimes of truncation selection! However unrealistic one might deem such a model, no one should miss the "bottom line": most populations don't include *any* viable individuals four standard deviations from the mean—and one can hardly imagine that the removal of such occasional misfits or anomalies could slowly move the mean value of a population to new adaptive heights over a million sustained years.

I do not mean to say that this paradox cannot be resolved to make gradualism intelligible once again, but I do hold that any revalidation demands a substantial reconceptualization for this venerable phenomenon. The obvious solution lies embedded in results such as Bell and Haglund (1982) on the fine-scale structure of stasis. Selection in the immediacy of ecological moments cannot be measured as either the net nontrending of stasis or the steady accumulations of changing means in anagenetic gradualism. Any local population constantly jiggles to and fro in selective accommodation to changing local environment (as when mean coloration for peppered moths becomes darker for a few centuries, but then lighter again, and back to previous values, when lichens return to trees after abatement of industrial pollution). The extent of selection in an anagenetic sequence must be cumulated through each and every one of these jiggles, not measured by calculating the coefficients needed simply to change one endpoint into another. (Such a tactic would lead to the evidently false conclusion that little or no selection had ever occurred in peppered moths.) In other words, perhaps we must construe gradualism itself as a "higher level" phenomenon of net accumulation through the jiggles, not as an expression of ordinary directional selection summed through the ages.

But such a conclusion then raises a different (and broader) question: what, then, *is* ordinary geological gradualism after all? How can such a minuscule directional effect persist through all the swings and jiggles? And what does such a phenomenon represent? Must we interpret such slow net change as caused by drift, as Lande's models made conceivable? Such a conclusion would seem unlikely given the common impression that certain features, size increase in particular, occur preferentially and nonrandomly in gradualistic sequences (but see Jablonski, 1997, and Gould, 1997a, on the apparent falsity, and status as a psychological artifact, of this venerable claim known as "Cope's Rule"). Can we even argue for natural selection as the primary cause of classical gradualism at all? I am confident that selection remains a good candidate, but of what sort, and at what level? The selective basis of gradualism surely cannot be ascribed to the extrapolated advantage at every given moment of traits so enhanced over the long run. Rather, the selective edge must lie in some form of more general benefit not consistently visible in ecological moments, but somehow skewed to a higher probability of immediate occurrence that can then cumulate to a consistent trend in macroevolutionary time.

One might be tempted to equate this skewing agent with some form of general biomechanical improvement that might hold cumulative sway above the jiggling of momentary advantage in any direction. But then the kinds of features that seem to prevail in gradual anagenesis do not stand out for potential membership in this category. Perhaps we need to consider selection on supraorganismal units, or perhaps we should entertain nonselectionist alternatives, especially in the light of Lande's modeling for drift. (Such hypotheses of random change would require a far better knowledge of relative frequencies, both for characters within a taxon and among taxa themselves, than we now possess or even know how to generate.) In any case, I do not think we have even begun to explore the range of potential explanations for the puzzling phenomenon of anagenetic gradualism. I, at least, find the subject very confusing and challenging.

Finally, once we recognize gradualism as an interesting puzzle rather than a dull expectation, we may be led to "dissect" the phylogenetic "anatomy" of such trends more carefully, thus adding an operational benefit to the renewed theoretical interest. In a striking example, Kucera and Malmgren (1998) published an elegant study of morphological change in the late Cretaceous *Contusotruncana* lineage of planktonic forams. After several million years of stasis, the defining feature of "mean shell conicity" increased in a gradualistic manner (see Fig. 4-3) for 3.5 million years, beginning 68.5 million years ago, in this anagenetic lineage.

The mean values of Fig. 4-3 record a conventional gradualistic sequence, but the greater detail of Fig. 4-4, illustrating the morphology of all specimens, not only the means for each level, reveals fascinating details that suggest novel interpretations. In short, the range of variation, after remaining stable during the preceding period of morphological stasis, increased rapidly during the half million year interval from 68.5 to 68.0 million years ago.

The subsequent gradual trend then developed *within* the envelope of this expanded range—a spread in variation that had already reached its full extent at the onset of the gradualistic interval. In other words, variation increased rapidly, and the gradual trend then unfurled into the enlarged morphospace of this new range. In fact, as Kucera and Malmgren point out, the upper endpoint of variation never expands after the initial surge, and the trend in mean values records a loss of variation by removal of flattened shells at the lower end.

I do not mention these details as a punctuational partisan trying to downgrade this example of gradualism, or to reinterpret the trend as a "mere" consequence of a punctuationally expanded range of variation. The gradual trend is both genuine and well documented—but the mapping of variation into its space gives us new insight into potential mechanisms of gradualism (while also imparting an important lesson about the significance of variation, the perils of not recording such data, and the potential for misreading patterns in expansion and contraction of variation as conventionally

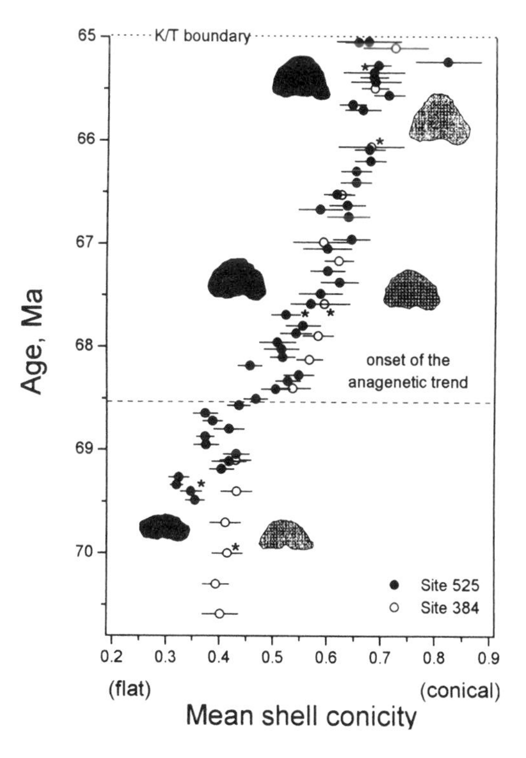

4-3. From Kucera and Malmgren (1998). A good example of gradualism for mean shell conicity in the planktonic foram *Contusotruncana.* After several million years of stasis, this trait increases in a gradualistic manner for 3.5 million years beginning 68.5 million years ago.

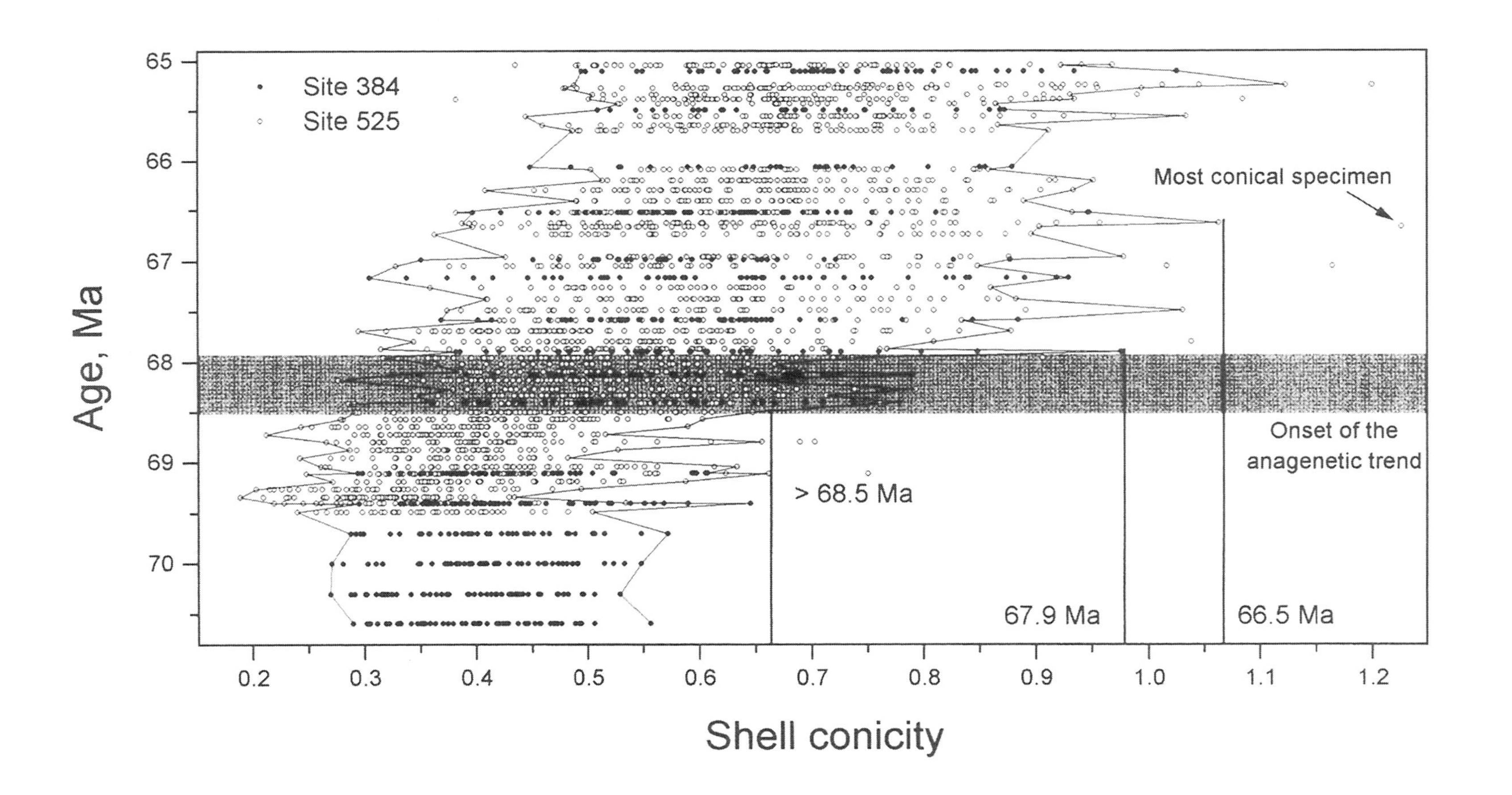
Site 384
Site 525
Most conical specimen
Onset of the
anagenetic trend
> 68.5 Ma
67.9 Ma
66.5 Ma
Age, Ma
65
66
67
68
69
70
Shell conicity
0.2
0.3
0.4
0.5
0.6
0.7
0.8
0.9
1.0
1.1
1.2

directed trends in mean values—see Gould, 1996). The gradual trend to greater conicity in *Contusotruncana* probably warrants a conventional selectionist explanation, in part—shifting means resulting from selective removal of disadvantaged flatter specimens. But we also need to understand the potentiating condition established by an initial (and geologically rapid) expansion in the range of variation. What mechanisms underlie such change in a variational spectrum? Evolution can't anticipate future needs for altered means, so the enlarged range can only be exaptive for the subsequent trend. What, then, lies behind such rarely documented (but eminently testable) expansions and contractions of variational ranges?

The Punctuations of Punctuated Equilibrium: Tempo and Mode in the Origin of Paleospecies

Stasis is data, and potentially documentable in any well-sampled series of persistently abundant fossils spanning a requisite range of time—i.e., most of the duration of an average species in a given taxon, ranging roughly from a million years or so for such rapidly evolving forms (or perhaps just more closely scrutinized, or richer in visibly complex characters) as ammonites and mammals to an average of 5 to 10 million years for "conventional" marine invertebrates. But punctuation may only record an absence of intermediary data. Thus, as noted several times before, the second word of our theory stands more open to general test than the first, and this operational constraint inevitably skews the relative abundances of published information.

If punctuational claims were truly untestable, or subject to empirical documentation only in the rarest of special circumstances, then the entire theory would be severely compromised. Fortunately, many cases at the upper end of a spectrum in richness of data (both in abundance of specimens and

4-4. When we plot the variation for all specimens abstracted as mean values in Figure 4-3, a fascinating pattern emerges (from Kucera and Malmgren, 1998); the variation remained stable during stasis in the ancestral lineage. Variation then increased rapidly during the half million year interval from 68.5 to 68.0 million years ago. The subsequent gradual trend then developed *within* the envelope of this expanded range—a spread in variation that had already reached its full extent at the outset of the gradualistic interval. The change in mean values remains gradualistic, but this addition of data on variation provides a new perspective upon the mechanism.

fine-scale temporal resolution) offer adequate materials for distinguishing the causes and modalities of punctuation. The testable cases do not nearly approach a majority of available species, but neither do they stand out as preciously unusual. Thus, we face a situation no different from most experimental testing in science, especially when we cannot construct ideal conditions in a laboratory, and must use nature's own "experiments" instead. That is, we must pick and choose cases with adequate information for resolution, and without inherent biases that falsely presuppose one solution over others. But experimentalists in "hard science" seek the same unusual resolvability when they "improve" upon the ordinary situations of nature by establishing fixed, simplified, and measurable circumstances in a laboratory. Natural historians proceed no differently, and with no more artificiality or rarity of acceptable conditions for testing—except that we must ask nature to set the controls (and must therefore live by her whims rather than our manipulations. In this sense, the naturalist's tactic of "choosing spots" selectively corresponds with the experimentalist's strategy of establishing controls in laboratories).

The testing of punctuations has generated two primary themes of research: the establishment of criteria for distinguishing among the potential causes of literal punctuations in the fossil record; and the specification of fine-scale "anatomy" (data of timing, mode, morphology, geography, etc.) for punctuational events.

THE INFERENCE OF CLADOGENESIS BY THE CRITERION OF ANCESTRAL SURVIVAL

I doubt that any professional paleontologist would dispute the statement that a great majority of paleospecies make a geologically abrupt first appearance in the fossil record. But this statement about an observed, literal pattern carries almost no interpretive weight because the phenomenon so described can be explained by such a wide variety of putative causes, including the following distinctly different proposals:

THE TRADITIONAL GRADUALIST VIEW. The species arose by geologically gradual transformation of an ancestral population, but our woefully imperfect fossil record did not preserve the intermediary stages.

THE CLADOGENETIC PROPOSAL OF PUNCTUATED EQUILIBRIUM. The literal record represents the expected geological scaling of biological processes responsible for the origin of species. New species arise by isolation and branching of a segment of the ancestral population. The branch evolves to a new species by continuous transformation, but at a rate that, however "slow" by the inappropriate standard of a human lifetime, runs to completion during the "geological moment" of a bedding plane in most cases.

PUNCTUATED ANAGENESIS. The new species arises by continuous transformation, *in toto* and without branching, of an ancestral species, but at a rate too rapid for geological resolution of intermediary stages.

SUDDEN APPEARANCE BY MIGRATION INTO A LOCAL SECTION. The new species arose in another region at a rate and mode that cannot be determined from local evidence. A punctuational first appearance in any particular geographic region records a process of migration from another area of earlier origin.

These four proposals have strikingly different implications for the validation of punctuated equilibrium. The first opposes punctuated equilibrium unambiguously and would disprove the theory, or at least consign it to irrelevancy as a cause of pattern in the history of life, if this mode of classical gradualism could be affirmed at dominant relative frequency for the origin of new species.

Punctuated equilibrium predicts that the second explanation must hold as the primary generator of the dominant empirical signal of punctuational origin for paleospecies. If most species did not arise by rapid cladogenesis at appropriate geological scales, then punctuated equilibrium would be disproven as a major cause of evolutionary pattern (and would be relegated to a status of marginality and insignificance in the history of life).

The third explanation may fall within the "spirit" of punctuated equilibrium, by identifying a genuine geological punctuation, rather than a false appearance based on missing data in gradualistic sequences, as the source for an empirical observation of abrupt origin. But if punctuated anagenesis could be validated at high relative frequency for the origin of paleospecies, then punctuated equilibrium—a theory about cladogenesis—would be demoted or negated, while the important ancillary concept of explaining trends, within a hierarchical model (see Chapter 8 of *SET*), as differential success of species within clades would also become marginalized.

The fourth, or migrational, alternative may resolve a local issue in a given section, but can only indicate, for questions about tempo and mode of speciation, a need for additional information of wider geographic scope. For we must still learn whether the new species, arriving as a punctuational migrant, arose by anagenesis or cladogenesis, and at either a gradual or punctuational tempo, in its natal area. But if the migrant invades the territory of a surviving ancestor—a common pattern in recorded literature—then, at least, we have documented the cladogenetic origin that punctuated equilibrium requires.

Against the charge that our theory cannot be adequately tested, participants in the empirical debate about punctuated equilibrium have long recognized, and generally utilized, an excellent criterion possessing the two cardinal virtues of a probing agent for scientific hypotheses: ready (and unambiguous) application in most cases, and an inherent bias against punctuated equilibrium by underrepresentation of actual cases. I presented this tool—ancestral survival following punctuational origin—in Chapter 3 (see pages 73–76), while leaving the primary documentation for this section.

The criterion of ancestral survival invokes paleontological data of the most conventional and easily acquired kind—specimens in local sections, forming samples of sufficient size for basic taxonomic identifications—and not distant inferences from models or from fossil data to unobservable correlates in behavior or physiology. Moreover, the criterion is properly biased against punctuated equilibrium in recognizing only the subset of legitimate cases with documented ancestral survival (therefore leaving in limbo all genuine cases where ancestors may have survived in other regions, or may not yet have been found). Any tabulation based on the criterion of ancestral survival must therefore underestimate the true relative frequency of punctuated equilibrium.

One potential biasing factor, however, might lead to an overestimate for punctuated equilibrium under this criterion, and must therefore be scrutinized and avoided. Under the fourth explanation presented above for literal observations of punctuation between a descendant and a surviving ancestor in a local section, migration of the descendant from a different region (where it might have originated gradually), rather than punctuational evolution *in situ,* could produce an artificial boost in frequency if falsely counted as a proven case of punctuated equilibrium. (I suspect that most cases in this mode do represented punctuated equilibrium, based on general arguments that most speciation events in unobserved regions of the geographic range will themselves be punctuational, but we obviously cannot count these examples favorably, because our entire case would then become circular by assuming the premise supposedly under test.)

The proper solution to such unresolvable cases lies in proper scrutiny, and in declining to count them as proven support for punctuated equilibrium. Most paleontologists recognize and follow this recommended practice. For example, to cite three titles from opposite ends of the conventional taxonomic spectrum, Sorhannus (1990) could not determine whether the punctuational origin of the diatom *Rhizosolenia praebergonii* from ancestral *R. bergonii* 2.9 million years ago in the Indian Ocean occurred *in situ* or by migration from the central Pacific. He entitled his article: "Punctuational morphological change in a Neogene diatom lineage: 'local' evolution or migration?" Schankler (1981), as previously reported, attributed punctuational patterns of Eocene condylarth *Phenacodus* to probable migration, and called his paper: "Local extinction and ecological re-entry of early Eocene mammals." And Flynn (1986) documented an excellent case of ancestral survival in Miocene rodents from Pakistan (in a group frequently cited for high relative frequencies of gradualism), but couldn't distinguish evolution *in situ* from migration as the cause of observed cladogenesis. He therefore only cited the literal pattern itself in his title: "Species longevity, stasis, and stairsteps in rhizomyid rodents."

Among affirmations of punctuated equilibrium by the criterion of ancestral survival, and ordering my discussion along a conventional taxonomic spectrum (for no reason beyond antiquated custom), Wei and Kennett's classic study (1988) illustrates how geographic data can be integrated with vertical sequences to resolve evolutionary modes not deducible from data of single sections. These authors showed that the upper Miocene planktonic foram *Globorotalia (Globoconella) conomiozea terminalis* evolved gradually into *G. (G.) sphericomiozea* during a 0.2 million year interval in central parts of its geographic range.

At the same time, intensification of the Tasman Front (Subtropical Divergence) separated peripheral populations of the warm subtropics from the central stock. The isolated population then branched rapidly into a new species, *G. (G.) pliozea,* in less than 0.01 million years, or 5 percent of the time taken for anagenetic transformation of the ancestral stock at the center of its range. The anagenetic trend proceeded in a direction (loss of keel and development of a more conical test) opposite to the morphological innovations (flattened test and more pronounced keel) of the allopatrically speciating peripheral form. The new species, following its punctuational origin, persisted in stasis for more than a million years. About halfway through this interval, a descendant of the central stock migrated into the warm subtropical region of *G. (G.) pliozea.* The two species then coexisted for half a million years without apparent intermixing, and with no interruption of stasis.

The rich data of microfossils from oceanic cores, often providing good resolution for both geographic and temporal variation, have also documented punctuational speciation (usually allopatric) with ancestral survival in several other cases. Cronin (1985) correlated the punctuational origin of six species in the ostracode *Puriana* with changes in oceanographic circulation engendered by the Pliocene rise of the Isthmus of Panama. Cronin comments (p. 60) that "since speciation occurred, ancestral species and their descendants have coexisted, in some cases sympatrically." In a study of Miocene deep sea ostracodes from the southwest Pacific, Whatley (1985, p. 109) documented two cases of allopatric and punctuational origin for new species followed by migration back to the parental range and subsequent coexistence with the ancestral species.

Alan Cheetham's work on American Cenozoic clades of the cheilostome bryozoan genera *Metrarabdotos* and *Stylopoma* (Cheetham, 1986, 1987; Jackson and Cheetham, 1994; Cheetham and Jackson, 1995) merits citation at several points in this chapter for its unparalleled documentation of all major tenets of punctuated equilibrium—both in clarity of conclusions and richness of empirical evidence. I present a general summary in the section on relative frequency (p. 164), but Cheetham's fruitful use of ancestral persistence should be noted here. Jackson and Cheetham (1995, p. 204) cite three primary empirical sources for documenting punctuated equilibrium from paleontological data: "The geologically abrupt appearance of species in the record, the static morphologies of species for millions of years, and the extensive temporal overlap between apparent ancestor-descendant species pairs."

Their summary of overwhelming support for punctuated equilibrium from the last source (Jackson and Cheetham, 1994, p. 407) states that "most well-sampled *Metrarabdotos* and *Stylopoma* species originated fully differentiated morphologically and persisted unchanged for > 1 to > 16 million years, typically alongside their putative ancestors."

On Cheetham's celebrated and frequently reprinted diagram of evolution and cladogenesis in the *Metrarabdotos* clade (Fig. 4-5, and redundant in citing "evolution and cladogenesis" because all phyletic change occurs by cladogenesis in this lineage), ancestors persist after the origin of descendants in 7 of the 9 cases where Cheetham felt confident enough to assert a phylogenetic claim for direct filiation. (Marshall's important challenge [1995] to assessments of stratigraphic range in several cases does not counter Cheetham's hypotheses about filiation, and certainly does not challenge the assertion of overlap, a claim based on direct observation of joint occurrence, not on inference.) The two cases where ancestral persistence has not been directly observed (see Fig. 4-5), but may well have occurred (the derivation of *M. tenue* from *sp. 10*, and of *M. unguiculatum* from *M. lacrymosum*), both fall "outside the interval of dense sampling" (Cheetham, 1986, p. 201), where Cheetham achieved a stratigraphic resolution by Sadler's (1981) criteria of 0.63. For *Stylopoma*, "eleven of the nineteen species originate fully formed at $p > 0.9$, with no evidence of morphologically intermediate forms,

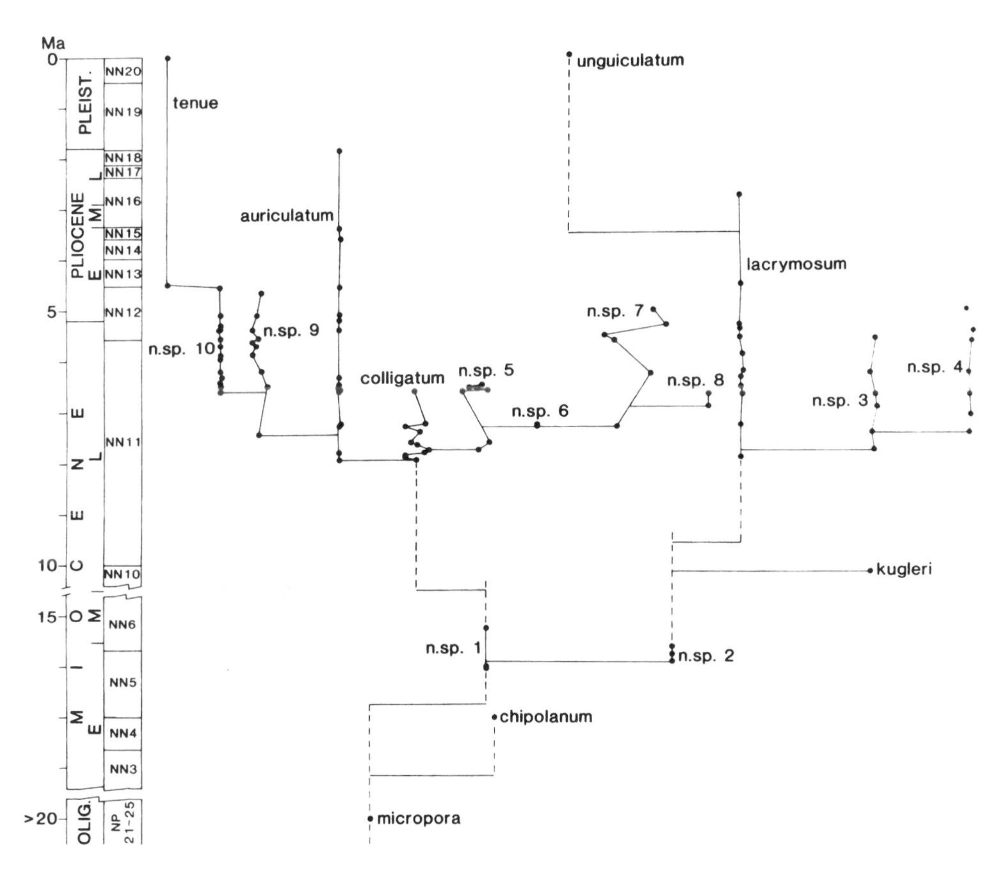

4-5. The best-documented, indeed already canonical, example of punctuated equilibrium as an invariant pattern for an entire clade across its full geographic range—the research of Cheetham on Tertiary and Quaternary Caribbean species of *Metrarabdotos*. Each point depicts a multivariate centroid based on all characters, not just a single feature. All species express stasis, several for extended periods and a large number of samples. Ancestors persist after the origin of descendants in 7 of 9 cases where Cheetham felt confident enough to assert a claim for direct filiation.

and all ancestral species but one survived unchanged along with their descendants" (Jackson and Cheetham, 1994, p. 420).

By dense sampling in both vertical sequence and geographic spread, Nehm and Geary (1994) demonstrated the punctuational origin of the gastropod *Prunum christineladdae* from its ancestor *P. coniforme* in a small part of its Caribbean range, during a short interval (0.6 to 2.5 percent) of ancestral persistence in stasis. Following the descendant's origin, ancestors continued to thrive in central areas of the range.

In a common pattern found in many taxa, punctuated equilibrium can be confirmed, even in local sections, and even when ancestors do not occur in the same strata as their descendants. Frequently, a population from an ancestral species of known and widespread geographic range branches punctuationally to a descendant that maintains exclusive occupancy of the range for a time, but then becomes extinct. The ancestor subsequently reinvades the range, thus establishing earlier coexistence during the descendant's geological tenure. For example, Bergstrom and Levi-Setti (1978) documented the threefold reappearance of the Middle Cambrian trilobite *Paradoxides davidis davidis* following local and allopatric origins of derived taxa that then become extinct at diastems, with the ancestor reappearing in strata just above.

Similarly, Ager (1983) traced the allopatric origin of late Pliensbachian brachiopod species from the central stock of *Homoeorhynchia acuta,* and the later Toarcian migration of the descendant *H. meridionalis* into the ancestral region. Williamson's (1981) celebrated and controverted study (see p. 44) of punctuational origin for several pulmonate snail species in African Pleistocene lakes invokes the same kind of evidence—as the ancestral species migrates back (in several separate episodes, moreover) after a coalescence of lakes and the extinction of descendant species that had originated in previous times of isolation.

When all evidence derives from a restricted region, the separation of punctuation *in situ* from migrational incursion (with origin elsewhere at an unspecified tempo) becomes more difficult, but some criteria of admittedly uncertain inference may still be useful. For example, Smith and Paul (1985) argue that the sudden appearance of the descendant echinoid *Discoides favrina* in strata still holding ancestral *D. subucula* may represent an event of punctuational speciation on morphological grounds—for the descendant species, though visually distinct in many features, can be easily derived, given allometric patterns shared by both forms, through a simple heterochronic process of hypermorphosis.

In graptolites, the pattern of ancestral survival after cladogenetic origin of a descendant taxon has been noted frequently enough to inspire its own terminology as the concept of "dithyrial populations" (Finney, 1986), or samples from the same stratum containing two directly filiated and nonintergrading species.

The widespread geographic distribution of many late Tertiary and Quaternary mammalian lineages provides several examples of geographically resolvable allopatric origin followed by later survival with the ancestral species. For example, *Mammuthus trogontherii,* the presumed ancestor of the woolly mammoth *M. primigenius,* first appears in northeastern Siberia while its presumed ancestor, *M. meridionalis,* continued to survive in Europe (Lister, 1993a, p. 209).

Other forms of evidence can lead to strong inferences from data of ancestral survival to origin of descendants by punctuated equilibrium, even in the absence of such firm geographic data. I previously mentioned the growing evidence for rapid cladogenesis as the primary pattern in hominid evolution, based on several criteria, including the high relative frequency of observed overlap, the limited time available for cladogenetic origin (even when place and geological moment have not been clearly specified), and our confidence that all events (at least preceding the origin of *Homo erectus*) occurred in Africa. In his review, McHenry (1994, p. 6785) stated that "ancestral species overlap in time with descendants in most cases in hominid evolution, which is not what would be expected from gradual transformations by anagenesis." McHenry's summary diagram (reproduced here as Fig. 4-6) shows a clear pattern of dominant relative frequency for rapid cladogenesis—a weight that has only increased in the light of discoveries since then (see Leakey et al., 2001), particularly for a vigorous phase of cladogenesis 2–3 million years ago, leading to at least half a dozen hominid species (see Johanson and Edgar, 1996).

On the same subject of punctuational and cladogenetic reformulations for classic evolutionary trends previously framed (and widely celebrated in both textbook and story) as exemplars of anagenetic gradualism, the phylogeny of horses has been rewritten as a copious cladogenetic bush replete with ancestral survival in the very parts of the sequence once most firmly read as a tale of linear progress. For example, Prothero and Shubin (1989) have shown that the Oligocene transition from *Mesohippus* to *Miohippus* conforms to punctuated equilibrium, with stasis in all species of both lines, transition by rapid branching rather than phyletic transformation, and stratigraphic overlap of both genera (one set of beds in Wyoming has yielded three species of *Mesohippus* and two of *Miohippus*, all contemporaries). Prothero and Shubin conclude: "This is contrary to the widely-held myth about horse species as gradualistically-varying parts of a continuum, with no real distinctions between species. Throughout the history of horses, the species are well-marked and static over millions of years. At high resolution,

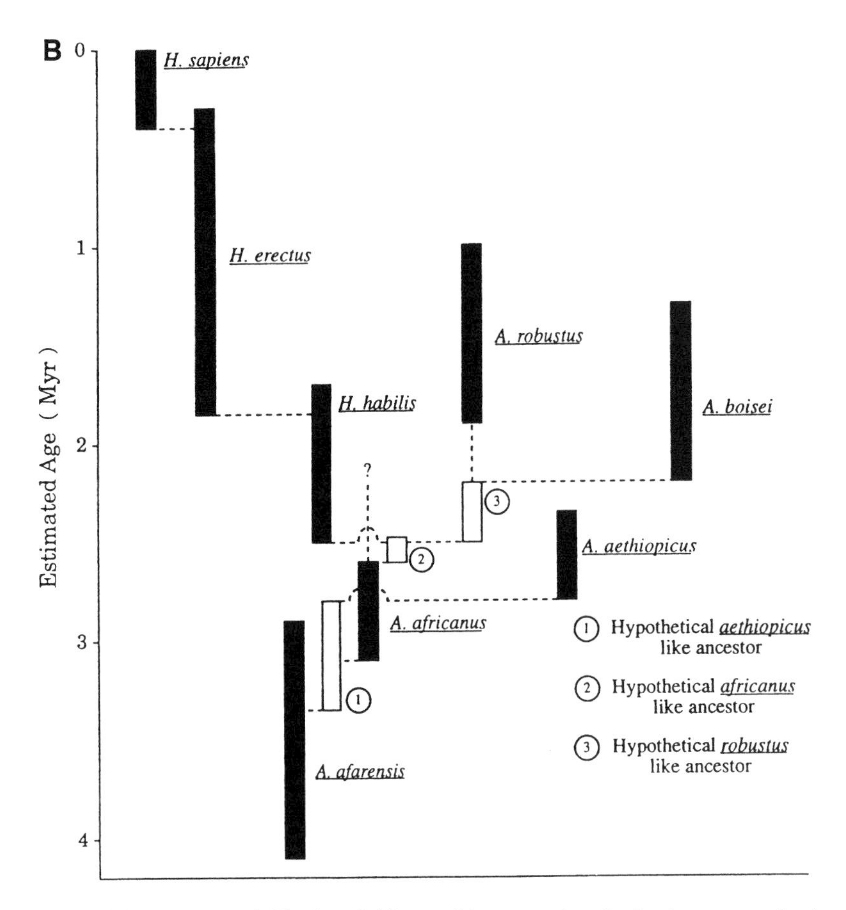

4-6. From McHenry (1994). The hominid record is spotty, but the basic pattern of substantial stasis within several species—particularly *A. afarensis*—and numerous branching points with persistence of putative ancestors lends support to the model of punctuated equilibrium.

the gradualistic picture of horse evolution becomes a complex bush of overlapping, closely related species."

To end this section with a particularly instructive example, punctuated equilibrium has frequently been saddled with the charge that inherent limitations of paleontological data yield biased results, artificially and superficially favorable to the theory—with Darwin's classic argument against a literal reading of punctuations as the conventional antidote. However, an opposite bias may also be significant, and may lead to serious underestimation of punctuated equilibrium in a circumstance likely to be quite common: when a descendant, fully distinct at its origin but initially rare, enters the ancestral area, and then increases steadily in relative abundance as the ancestor declines to extinction. The true evolutionary pattern will be fully punctuational, with stasis in both ancestor and descendant throughout, and with abrupt geological origin of the descendant. But if we misread the event as a tale of anagenetic transformation, and if the two species overlap extensively in ranges of variation, then we will misinterpret the full pattern as transformation by anagenesis, rather than replacement with steadily increasing relative abundance of the descendant species.

The important distinction between these interpretations can be made with appropriate statistical tools applied to samples of sufficient size—but the punctuational alternative must be conceptually available to suggest such a test. In this subtle sense, among so many other more overt reasons explored in this book, expectations of gradualism seriously restrict our range of potential explanations for evolutionary modes and tempos—and punctuated equilibrium therefore becomes both suggestive and expansive, whether or not the hypothesis holds in any particular case.

In an elegant demonstration of this principle, Heaton (1993 and 1996 for data *in extenso*) showed that a classic case of supposedly gradualistic anagenesis in Oligocene rodents from the western United States really represents a case of replacement. Heaton writes (1993, p. 297): "Statistical investigation of large samples suggests instead that two closely related species coexisted, and the shift in mean size that was thought to represent anagenesis actually represents replacement."

Heaton demonstrated the distinct character of the two taxa both by bimodality in their joint occurrences (Fig. 4-7), and by showing that the two species maintained distinctly different geographic ranges (with overlap in Nebraska and eastern Wyoming, but only the descendant taxon living at the same time in the Dakotas—see Figs. 4-8 and 4-9). The small spe-

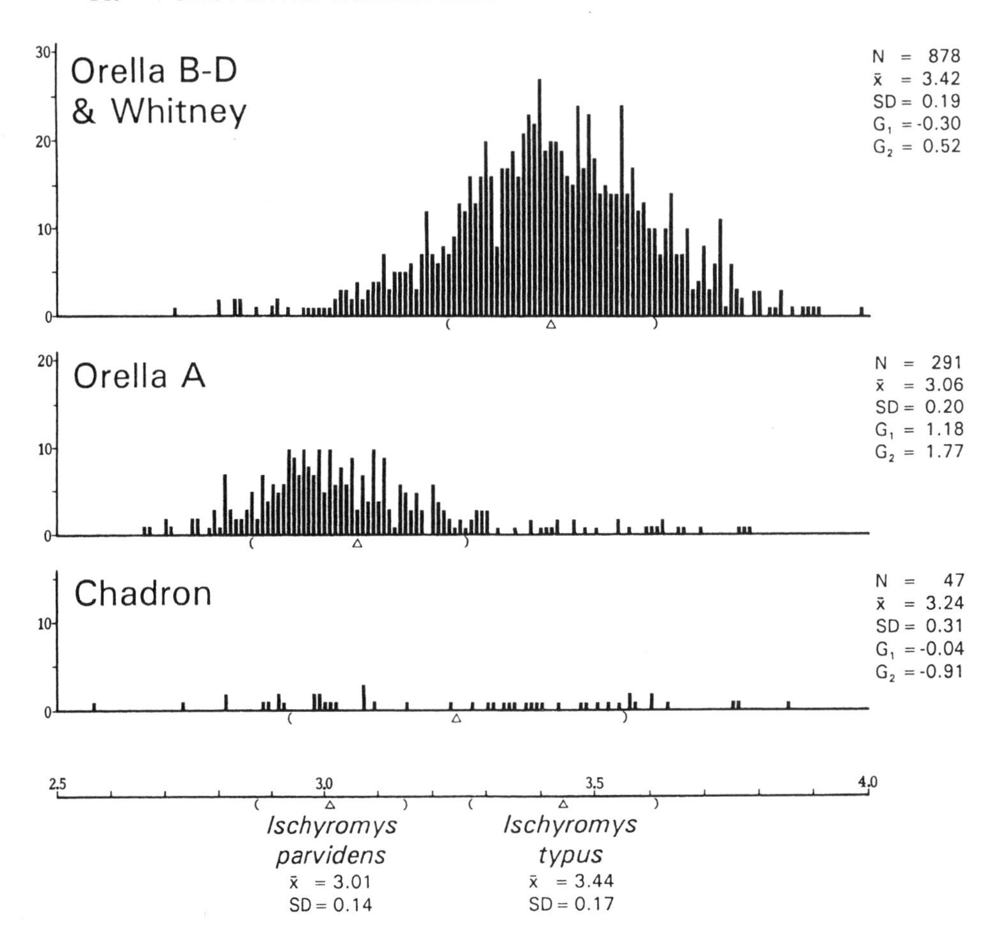

4-7. From Heaton (1993). Data that had, in the past, been interpreted as a gradualistic evolution of increasing size within a single species actually represent a change in relative abundance of two species, each stable throughout its interval—with the species of larger body size gradually becoming more common in the local section.

cies, *Ischyromys parvidens,* predominates in the early Orellan, although the larger *I. typus* already occurs low frequency in the same strata. *I. typus* then increases, as ancestral *I. parvidens* declines, throughout the remainder of Orellan times.[1] Interestingly, *I. typus* does undergo a small anagenetic increase following the extinction of *I. parvidens,* "but this change is minor and not deserving of chronospecies recognition" (Heaton, 1993, p. 297), and the species, in any case, becomes extinct soon thereafter—a common pattern, also strongly implicating punctuated equilibrium as the major generator of

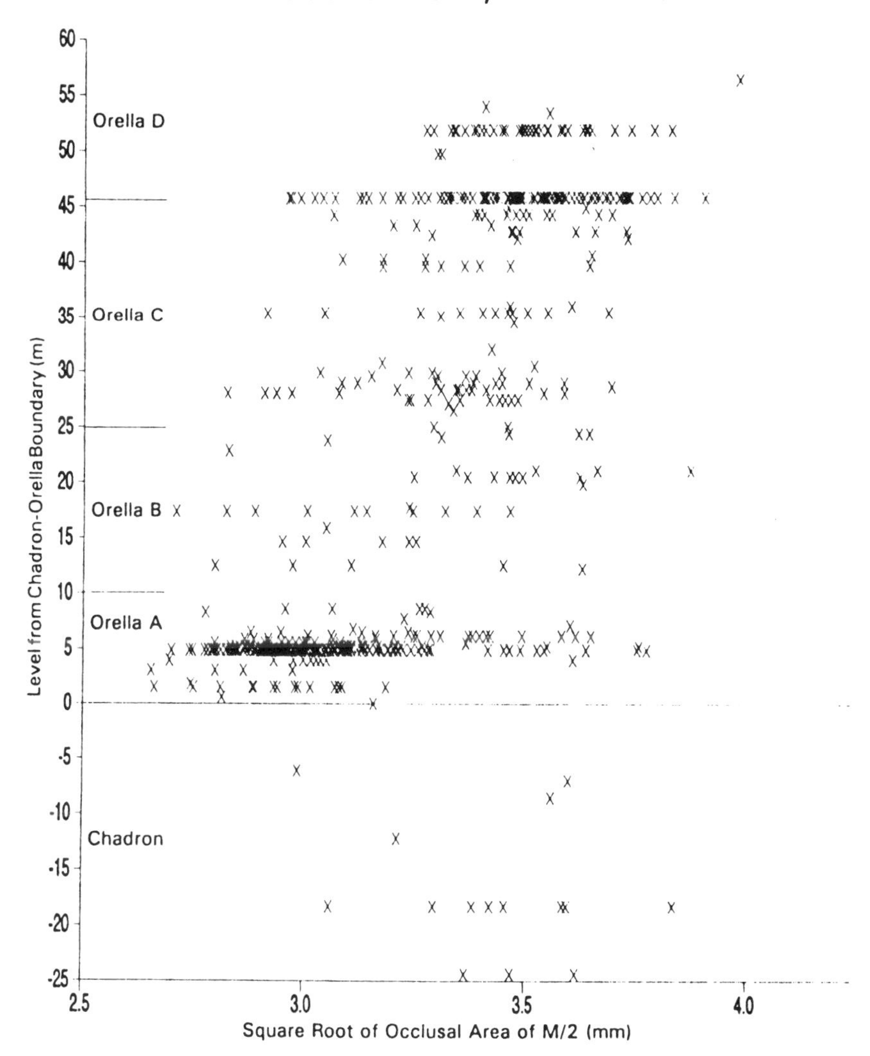

4-8. Note that both species of *Ischyromys* live sympatrically and remain in stasis in some parts of their range, particularly in Nebraska and Wyoming.

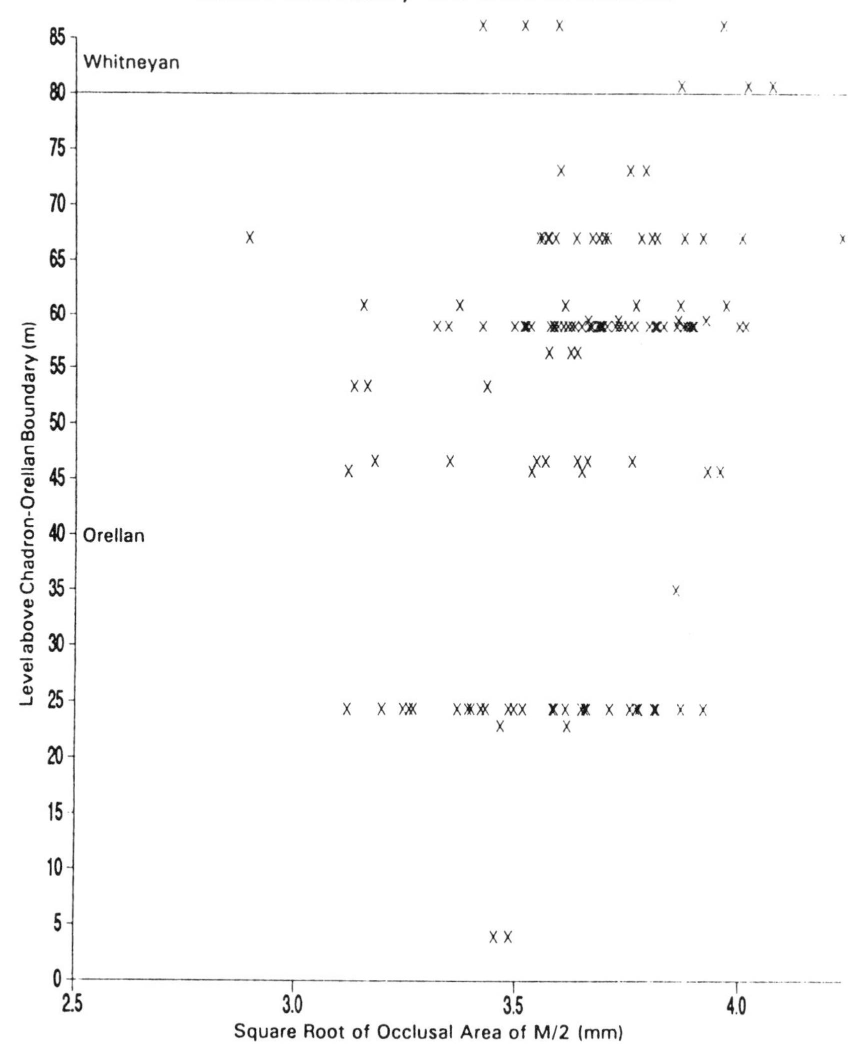

4-9. In other areas of the range, as here in South Dakota, only the descendant species lives during the entire interval.

larger trends, if only because the fruits of anagenesis get plucked so quickly unless they can be "locked up" in cladogenetic iterations.

THE "DISSECTION" OF PUNCTUATIONS TO INFER BOTH EXISTENCE AND MODALITY

Once a literal punctuation has been noted, and a cladogenetic origin inferred by such criteria as the documentation of ancestral survival, further testing of punctuated equilibrium as the mode of origin for the new species may be achieved by several standards that might be characterized (somewhat metaphorically) as devices for "dissecting" the punctuation by revealing an internal "fine structure" with probative value for inferences about evolutionary causes. Three major modes of dissection have been featured in the existing literature (although the theme has not been organized in this manner before), each explicitly invoked as a tool for the potential validation or refutation of punctuated equilibrium.

Time

I discussed the operational definition of punctuation as scaled to periods of stasis (see pp. 39–49). The obvious barrier to testing this primary requirement of the theory lies in our inability to specify requisite information about time in "standard" paleontological situations, where the duration of speciation lies beneath the resolving power of our basic operational "moment"—the bedding plane. Therefore, to achieve a proper dissection, we must search for unusual situations that permit an adequate resolution of time in one of two manners dictated by the logic of the problem: either by finding a way to date individual specimens compressed on a single bedding plane, or by locating situations of unusually rapid sedimentation, where a sequence of events usually collapsed onto a single bedding plane can be expressed in true temporal order through a vertical sequence.

The first tactic can be applied only in highly unusual circumstances effectively limited to nearly modern bedding planes with specimens that can be dated individually, for the error bars associated with most radiometric techniques exceed the entire duration of most bedding planes (except for isotopes with very short half lives, which can then only be applied to Pleistocene or Holocene specimens). However, in a recent example (discussed more fully on p. 47 and Fig. 2-2), Goodfriend and Gould (1996) traced a species transition by hybridization in the land snail *Cerion* on the Bahamian island of Great Inagua. We found all specimens jumbled together on a modern mudflat (a bedding-plane-to-be, if you will), and we then used a combination of radiocarbon dating and amino acid racemization to determine that the smooth and complete species transition occupied 15,000 to 20,000 years—a reasonable figure for a punctuational event (here compressed, as usual, into a single geological "moment," which we, thanks to the rare combination of recent occurrence and availability of dating techniques, were able to disaggregate and resolve).

In the far more common situation of sedimentation rates high enough to spread the usual compressions of single bedding planes into resolvable vertical sequences, assessments have been made in both relative and absolute terms. I previously cited Fortey's (1985) conclusion in the relative mode (see p. 44), based on calibrating punctuational origins against gradual transitions observed for other taxa in the same strata (thus obviating the usual claim that literal punctuations probably represent a geologically slow gradualism that extremely spotty sedimentation cannot record). With this technique, Fortey's found about a 10:1 ratio for punctuated vs. gradual origins of species in Ordovician trilobites from Spitzbergen.

The best examples in the more satisfactory absolute mode do not arise from direct paleontological records, but as firm inferences based on species flocks in lakes or on islands of known and recent origin (with African cichlid fishes as the classic case of modern evolutionary biology). These evolutionary "explosions" often produce several hundred species in just a few thousand years, and must be ranked as punctuational with a luxurious vengeance! But such circumstances do not represent a norm for most speciation in most clades, and such an unusual phenomenon, however stunning and however well documented, cannot suffice to validate a proposed generality.

The punctuational origin of many species can be accurately timed with direct paleontological data. Lister (1996) calculated a maximum of only 5000 years for the Quaternary evolution of dwarfed woolly mammoths on Wrangel Island, and 6000 years for the dwarfed red deer of Jersey. The punctuational origin of the marginellid gastropod *Prunum christineladdae*, based on the study of Nehm and Geary (1994), took 73,000 to 275,000 years, and spanned 0.6 to 2.5 percent of the full duration of the ancestral species. Reyment (1982) calculated outside limits of 100,000 to 200,000 years (perhaps a good deal less) for origin of the Cretaceous ostracode *Oertiella chouberti* from its ancestor, *O. tarfayaensis*.

Geography

I have already discussed this important tool for validating punctuated equilibrium by gathering data at a more inclusive and finer scale than the local documentation of a literal punctuation. On pp. 130–135, I described cases where geographic data affirmed an allopatric and punctuational event of cladogenesis, thus demonstrating that the abrupt appearance of a descendant species truly represents punctuated equilibrium, and not just a migrational incursion of a species that originated by an uncertain mode in an unknown place (Wei and Kennett, 1988, for protistans; Williamson, 1981, for mollusks; Lister, 1996, and Heaton, 1993, for mammals, among many others. Albanesi and Barnes (2000) present a particularly well documented case both for allopatric and punctuational origin of new taxa and survival of ancestors in their original regions for a lineage of Ordovician conodonts).

Morphometric Mode

By quantitative study of patterns in morphological transition between ancestral and descendant species, several criteria of inference can increase our confidence in the identification of punctuated equilibrium, both by establishing a case for direct filiation rather than simple replacement by a taxon evolved elsewhere, and by indicating a punctuational mode for the cladogenetic event. As illustrations of this approach, consider three effective morphometric arguments:

1. Visually extensive change (supposedly requiring many independent inputs expressed over substantial time) can arise as coordinated consequences of one, or few, generating factors, and can therefore readily be accomplished at a punctuational tempo. This "standard" argument has a long pedigree, and serves many purposes, in evolutionary theory (see *SET,* Chapter 10, for example, on "positive" constraints). In the context of punctuated equilibrium, this theme establishes plausibility for temporal compression of visually substantial change into a single cladogenetic event at a punctuational tempo. The argument also proceeds in several modes, including inductive approaches based on "covariance sets" (Gould, 1984a) of correlated characters transformed together along a single multivariate axis. For example, the transition documented by Goodfriend and Gould (1996) involved several measures coordinated by a single change in direction and rate of growth along the axis of coiling—a coherently correlated pattern running orthogonal to, and therefore independent of, the standard covariance set represent-

ing shell size alone. Since the cladogenetic event altered shape (as expressed along the axis of coiling), but not size, this multivariate separation established the source of morphometric change and also revealed its unitary nature in modified growth.

Among examples of the opposite deductive approach, based on fitting an apparent complexity of observed changes to a simple model of underlying generation, Smith and Paul (1985) recognized a suite of alterations through a punctuational event in Cretaceous echinoids as coordinated consequences of a single heterochronic change; and Benson (1983) explained a "punctuational event" (1983, p. 398) in the ostracode *Poseidonamicus* as a set of secondary, and mechanically automatic, accommodations of the carapace to a primary change in shape.

2. Patterns of changing variation through a "dissected" punctuation may reveal a cladogenetic mode of direct filiation. Empirical patterns of variation may permit distinction between punctuational incursion from elsewhere (a migrational event of uncertain interpretation) and cladogenesis *in situ*. To contrast two studies previously reported, Heaton (1993) found consistent and unaltered bimodality in a vertical sequence of Oligocene rodents, with an appearance of overall gradualism arising as an artifact of directionally changing relative abundances. Such a pattern—while confirming stasis and showing the utility of punctuated equilibrium as a generator of alternative hypotheses—cannot resolve the descendant's mode of origin, for the new species enters the fossil record in full and complete distinction. In Williamson's study of African lake mollusks (1981), however, a distinctive pattern of variation (see previous discussion on page 44) implicates cladogenesis *in situ*—for Williamson's fine-scale vertical resolution allowed him to discern an initial period of expanded intrapopulational variation followed by a reduction back to ancestral levels, but now centered about the altered mean of the new species.

3. Comparison of within- and between-species variation as a test of extrapolationism. If new species arise by gradualistic anagenesis, then the direction of selection within populations and the pattern of temporal variation during the life of a species should mirror the morphological changes between species in a geological trend. But Shapiro (1978) estimated natural selection in the Miocene scallop *Chesapecten coccynelus* by comparing specimens that died as juveniles to those that survived to adulthood. He measured the direction of implied change as not only different from, but actually orthogonal to, the distinction between this species and its descendant,

Chesapecten nefrens. Shapiro could not resolve the tempo or mode of the cladogenetic event itself, but he showed that its direction cannot be extrapolated from an inferred pattern of change by natural selection within the ancestral species.

Kelley (1983, 1984) then conducted a more extensive study of all molluscan species with sufficiently rich vertical records in these classical and well-studied Miocene beds of Maryland. She found trending within species for only 16 of 90 cases (17.8%) as defined by well-determined rank correlation coefficients between a shell measurement and stratigraphic position. But for the majority of positive coefficients, she found that the trend within a species is "oriented opposite to the direction of the succeeding species's morphology, indicating a decoupling of macroevolution from microevolution in those cases" (1983, p. 581).

Proper and Adequate Tests of Relative Frequencies: The Strong Empirical Validation of Punctuated Equilibrium

THE INDISPENSABILITY OF DATA ON RELATIVE FREQUENCIES

As stated at the beginning of Chapter 3, proponents of punctuated equilibrium have always recognized that the theory cannot be proven, and can win only two minimal validations—proof of plausibility and promise of testability—from documentations, however rigorous and complete, of individual cases (as presented in the last two sections). As its primary claim, therefore, punctuated equilibrium must assert a dominant role for stasis within species and rapid cladogenesis between species in the construction of macroevolutionary patterns at the appropriate scale of speciation and trends across species within clades. This assertion requires that punctuated equilibrium maintain a *dominant relative frequency* in the origin of new paleospecies. Tests of the theory must therefore focus upon percentages of occurrence in exhaustive, or at least statistically definitive, surveys of particular taxa, faunas, and times.

Species cannot be conceptualized as indistinguishable beans in the conventional bag of our standard metaphor for problems in probability—for the nature of history grants uniqueness to times and taxa, and therefore precludes any simple tabulation by global enumerative induction. We may, however, assess relative frequencies for well-bounded situations restricted to taxa of a given fauna, species within a monophyletic clade, or representatives of a particular time or geological formation. Several such studies have been carried out, and effectively all have found the clear signal of a dominant relative frequency for stasis and punctuation, as predicted by the theory of punctuated equilibrium. I regard these data as our most convincing indication of the validity and importance of punctuated equilibrium as a primary generator of pattern in the history of life. I am also surprised that this clear signal has not been more widely appreciated as the most decisive result in a quarter century of research and debate about punctuated equilibrium.

For reasons previously discussed under the heading of "publication bias" or "Cordelia's dilemma" (see the end of Chapter 1), proper tests of relative frequencies cannot be made by a "catch as catch can" style of simple enumeration based on previously published studies done for other reasons. Until quite recently in paleontology, strong and pervasive biases equated evolution with gradual change, and regarded stasis as "no data," and therefore not worth recording. Tabulations of older literature will inevitably favor gradualism both because no other style of evolution attracted study, and (even more problematically) because paleontologists, expecting only gradualism, tended to misread other patterns in this conventional light. Proper (and noncircular) testing—as in any statistical study—requires that the items chosen for sampling display no bias (imposed by human choice or preference) away from their relative frequencies in nature. When this ideal cannot be realized in natural experiments, which necessarily lack the rigor of laboratory controls, we should at least insist that unavoidable biases be directed against the hypothesis under test.

Thus, one cannot achieve a reliable relative frequency for punctuated equilibrium by tabulating cases from an existing literature, where strong biases in favor of gradualism may reasonably be suspected (or, to put the issue more accurately, virtually guaranteed). May I simply restate Tony Hallam's comment to me on why evolutionary studies of mollusks in English Liassic beds have concentrated with near exclusivity on *Gryphaea* (which, ironically, does not, after all, display the kind or direction of gradualism that initiated this literature in Trueman's famous (1922) paper—see Hallam, 1968; Gould, 1972; Jones and Gould, 1999): "Why hasn't anyone ever examined any of the 100 or so *other* molluscan species, many with equally good records, in the same strata?" Hallam then answered his own rhetorical question: "Because they seem to show stasis, and were therefore regarded as uninteresting"(see Johnson's (1985) affirmation of this stasis).

As an example of major differences between adequate and biased modes of sampling, two contrasting studies were presented at the North American Paleontological Convention, Boulder, Colorado, 1986. Barnovsky calculated the relative frequency of punctuated equilibrium vs. anagenetic transformation for Pleistocene mammals based exclusively on previously published reports in the literature. The two modes were supported at close to equal frequency.[2] Prothero then reported his field study for *all* mammalian lineages in Oligocene rocks of the Big Badlands of South Dakota. (See pages 156–160 for a full discussion of Prothero's refined and extended results—an even more impressive validation of punctuation equilibrium by well-established relative frequencies.) Nearly all lineages remained in stasis, and all new forms entered the record with geological abruptness. Prothero found very few cases of gradual anagenesis. Of course the differences might be real; perhaps the Pleistocene did witness a much higher frequency of gradualism. But I suspect that Barnovsky's result records a bias in older literature, when paleontologists tended to publish only when they found "interesting" lineages in the midst of change. But Prothero studied *all* lineages for a time and place, without preconception about modes or tempos—and his relative frequencies matched the predictions of punctuated equilibrium.

Proper empirical tests of relative frequencies impose two crucial requirements: *first,* that cases be sampled without any preselection in favor of one outcome or the other; and *second,* that cases be sufficiently numerous to establish a statistically significant relative frequency for a totality. The "totalities," "universes," or "populations" that inspire studies of relative frequencies for testing punctuated equilibrium constitute the "usual suspects" of evolutionary research: all species in a monophyletic taxon (genealogical criterion), or all species (perhaps of restricted taxonomic scope) in a given biota over a specified time and area (temporal and geographic criteria).

RELATIVE FREQUENCIES FOR HIGHER TAXA IN ENTIRE BIOTAS

I previously cited the admittedly subjective testimony of many leading experts about the overwhelming predominance of punctuated equilibrium among all lineages in the group of their lifelong expertise and specialization (not just those featured in published studies)—see pages 22–26. Some paleontologists have tried to provide a rough quantification for this "feel." For example, Fortey (1985) states that, for graptolites and trilobiles, "the gradualistic mode does occur especially in pelagic or planktonic forms, but accounts for 10% or less of observations of phyletic change, and is relatively slow." J. Jackson (cited in Kerr, 1995, p. 1422) attempted to separate out only the most persuasive cases of unbiased sampling in faunal studies of relative frequencies. Of this subset, he remarked: "I'm imposing pretty strict criteria, but in the few cases I know [that meet these criteria], it's perhaps 10 to 1 punctuated." Later, and after a more rigorous attempt to compile best documented cases for the time and general environment best suited for supplying the requisite density of data—Neogene benthonic species of macroinvertebrates—Jackson concluded (in Jackson and Cheetham, 1999, p. 75): "Overall, 29 out of 31 species of Neogene benthos for which phylogenetic data are available exhibited punctuated morphological change at cladogenesis that is consistent with the theory of punctuated equilibria. Cases of punctuation more than double if we include extended morphological stasis . . . Thus, most but not all cases of speciation in the sea are punctuational."

The most persuasive studies have applied morphometric methods to large numbers of species in exhaustive (or at least statistically well validated) tabulations for the full diversity of higher taxa within particular faunas or spans of time. Hallam (1978), for example, tabulated data for all adequately defined European Jurassic bivalve species, forming a compendium of 329 taxa. He found "overwhelming support" (p. 17) for punctuated equilibrium, with the single exception that 15 to 20 percent of his species showed phyletic size increase—but no changes in shape—during their geological tenure. Only one lineage, the famous oyster *Gryphaea,* showed a corresponding gradual change in shape as well—a consequence of heterochronic linkage to phyletic variation in size (see affirmation of Jones and Gould, 1999). Hallam concluded, in persuasive support of punctuated equilibrium by the proper criterion of relative frequency, and with explicit attention to important and potentially confounding issues of geographic variation and missing data due to gaps in the geological record (1978, p. 17):

> The results of my analysis of 329 European Jurassic species provide, with an important exception, overwhelming support for the punctuated equilibria model. Species whose morphology appears to persist unchanged for long periods are abruptly terminated usually with one or more species of the same genus succeeding the older species with marked morphological discontinuity. The species ranges are long compared with the ammonites that allow fine stratigraphic subdivision and can be used to eliminate the possibility of significant stratigraphic gaps in the rock succession. Geographic variation within Europe is negligible, and more cursory examination of data from

> other continents provides no encouragement for the view that gradualistic events linking the "punctuated equilibria" in time took place outside Europe.

I trust, however, that Tony Hallam, one of my best friends in science, will not think me fractious or ungrateful if I point out that he then devoted the empirical content of his paper to documenting phyletic size increase in several species and, especially, to tracing gradual evolutionary changes within *Gryphaea*—in other words to the 1 lineage among 329 that illustrated phyletic gradualism. He presented no morphometric data for the overwhelming majority of species that remained in stasis throughout their existence. He wrote (1978, p. 17): "The succeeding sections of this paper are devoted primarily to this aspect of phyletic gradualism [size increase] and its implications in the broader context of environmental control of speciation, starting with the detailed analysis of *Gryphaea*."

The unconsciously imbibed power of gradualism thus remained so strong during these early years of the punctuated equilibrium debate, that Hallam could declare "overwhelming support for the punctuated equilibria model" as his primary conclusion and the focus of his study—and then follow conventional practice in applying morphometric methods only to rare examples of gradualism within his sample, even though the predominant signal of stasis could be validated just as rigorously by the same methods. For all the theoretical uncertainties that still animate the punctuated equilibrium debate, at least we have made substantial headway on this operational issue since the 1970s. Any similar study, done now, would almost surely include the documentation of stasis.

Kelley (1983, 1984) studied all molluscan lineages with adequate samples over sufficiently long ranges in one of the most famous and widely studied of all fossil faunas: the Miocene deposits of the Chesapeake Group in Maryland (Shattuck, 1904, for the classic statement; Schoonover, 1941, for the standard stratigraphic study). In an initial study of rank correlations between stratigraphic position and values of unit characters at a standard shell length estimated from bivariate regressions, she found no directional change for 82 percent of characters within species of 8 lineages. Of the 18 percent showing significant rank correlations with time, most directional changes either become reversed later in the same sequence, or run in a direction opposite to the net transformation between the measured species and its descendant. In other words, such changes, however genuine, should be read either as mild fluctuations within a pattern of stasis, or as intraspecific temporal variation unrelated to the trend of the larger lineage. For example, shells of the bivalve *Lucina anodonta* become gradually less inflated from the Calvert into the overlying Choptank Formation. But the same species then regains its ancestral degree of inflation in the succeeding St. Mary's Formation (*SET,* p. 587). Kelley (1983, p. 596) concluded with both substantive and methodological comments: "Within these middle Miocene mollusc species, then, changes are more commonly oscillatory than unidirectional . . . Most variables follow a pattern of fluctuation within a narrow range of values through time . . . In order to approach the goal of unbiased assessment of entire faunas, I examined all taxa of the mollusc faunas which were abundant enough for statistical analysis. Because no other bias controls the taxa chosen for study, these data provide strong evidence for punctuated equilibria."

Kelley's subsequent study (1984) affirms these patterns from a multivariate perspective based on discriminant analysis of 10 characters through 14 to 20 stratigraphic levels. Figure 4-10 (from Kelley, 1984, p. 1247) shows the stratigraphic distribution of centroids for each lineage at each level, as projected upon the first discriminant axis. Stasis prevails within most species (shown as unbroken vertical plots), while the four lineages composed of two or more successional species through the sequence generally show a stairstep pattern across transitions, and stasis within the bounds of species. In a very few cases, notably the transition from the lower to the middle species of *Anadara,* a trend within an ancestral species does move gradually towards the descendant's mean value. But even in this case, the third and uppermost species of the lineage then reverses the trend and moves back towards the beginning value.

Kelley (1984) also used patterns of misclassification for individual specimens to illustrate the character of predominant stasis. In the three successive species of *Astarte,* for example, 96.7 percent of specimens fall nearest the centroid of their own species—thus indicating sharp and clear division between successive species. But variation within species showed the opposite pattern. Only 42.1 percent of specimens fell nearest the centroid for their own stratigraphic level. Most remarkably, only 36.7 percent of misidentified specimens fall closest to centroids for samples of either the same or an immediately adjacent stratigraphic level. In other words, nearly of misidentified specimens stood closer to the centroids of stratigraphically distant populations than to the centroids for samples adjacent to their own time. This pattern of nondirectional distribution throughout the full vertical range of species—compared with sharp divisions between species—illustrates the strength and character of stasis in these well-known fossil lineages.

Perhaps the most impressive and definitive study of pervasive stasis in molluscan faunas has been presented by Stanley and Yang (1987) for Neogene bivalves from the Western Atlantic region. They studied 24 variables (normalized for shell size) in 19 lineages, for a total of more than 43,000 measurements. Stanley and Yang followed a comprehensive sampling method, unbiased with respect to likelihood of punctuation and stasis, and including

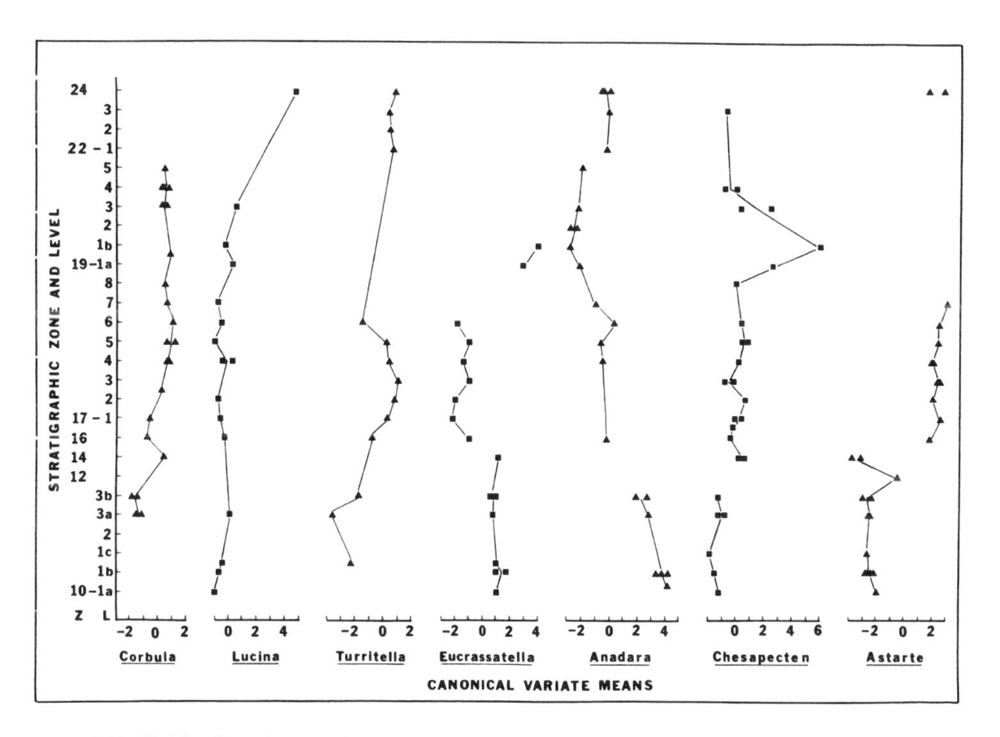

4-10. Multivariate changes based on discriminant analysis of 10 characters throughout 14 to 20 temporal units in the evolution of seven molluscan genera in Miocene strata of Maryland. From Kelley, 1984. Stasis prevails within a large majority of species. For most lineages where a descendant replaces an ancestor, a stair-step punctuation characterizes the transition. In a particularly interesting case of three successional species, ancestral *Anadara* does seem to move anagenetically towards the morphology of its descendant, which then remains quite stable. But the third and uppermost species, arising punctuationally, returns virtually to the morphology of the initial form.

all species within four bivalve taxa (Lucinidae, Tellinacea, Veneridae and Arcticacea) with shells sufficiently large and geometrically tractable (flat to only weakly convex) for their measurement protocol, and with adequate numbers of well-preserved specimens (almost always more than 20 per sample, with a minimum of 16) over a sufficient range of time (at least 4 million years from early Pliocene to Recent).

Two additional features enhance the methodological value of this study: first, these species belong to the best known and most intensely studied of all molluscan faunas; secondly, all species either still exist (12 of 19 cases) or can be compared with a close living relative (almost surely the immediate descendant in 4 cases, and perhaps directly filiated in the other 3). Thus, in the most important innovation of this study, temporal variation can be directly scaled against current geographic variation of the same species, or of a close relative. In testing whether temporal fluctuations exceed the limits of stasis, comparison with the range of geographic variation among current populations of the same species should serve as our best anchor and standard.

Using eigenshape analysis for multivariate representation of shell form, Stanley and Yang first compared variation among modern populations for each species with differences between these modern populations and early Pliocene (circa 4 million years old) samples of the same species. In a convincing demonstration of stasis properly scaled to realized intraspecific variation, Fig. 4-11 (from Stanley and Yang, 1987, p. 124) shows histograms for overlap of eigenshape areas in comparing modern geographic variation with

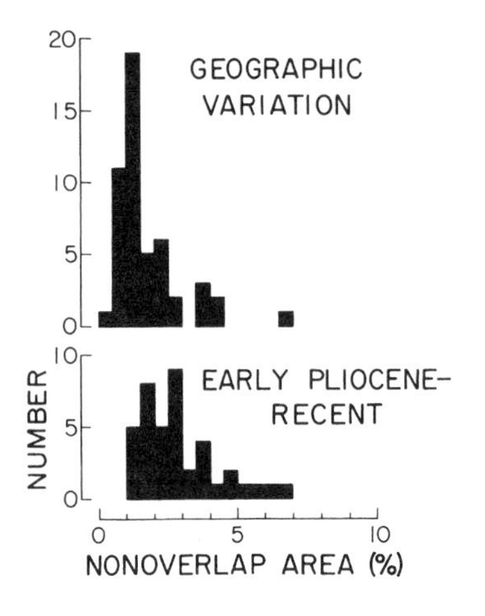

4-11. Non-overlap percentages based on comparisons of eigenshapes—from Stanley and Yang, 1987. The top histogram shows all conspecific pairs for the geographic variation of Recent populations. The bottom histogram shows differences between Recent populations and their presumed early Pleistocene ancestors. Note that the spread for the full geological range barely exceeds the spread for differences among geographic variants of living populations.

4 million year distances between early Pliocene and modern samples. The temporal mode slightly exceeds the geographic value, but the ranges overlap completely, and the difference in central tendency is very small. The authors conclude (p. 113) that "with minor exceptions, the distribution of morphologic distances between 4 million year old and Recent populations resembled the distribution of distances between conspecific Recent populations." "Approximate morphological stasis has been the rule for the taxa considered" (p. 124).

Stanley and Yang then extended their study (for species with available data) back to Miocene samples up to 17 million years old. Even for this extended duration, they found the same pattern of mild fluctuation, rarely extending outside the range of modern geographic variation, and with no accumulative directional effect. For example, Fig. 4-12 (from Stanley and Yang, 1987, p. 132) shows the temporal distribution of mean values for each of the 24 characters over 17 million years in the venerid bivalve *Macrocallista maculata*. For most characters, the full temporal range lies within the variational scope of living populations (noted by the "forks" for separate geographic samples at the top of the trajectory). They conclude (p. 113): "We calculated net rates of evolution separating pairs of populations that belong to single lineages. For all intervals of time, the distribution of differences between population means for individual variables is remarkably similar to a comparable distribution representing the comparison of pairs of conspecific Recent populations from separate geographic regions . . . Evolution has followed a weak zigzag course, yielding only trivial net trends."

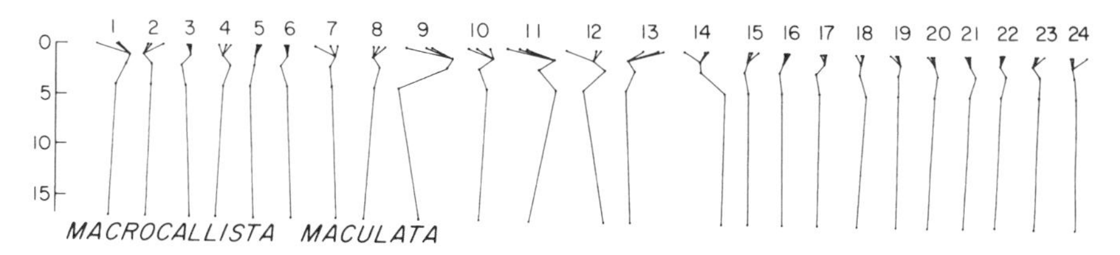

4-12. From Stanley and Yang, 1987. A history of change during 17 million years for each of 24 measured characters in the bivalve *Macrocallista maculata*. For the great majority of characters, the entire temporal spread lies within the scope of variation in the geographic range of living populations (represented by the "forks" for separate samples at the top of the trajectory). This form of comparison provides an excellent documentation of stasis by the criterion of scaling to the full range of geographic variation at a single time within the same taxon.

A particularly impressive study by Prothero and Heaton (1996) documents the overwhelming dominance of punctuated equilibrium in a full tabulation of one of the most prominent fossil faunas—a study that also gives us good insight into how biased reporting in general, and Cordelia's dilemma in particular (see end of Chapter 1), can so strongly skew tabulated results to appearances of equal frequency or only mild domination by punctuated equilibrium. These authors studied one of the world's richest and best known mammalian sequences—the upper Eocene and Oligocene White River Group of the American High Plains, particularly as exposed in the Big Badlands of South Dakota—"one of the densest and most complete records of mammalian evolution anywhere in the world . . . The spectacularly stark and beautiful outcrops . . . have been a Mecca for fossil collectors ever since the first fossils were described in 1846 . . . Enormous collections have accumulated, and White River fossils are found in nearly every rock shop and mineral show across the country" (p. 259). This large mammalian assemblage seems to possess sufficient long-term coherence (from Duchesnean strata of the late middle Eocene into Arikareean strata of late Oligocene times) for designation as the White River Chronofauna (Emry, 1981).

The authors spent more than a decade conducting "an unbiased survey of all fossil mammal lineages . . . which have large enough sample sizes and recent systematic revision" (p. 258) for the 7 million year period (37–30 million years ago) across the Eocene-Oligocene transition (Chadronian to Whitneyan North American land mammal "ages"). The protocol also includes the two other factors—good geographic spread and temporal resolution—most essential for proper studies of punctuated equilibrium, but all too rarely realized: "This study considers geographic variation over a wide area (from western Montana and North Dakota in the north and west, to Colorado in the south), with very fine-scale chronostratigraphic control from magnetic stratigraphy and ^{40}Ar/^{39}Ar dating" (p. 258). Finally, and fortunately, this interval includes a major global climatic change (with no disruption of continuity in sedimentation), thus permitting researchers to study how such an external input influences rates of speciation and styles of phyletic change.

Prothero and Heaton found near exclusivity for punctuated equilibrium in the 177 well-documented mammalian species of this fauna. "Most species are static for 2–4 million years on average, and some persist much longer" (p. 257). "Only three examples of gradualism can be documented in the entire fauna, and these are mostly size changes" (p. 257). The details of these three cases also illustrate the exceptional status of gradualism, even at the smaller scale of their own taxonomic context:

1. The lagomorph *Palaeolagus* undergoes reduction in size of upper molars, accompanied by loss of their roots, during the early Orellan, but maintains stasis for much longer invervals both above and below: "Chadronian *Palaeolagus* shows about 2 m.y. of stasis, followed by gradual reduction in size and development of rootless upper molars during the early Orellan. From Orella B onward, several species of *Palaeolagus* are present, and except for slight changes, they are static for several million years" (p. 273).

2. The artiodactyl *Leptomeryx* experiences "subtle, gradual change in a number of characters" (p. 263) in the transition from *L. speciosus* to *L. evansi,* but both species show stasis throughout most of their substantial history—so we do not here witness the "classic" continuous anagenesis that supposedly makes the definition of species so arbitrary in temporal sequences: "While the transition from *L. speciosus* to *L. evansi* is not stratigraphically instantaneous, it occurs in a relatively short time compared to the long durations of both species."

3. The oreodont *Merycoidodon* does seem to undergo extensive and gradual dwarfing (30 percent size reduction) over a one million year interval in the early Orellan. I accept this case as a good example of extended gradualism (see Fig. 4-13 taken from Prothero and Heaton, 1996, p. 262), but also note that the trend occurs within a common genus, including several species otherwise showing predominant stasis—and that the dwarfing trend only involved the labile character of size, without concomitant changes of shape, a common finding among exceptional cases of gradualism in faunas dominated by punctuated equilibrium (see previous discussion of Jurassic bivalves on page 150).

Such exhaustive and unprejudiced tabulations can give us insight into the limited value—that is, for establishing proper relative frequencies, not for resolutions of particular cases—of trying to infer the quantitative distribution of rates and modes for all taxa from previously published research carried out within the "best case" tradition, usually with strong (and unacknowledged) preference for defining evolution only as geologically gradual change. Before the punctuated equilibrium debate began, how would an evolutionary paleontologist have treated the White River Fauna? Almost surely, any expert on these strata would have selected the cases of apparent gradualism for study and publication, while ignoring the others as negative

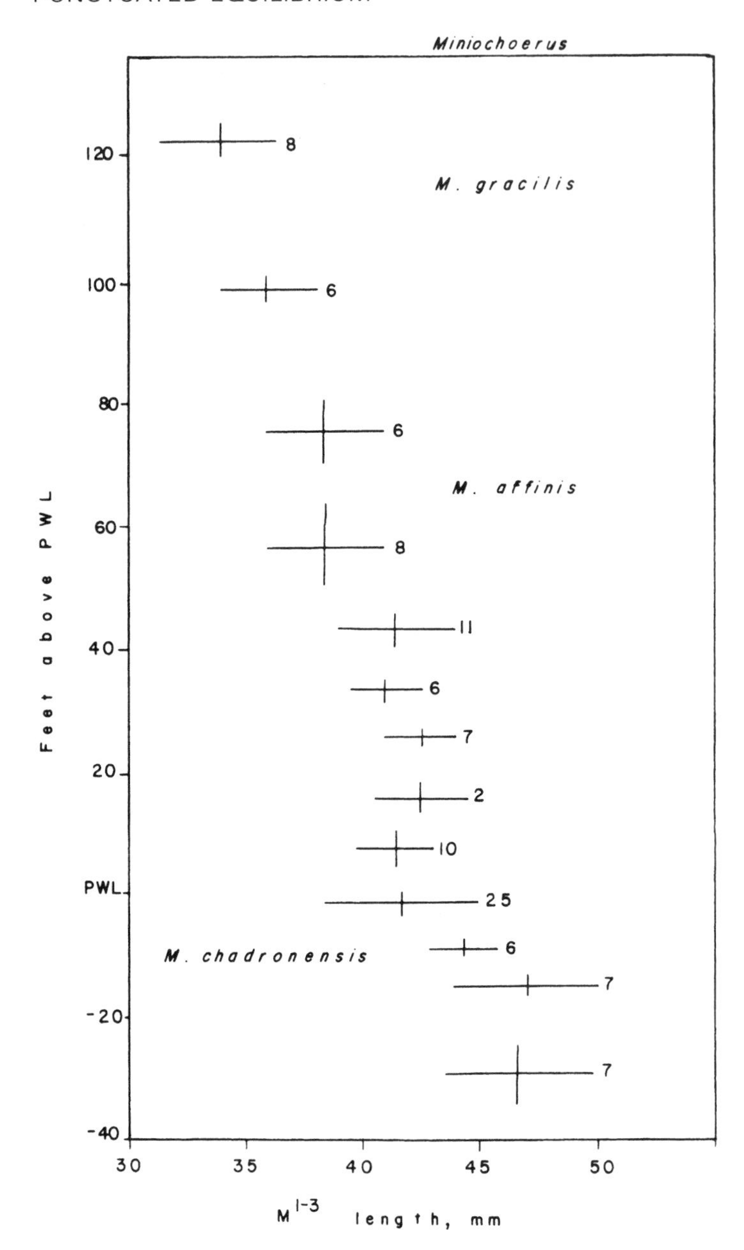

4-13. A rare case of gradualism amongst the overwhelming domination in relative frequency for stasis in 177 well-documented mammalian species of the Big Badlands fauna of South Dakota. This species of *Miniochoerus* (previously known as *Merycoidodon*) does undergo gradual dwarfing to 30% size reduction over a one million year interval. From Prothero and Heaton, 1996.

instances of no evolution, worth only a side comment at best, if noted at all, and suited for explicit mention (but without any quantitative analysis) only in formal taxonomic treatises. Thus, readers with no personal knowledge of the entire fauna—especially non-paleontological readers unaware of strong signals for stasis and punctuation in virtually all faunas—would almost surely assume that the three reported studies characterized the usual situation for the history of fossil species, rather than representing the only examples of a rare phenomenon.

Prothero and Heaton (1996, p. 258) raise the important point that the examples of gradualism most widely featured, and most frequently cited to urge the general case against punctuated equilibrium, derive from such faunas—where they stand as unusual examples against an unstudied (or simply non-discussed) but overwhelming prevalence for stasis and punctuation among all species. They remind us, for example, that Gingerich's most famous half dozen or so cases from lower Eocene beds of the northern Bighorn Basin "are just part of a fauna of over a hundred genera. Detailed monographs by Bown (1979), Schankler (1980), and Gingerich (1989) [in his very own taxonomic work] have shown that stasis is prevalent among most of the taxa not featured by Gingerich (1976, 1980, 1987)" (p. 258). Of another famous claim for gradualism (one that I do not challenge as a single case, while asking that relative frequencies also be acknowledged), Prothero and Heaton (1996, p. 258) write: "Krishtalka and Stucky (1985, 1986) reported a gradualistic transformation in the early Eocene artiodactyl *Diacodexis.* However, this is a single lineage from the same faunas described by Schankler, Gingerich, and Bown, so these studies do not address the overall prevalence of gradualism vs. stasis."

Finally, although this issue belongs more to the forthcoming discussion of faunal stasis as an extension of punctuated equilibrium (see pp. 222–229), Prothero and Heaton's (natural) experimental design in choosing the White River Chronofauna for such intensive study included the existence, in the midst of the fauna's duration, of one of the most profound and rapid climatic changes in Tertiary North America—"the earliest Oligocene climatic crash" (p. 257) at 33.2 million years ago, where "vegetation changed from dense forests to open forested grassland, mean annual temperatures dropped 13°C, and conditions got much drier and more seasonal" (p. 257). The nondisturbance of stasis—indeed, the virtual "ignorance" of this event by most species, at least by observable changes in diversity or skeletal anatomy—also illustrates the strength of stasis and the apparently active (rather than merely passive) sources of its maintenance. Prothero and Heaton write (1996, p. 257): "Only a few mammalian lineages speciated, a few more went extinct, and the vast majority (62 out of 70) persisted through this climatic event with no observable response whatsoever." The authors then end their paper by throwing down the gauntlet to supporters of traditional evolutionary views about response to environmental change: "Evolutionary theory has come a long way since Eldredge and Gould (1972) first pointed out that stasis is the norm in the fossil record, and the data cannot be simply dismissed or explained away . . . In fact, stasis and resistance to change is so ingrained that species can actually pass through the most significant climatic change of the last 65 million years as if nothing happened."

In recent years, studies of stasis and punctuation in entire faunas have blossomed, especially with the introduction and testing of two partly complementary, but partly dissonant, explanations for apparently concerted stability of entire faunas over substantial intervals—the *turnover-pulse hypothesis* of Vrba (1985), and the theory of *coordinated stasis,* developed by Brett and Baird (1995 and several other works), and extensively (often contentiously) treated in the symposium of Ivany and Schopf, 1996. I shall treat the theory itself on pages 222–229), but will record here the convincing documentation of extensive faunal stasis that established the evidentiary base for these ideas.

The famous Middle Devonian (Givetian) Hamilton fauna of the Appalachian Basin has provided a "type" case for coordinated stasis. The Hamilton fauna includes more than 330 species of mostly typical Paleozoic invertebrates, ranging through about 9 million years of strata in a series of about twenty identifiable "communities" or biofacies. About 80 percent of these species persist throughout the entire Hamilton, while fewer than 20 percent carry over from the fauna just below. Ever since Cleland's original and wistful comment in 1903 (quoted on pp. 20–21), students of the Hamilton fauna have recognized the overwhelming signal of stasis presented by almost every species of the assemblage—although this original wistfulness has now ceded to considerable positive interest! Brett and Baird write (1995, p. 301): "Individual lineages within particular biofacies of the Hamilton biotas appear to display very little morphological change, and that which is observed is neither progressive nor directional."

Several taxa of this fauna have now been analyzed in great morphometric detail, beginning with Eldredge's classic study of the trilobite *Phacops rana,* one of the "founding" examples of punctuated equilibrium (see Eldredge,

1971; Eldredge and Gould, 1972). Eldredge found stasis in more than 50 characters, and directional, but punctuational, change only in one—reduction in rows of eye facets in two punctuational steps during a 5–6 million year period otherwise marked by stasis for this feature as well. Other quantitative studies of stasis in Hamilton species include Pandolfi and Burke (1989) for tabulate corals, Lieberman (1995) on trilobites, and Lieberman, Brett, and Eldredge (1994) on brachiopods. The last study considered 8 characters in two species using principal components and canonical discriminant analysis. The authors found fluctuating variation, correlated neither with age nor facies, throughout the interval. However, and ironically given past expectations of gradualism, the uppermost samples plotted closer to the lowermost than to any intervening population. For the entire fauna, Brett and Baird conclude (1995, p. 303): "Taken together, these studies indicate that a majority of Hamilton lineages display virtual stasis from oldest to youngest samples. Slight nondirectional change is observed in some cases. Such variation seemingly records very minor evolutionary fluctuation . . . However, it clearly does not lead to development of major new grades of morphological development . . . Several of the species appear abruptly in the Appalachian Basin near the beginning of the Hamilton fauna or become locally extinct at its end."

RELATIVE FREQUENCIES FOR ENTIRE CLADES

We add another component to studies of relative frequencies when, in addition to the thoroughness provided by assessing all lineages within a given time or region, we add the phylogenetic component of complete coverage for clades (preferably monophyletic of course, but sometimes paraphyletic in the existing literature). Obviously, we feel most secure about such phylogenetic assertions when truly cladistic, or at least stratophenetic, criteria have been used for definitions, but many studies in this mode employ a standard that, albeit and admittedly less preferable, probably provides as much confidence in practical utility: investigations of distinctive taxa known on reliable biogeographic grounds to be restricted to a region exhaustively studied. Clades confined to isolated islands, lakes, or other such distinct and coherent places and environments constitute our best cases under this criterion.

I have discussed nearly all the best examples in this mode in other chapters, and will only make brief reference here. Several "classic" mammalian lineages fall into this category of excellent cladistic definition and overwhelming domination by the punctuational pattern of stasis within species and geologically abrupt transitions between—all despite (or rather, in a punctuational reformulation, because of!) such celebrated evidence for sustained and important trends. I include here the excellent evidence for horses (see p. 209) and the spottier but still persuasive data on hominid evolution (see pp. 212–222)—in each case, for clades well delimited both by morphology and geography.

Among such geographically confined clades, Vrba's classic studies (1984a and b) of African antelopes stand out for detailed data on one of the most successful and speciose of vertebrate higher taxa. In the maximally diverse tribe Alcelaphini (including blesbucks, hartebeests, and wildebeests), the Quaternary record includes 25 species, all with a geologically sudden origin in recorded data, and with cladogenesis as a reliably inferred mode of origin for at least 18 species. Several species lived for 2 million years or longer in stasis, and no ancestors with incrementally transitional morphologies have been found for any of these forms.

I continue to be amazed by the skewed interpretation often imposed by gradualistic expectations upon data for clades that seem, at least in my partisan judgement, clearly dominated by punctuated equilibrium in overall relative frequency. For example, in a well-known work, White and Harris (1977) used the Plio-Pleistocene record of African pigs for supposed validation of gradualism as a primary guide in biostratigraphic resolution (particularly of some important hominid-bearing strata). They did document one or two cases of gradual change, notably an increase in third molar length for *Mesochoerus limnetes.* But the clade includes 16 species during this short period of no more than 4 million years, 8 of which arise by punctuational cladogenesis even in White and Harris's own diagram (1977, p. 14). The authors' comments, unwittingly I suspect, frequently point to the domination of evolutionary history in this clade by cladogenetic events and their consequences. They write (1977, p. 14), for example, that "*Metridochoerus* underwent a substantial adaptive radiation during the early Pleistocene, and at one point four distinct metridochoere species existed contemporaneously."

Many of the best invertebrate examples fall into the same category of unique and endemic taxa confined to isolated places, and therefore forming, by strong inference, a complete and coherent phylogenetic unit—as in Williamson's study (1981), cited several times previously, of speciation in pulmonate snails from separated African lakes. In another example from a famous sequence of much greater temporal extent, Geary (1990, 1995) studied the evolution of melanopsid gastropods in the Middle to Late Miocene beds (spanning 5 to 10 million years) of the Pannonian Basin in eastern Europe. In one case of gradualism, following a much longer interval of at least 7 million years in stasis for the ancestral form, *Melanopsis impressa* transformed to *M. fossilis* by directional increases in shell size and shouldering over a two million year interval. However, within the same Pannonian Stage, at least six new melanopsid species arose by punctuation: "their first appearances are abrupt, and preceded by no intermediate forms" (Geary, 1995, p. 68). Geary (p. 69) regards the stratigraphic resolution as "not particularly good," but still fixes the origin of these species to within "tens of thousands of years"—a clear punctuation by the criterion of scaling against average species duration in stasis within the clade. Figure 4-14 (from Geary, 1995, p. 68) depicts Geary's results for this geographically isolated evolutionary radiation.

As mentioned many times in several contexts within this chapter, Alan

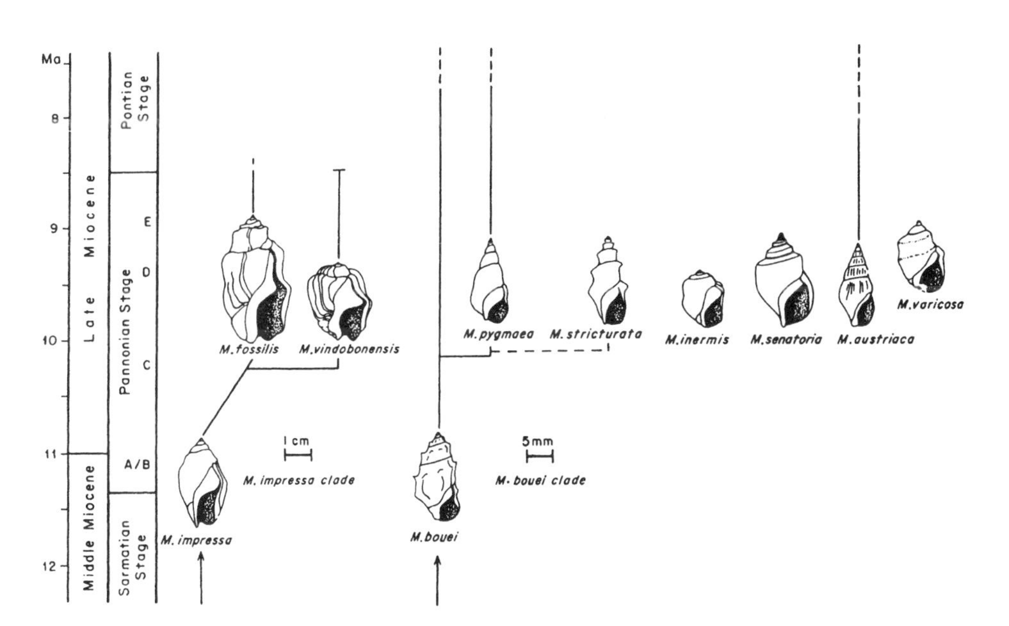

4-14. From Geary, 1995. In the radiation of melanopsid gastropods in Middle to Late Miocene beds (spanning 5–10 million years) of the isolated Pannonian Basin, Geary found one case of gradualism where, after at least 7 million years of stasis, ancestral *M. impressa* transformed to *M. fossilis* by directional increase in shell size and shouldering over a million year interval. However, during the same time, at least six new melanopsid species arose by punctuation, as also shown.

Cheetham's studies of the bryozoan *Metrarabdotos* (1986, 1987), now supplemented with the work of Jackson and Cheetham (1994, and Cheetham and Jackson, 1995) on *Stylopoma*, have set a standard of excellence and confidence for empirical studies of relative frequency. All major desiderata for such research have been realized in these genera—a group with a well-resolved phylogeny, in a clade restricted to a geographic region, and exhaustively sampled in strata of unusually complete resolution over a long period. Moreover, Cheetham's multivariate morphometrics permit us to assess stasis and punctuation as a morphological totality, not only as a potentially biased impression based on a few preselected characters. Finally, studies with Jackson on the ecology and genetics of extant species demonstrate (see pages 64–66) that the morphology of paleospecies almost surely provides a good surrogate and identifier for true biospecies in this clade. (As a personal note, I am also gratified that Cheetham began these studies with the intention of proving his suspicions for gradualism in the context of the developing debate about punctuated equilibrium—and ended up with the finest data-driven evidence ever gathered for the domination of a total evolutionary pattern by punctuated equilibrium.)

To recapitulate the major conclusions of these studies (see also Fig. 4-5 for the phylogeny of *Metrarabdotos*, with morphology expressed as multivariate Euclidian distances between samples based on all canonical scores of a discriminant analysis, and connecting nearest morphological neighbors in stratigraphic sequence), Cheetham measured 46 characters in 17 species of *Metrarabdotos* over a duration of 15 million years, with intense sampling for a 4.5 million year interval of Upper Miocene to Lower Pliocene sediments (3.5 to 8.0 million years ago) in the Dominican Republic. Cheetham (1986, p. 195) specified the favorable features of *Metrarabdotos* on both geographic and phylogenetic grounds: "The ascophoran genus *Metrarabdotos* is a favorable subject for detailed analysis of evolutionary pattern because of its diversity and wide distribution during much of Miocene and Pliocene time. Caribbean species . . . form an apparently monophyletic subset within which phylogenetic relationships can be inferred independently of evolutionary events in eastern Atlantic-Mediterranean congeneric species groups."

Moreover, the unusual resolution for the detailed sampling interval permits "a fine-scale comparison of successive populations similar to those made with oceanic planktonic groups in deep-sea cores" (1986, p. 195), thus dispelling the common argument of gradualists that stasis in metazoans from conventional sediments must arise as an artifact of coarseness of resolution, while gradualism of microfossils in deep-sea cores must record a general pattern that would be seen wherever stratigraphic sampling could attain such completeness. Cheetham (1986) calculated 160,000 years for average spacing between successively sampled populations in the intensely collected interval—for a stratigraphic completeness of 0.63 by Sadler's (1981) criteria.

In *Metrarabdotos* (again see Fig. 4-5), 11 of the 17 species persist in stasis for 2–6 million years, and all originate punctuationally within the limit of resolution (at least in the intensely sampled interval) of 160,000 years—undoubtedly in far less time for many branching events, since 160,000 years represents a maximum figure based on the available unit of measurement. Again for the intensely sampled interval, Cheetham writes (1986, p. 190) that "nine comparisons of ancestor-descendant species pairs all show within-species rates of morphologic change that do not vary significantly from zero, hence accounting for none of the across-species difference. In all cases, the ratio of within-species fluctuation to across-species difference is low enough to allow the punctuated pattern to be distinguished with virtual certainty. In at least seven of the cases, ancestor species persisted after giving rise to descendants, in conformity with the punctuated equilibrium mode of evolution."

The morphometric details can only increase confidence in "the remarkably clear-cut evidence for a punctuated evolutionary pattern in these *Metrarabdotos* species" (1986, p. 201). The reported central tendencies of samples integrate data from 46 measurements, providing a good assessment of general anatomical distance (based on characters considered important in the taxonomy and functional morphology of these organisms), and not on selected single characters (see Cheetham, 1987, for affirmation of punctuated equilibrium from analyses of temporal trends in individual characters as well). In supplementary affirmation, Cheetham studied the fine-scale pattern of temporal variation by computing autocorrelations between mean scores of stratigraphically successive pairs of populations: "In all cases, the autocorrelations of mean scores of successive populations are nonsignificant and near zero, and the autocorrelations of rate deviations are negative and (except in one case) nonsignificant. These autocorrelations clearly indicate that changes within species are fluctuations around a near-zero, otherwise unchanging rate" (1986, p. 201).

Finally, some authors (see Marshall, 1995) have challenged Cheetham's phylogeny for its stratophenetic basis. But a purely cladistic analysis, as now preferred by many researchers, not only changes the previous scheme in only minor ways (Cheetham and Jackson, 1995, p. 192), but also—and the point becomes almost amusingly obvious once one grasps the different criteria used by the two methods—leads to an even stronger pattern of punctuated equilibrium, for the stratophenetic phylogeny minimizes the mean morphologic distances between putative ancestor-descendant pairs, while the cladistic phylogeny makes no such assumption and must therefore yield a larger mean difference between species. Since the documented stasis *within* species is not affected in either case, the cladistic scheme must increase the average magnitude of punctuational events, thus only decreasing the likelihood that between-species differences could be extrapolated from temporal variation within species (see Cheetham and Jackson, 1995, p. 192, for an elaboration of this argument with appropriate data).

The corroborative study of a second bryozoan genus, *Stylopoma*, from the same beds yields an identical conclusion of overwhelming predominance—indeed, exclusivity—for punctuated equilibrium. Jackson and Cheetham (1994) used 12 morphological features to identify 19 species in this rarer genus. Despite their more limited information, Cheetham and Jackson (1995, p. 195) found that "temporal overlap between putative ancestor-descendant species pairs is even greater than for *Metrarabdotos*, with 10 species surviving beyond the detailed sampling interval more than 6 million years to the Holocene." Moreover, "no evidence of morphologically intermediate forms" (Jackson and Cheetham, 1994, p. 420) has been found for any transition; all species origins are fully punctuational at the scale of detailed sampling.

Finally, since *Stylopoma* provided Jackson and Cheetham's principal data for the correspondence of genetically defined biospecies with morphologically designated paleospecies (modern specimens of *Metrarabdotos* are much less common and not so well suited for genetic work), this second study provides strong additional support for punctuated equilibrium by coordinating several potentially independent indicators of evolutionary change with rapid events of branching speciation: "Moreover, the tight correlation between phenetic, cladistic, and genetic distances among living *Stylopoma* species suggests that changes in all three variables occurred together during speciation. All of these observations support the punctuated equilibrium model of speciation."

I regard these empirical studies of relative frequencies as the strongest evidence now available for the most important and revisionary claim made by the theory of punctuated equilibrium: the overwhelming domination of evolutionary patterns in geological time by events at the species level (or higher), and the consequent need to explain macroevolution by patterns of sorting among species rather than by extrapolated trends of anagenetic transformation within continuous lineages.

CAUSAL CLUES FROM DIFFERENTIAL PATTERNS OF RELATIVE FREQUENCIES

Once we set our focus of inquiry on determining the relative frequencies of punctuated equilibrium in different times, places, environments, and taxa, we can ask the classic question of natural history, a subject rooted in the concept of variation: do we note characteristic differences in relative frequencies based on any of these factors and, if so, can we draw any causal inferences (useful to evolutionary theory) from these patterns. I have already raised this question in a number of preceding contexts in this chapter, most extensively for the observed higher frequency of gradualism in predominantly asexual oceanic protistan lineages, where I argued that this unusual result may not record the greater completeness of strata in oceanic cores (as traditional views have assumed), but probably arises from interesting biological differences that have led us to look for a truly underlying punctuational pattern at the wrong scale in this case (see pp. 84–93).

Most discussion on the linkage of differences in frequencies to distinctive structural and functional characteristics of organisms (rather than to types of environments *per se*) has focused on claims for variation among broad taxonomic groups (high frequency for gradualism in rodents vs. rarity in bovids, for example), but I strongly suspect that this genealogical emphasis reflects traditions of specialization in research more than any inherent preference for taxonomic parsing in such a search. We should also consider more general features of organisms that cut across taxonomic lines, and we should therefore examine broader differentia potentially related to chosen environments or tendencies to speciate. Schoch (1984), for example, suggested a link between high frequencies for punctuational speciation and intense social competition, arguing that selection on such features tends to proceed so rapidly, even in ecological time, that speciation would almost surely occur in a geological instant. Breton (1996) linked punctuational modes to evolution of "pioneer" structures (evolutionary novelties tied to morphological reorganizations), and gradualism to "stabilization" and "settlement" structures (refinements and improvements in local adaptations). I suspect that such arguments may apply better to the different issue of average amounts of change per speciation event, than to questions about the relative frequency of punctuational events (at whatever degree of alteration) *per se.*

Most arguments about patterns of differences in relative frequencies have invoked "externalist" claims about characteristic environments, rather than "internalist" correlations with structural features of organisms (although the two subjects may, of course, be correlated and need not stand in antithesis). In a first attempt, Johnson (1975, 1982), working with Devonian brachiopods and conodonts but generalizing more widely in an important set of papers, linked higher frequencies of gradualism to pelagic environments and greater prevalence of punctuated equilibrium to benthic habitats. He then justified the ecological correlations by linking characteristic evolutionary modes with relative stability of environments: "Among marine invertebrates, pelagic organisms are the most likely to have inhabited extensive, gradually changing environments and are therefore the most likely to have evolved by a rate and pattern that can be described as phyletic gradualism . . . Post-larval, attached and stationary benthic organisms are the most likely to have inhabited environments that are subject to relatively abrupt changes and are therefore the most likely to have evolved by a rate and pattern that could be described as punctuated equilibria."

In a series of papers, Parsons (well summarized in 1993) suggested a similar linkage, while proposing a different, but generally concordant, explanation based on a putative correlation between environmental "stress" and patterns of genetic variation and available "metabolic energy." (I put Parsons's last factor in quotation marks because I have trouble grasping both the definition and operationality of such a concept.) Parsons writes (1993, p. 328): "In moderately stressed and narrowly fluctuating environments, sufficient genetic variability and metabolic energy should be available to permit adaptation. In these environments, phyletic gradualism is expected. In highly stressed and widely fluctuating environments, a punctuated evolutionary pattern is expected whereby stasis occurs most of the time."

In one of the most important recent papers on punctuated equilibrium, Sheldon (1996) has generalized a superficially paradoxical link of morphological stasis to highly fluctuating environments, and gradualism to more stable and narrowly fluctuating environments, as the "*plus ça change*" model—citing the sardonic French motto that "the more things change, the more everything's the same," a reference to the proposed link of morphological stasis with highly variable environments. Sheldon (1996, p. 772) explained the basis and resolution of the paradox: "One might expect a changing environment to lead to changing morphology, and a stable environment to stable morphology. But over long intervals the opposite may often occur . . . Perhaps gradual phyletic evolution can only be sustained by organisms living in or able to track narrowly fluctuating, slowly changing environments, whereas stasis, almost paradoxically, seems to prevail in more widely-fluctuating, rapidly changing environments." Sheldon's figure (reproduced here as Fig. 4-15) will make his argument clear. Species in highly variable habitats must adapt to pervasive and rapid fluctuations, and generally do so by evolving a stable and generalized morphology suited to the full environmental range. But when external fluctuation exceeds a certain limit of internal toleration, rapid speciation may be the only viable response. On the other hand, mildly fluctuating environments may enhance selection for more precisely tuned adaptations capable of tracking long-term climatic trends by gradual adjustment.

Sheldon (1987) began his work by publishing one of the most widely discussed empirical defenses of gradualism, based on several lineages of Ordovician trilobites from the Builth Inlier, an environment interpreted as generally stable and only narrowly fluctuating. (I appreciate the richness of Sheldon's data, but regard his interpretations as ambiguous, for most of his published trajectories seem to me—from my partisan standpoint [as I keep repeating to remind readers to be especially critical]—more consistent with expectations of fluctuating stasis with little or no net change, see Eldredge

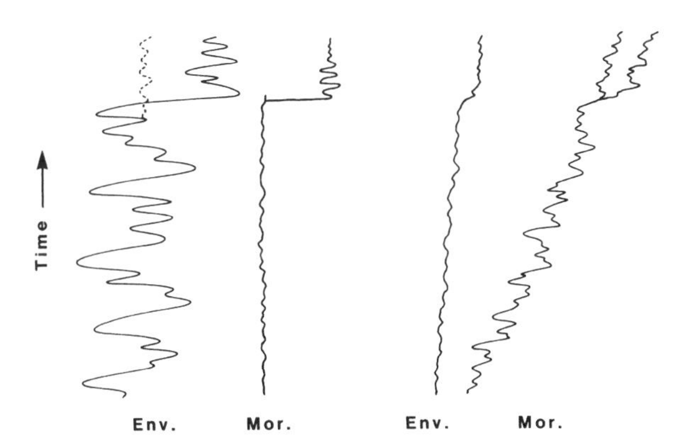

4-15. An epitome of Sheldon's argument (1996) that, paradoxically, highly fluctuating environments may induce stasis and punctuation, with gradualism more commonly found in environments undergoing slower but more steady change.

and Gould, 1988.) I am particularly grateful that Sheldon, even while developing one of the most famous data sets against the theory, has always accepted the importance of establishing relative frequencies for different groups and situations, and has consistently regarded punctuated equilibrium as a valuable theory (to which he has made major contributions), with important implications for our understanding of macroevolution.

Such broad arguments about environmental correlations have been notoriously difficult to document because, even when the effect can be validated as both real and pervasive, so many other factors will be operating in any particular case (including immediate and local influences able to overwhelm the smaller impact of the generality under test) that the signal may be lost in surrounding noise. But I am strongly attracted to Sheldon's *plus ça change* hypothesis for two primary reasons. First, the concept makes good sense of patterns that have often been noted empirically, but regarded as confusing in interpretation—particularly the common finding of pronounced stasis through major climatic fluctuations, including Pleistocene ice age cycles (Cronin, 1985, on ostracodes; Coope, 1980, 1994 on beetles), and the largest climatic crash in Tertiary North America (Prothero and Heaton, 1996). The presentation of a hypothesis like Sheldon's prompts researchers to focus studies on interesting issues, and to seek wider implications. For example, Wei (1994) used Sheldon's hypothesis to explain the link of stasis to intensification of ice-age climatic fluctuations in the planktonic foram *Globoconella inflata.* Wei (1994, p. 81) suggests that stasis may represent "a compromise for the species as an attempt to meet with both glacial and interglacial extremes."

Second, Sheldon's hypothesis predicts a large suite of definite correlations subject to empirical test. *Plus ça change* predicts linkages of different relative frequencies for punctuated equilibrium and gradualism to geographic gradients (with more punctuated equilibrium expected in temperate areas, and more gradualism in the topics), environmental distinctions (with more punctuated equilibrium in nearshore shallow-water strata and more gradualism in offshore regions, as Johnson had earlier predicted), and evolved responses of organisms and populations (with, *ceteris paribus,* more punctuated equilibrium in eurytopes and *r*-strategists, and more gradualism for stenotopes and *K*-strategists). Needless to say, *ceteris paribus* does not always hold—but with so many expected consequences, the probability of finding patterns (if they exist) does rise substantially.

Finally, as for any good hypothesis in science, Sheldon's *plus ça change* suggests several interesting extensions. For example, Sheldon raises an intriguing argument for linking these putative correlations with patterns of genuine selection at the species level or above:

> Perhaps the most important (and perhaps the most controversial) mechanism I am suggesting here is a type of lineage selection with two stages: (1) if an established or an incipient species experiences a widely fluctuating environment on geological timescales, the evolutionary response (morphological change) tends to become damped with time, and (2) those species that are least sensitive to environmental change (the most "generalized" in a long-term sense) are the ones that tend to persist, remaining in morphological stasis until a threshold is reached (Sheldon, 1996, p. 218).

In a second extension, Sheldon makes an almost quizzical, but oddly compelling, argument based on another important source of potential correlations, previously unaddressed here but perhaps quite important: time itself, expressed either as the absolute time of particular intervals in the earth's history, or as the relative time of distinctive segments in the general "ontogeny" of a species's duration. Many biologists have noted the apparent paradox that so little sustained and directional evolution (as opposed to abundant evidence for rapid and adaptive fluctuations in such characters as bill form in Darwin's finches or wing colors in peppered moths) has been noted for species in historic, and recent prehistoric, times during the tenure of modern humans (who have also remained in stasis) on earth. I would, of course, attribute this phenomenon mostly to a general prediction for stasis in the vast majority of lineages at any time (while charging our puzzlement only to the false equation of evolution with gradual change). But Sheldon raises the interesting ancillary argument that this general expectation may now be enhanced by special advantages for stasis in the regimes of strong and rapid worldwide climatic fluctuation that our earth has been experiencing in these geologically unusual times: "Given the Quaternary climate upheavals, relatively little evolution may be occurring worldwide at present (except for evolution induced by humans)" (Sheldon, 1996, p. 209). I can only hope that the more punctuated equilibrium induces change in our evolutionary views, the more things will *not* be the same in our interpretations of the history of life.

5 THE BROADER IMPLICATIONS OF PUNCTUATED EQUILIBRIUM FOR EVOLUTIONARY THEORY AND GENERAL NOTIONS OF CHANGE

What Changes May Punctuated Equilibrium Instigate in Our Views about Evolutionary Mechanisms and the History of Life?

THE EXPLANATION AND BROADER MEANING OF STASIS

As emphasized throughout this chapter, the stress placed by punctuated equilibrium upon the phenomenon of stasis may emerge as the theory's most important contribution to evolutionary science. The material world does not impact our senses as naturally and objectively parsed categories. We can make accurate observations and measures of particular "things," but the ordering of "things" into categories must be construed largely as a mental operation based on our theories and attitudes towards "reality." Moreover, we must also apply mental screening to select "things" meriting our attention within nature's potential infinity, and even to recognize a configuration of matter as a "thing" in the first place. Therefore, phenomena without names, and without theories marking them as worthy of notice, will probably not be recognized at all.

The phenomenon always existed "out there" in nature, of course, but punctuated equilibrium largely "created" the category of stasis as an important item in evolutionary theory through a four-step process of (1) defining stasis as a positive "thing" with properties and boundaries, a phenomenon rather than an unnamed and unrecorded absence of evolution; (2) bringing stasis to visibility as the expectation of a particular theory of evolutionary modalities; (3) suggesting methods for the active and rigorous study of stasis, so that the concept could be operationalized as a subject for empirical research; and (4) granting interest and importance to stasis as a controversial topic with broad implications for revising traditional modes of thought in evolutionary biology.

Before Eldredge and I published our first paper in 1972, most paleontologists treated stasis as an embarrassment, imposed by the poverty of the fossil record upon hopes for recording evolution (defined as gradualistic anagenesis), and therefore as not meriting active study, or even explicit recognition as a discrete phenomenon. Just a decade later, the situation had changed so dramatically that Wake, Roth and Wake (1983) could write, "perhaps no phenomenon is as challenging to evolutionary biologists as what has been termed 'stasis'" (p. 212), defined by them as "the maintenance of a standard morphology over vast periods of time during which much environmental change has taken place" (p. 211). Illustrating my claim that a phenomenon becomes interesting only in the light of defining theories, Wake *et al.* (1983, p. 212) then stated: "With natural selection operating in a changing environment as an agent of adaptation, we expect to see changes at the organismal, ultimately physiological and morphological, level. How, though, can we explain the paradoxical situation in which environments change, even dramatically, but organisms do not?"

As I now survey the subject, a quarter century after our initial presentation and definition, stasis has become an even more general and important issue in evolutionary theory for three principal reasons:

Frequency

Once the phenomenon had been named, and criteria established for recognition and study, researchers documented stasis at far too high a relative frequency to represent anything other than an evolutionary norm and expectation. Such predominance also implicates stasis as a property *actively* maintained by species—thus leading to a substantial literature (discussed at the end of this subsection) on the causes of non-change. Several authors, notably Paul (1985) and Jackson and Cheetham (1994, also Cheetham and Jackson, 1995), developed models and data sets to prove that stasis occurs too frequently for explanation under random models (including pure neutralism with no natural selection), and therefore must be caused by active forces promoting such a result, either directly, or as a consequence of some important linked property of organisms or populations.

This growth in emphasis has been so vigorous since 1972 that geological gradualism, once the unquestioned expectation of evolution itself, is now generally regarded as an infrequent, if not anomalous, phenomenon requiring a special explanation in the light of anticipated stasis. Geary (1990, p. 507) after documenting a case of gradualism within a clade showing a much higher frequency of stasis, wryly noted: "Given that past studies were assumed complete only if gradual change was apparent, it seems somewhat ironic that unseen mechanisms or events, however realistic, must now be invoked in order to explain an instance of gradual change!" Gradualism, in short, has become both a rarity and a puzzle. (Much as I take a rather wickedly and secret personal pleasure in this sea change, I'm not sure that I can, in good scientific conscience, regard such *a priori* mental downgrading of gradualism as a "good thing." *A posteriori* downgrading based on documented rarity represents nature's chief signal in my view, but I do think that any study should begin with equal potential welcome for either result!)

Generality

The interest in stasis, originally generated by punctuated equilibrium for inquiries at the appropriate level of species durations through time, has since expanded to other domains of size and time, and to more comprehensive questions about the nature of change itself. Causes operating at punctuated equilibrium's proper scale will not explain other forms of stasis, but the generalized definition and inquiry did arise by expansion from our theory (at least as a sociological phenomenon), while we may also anticipate the identification of some common causes or constraints (see further discussion on conceptual "homology," pp. 236–240)—that is, in evolutionary parlance, causal parallelisms, based on structural homologies, rather than convergences or mere analogies of appearance—behind the deeper generality (with different immediate forces producing similar and partly homologous results at various levels). I shall discuss some of these other scales later in this subsection. These extensions include: punctuational anagenesis for directional changes in lineages of asexual organisms by clonal sorting (in a domain below punctuated equilibrium, which, *sensu stricto,* only operates at the level of speciation to explain trends in multicellular sexual lineages by species sorting); longterm morphological stability for basic anatomical features of larger clades (at a level above punctuated equilibrium and within monophyletic lineages—see *SET,* Chapter 10); putative "lock-step" stasis for the great majority of defining species within larger faunas through significant geological intervals (at a still higher level above punctuated equilibrium and across genealogical lineages to a consideration of faunal dynamics—see pp. 222–229). Interest has also extended beyond evolutionary systems to the meaning and causes of stasis in stairstep patterns of ontogenetic growth, stubbornly persistent plateaus followed by thresholds of rapid change in response to continuous input in human learning, and active stasis followed by punctuational breakdown in the history of human ideas and social organization (see pp. 265–289).

Causality

Fruitful debate about the causes of stasis must first specify the level manifesting the common phenomenology. (Obviously, causes of learning plateaus in piano playing cannot be strict homologs of ecological reasons for joint stability of species in coordinated stasis, even though the graphed pattern of change may manifest the same geometrical form.) In this section, I confine my discussion to punctuated equilibrium *sensu stricto,* and not to the general pattern of punctuational change at any level—that is, to proposed reasons for the observed high relative frequency of stasis during the full geological range of metazoan species as preserved in the fossil record.

But first, and as an example of how discussion can proceed at cross purposes when proper scales have not been specified, I must note that most of the literature proclaiming punctuated equilibrium as "old hat" (Lewin, 1986) or something long known and merely hyped by ill-informed paleontologists, has only analyzed ecologically rapid anagenesis in populations rather than the relevant phenomenon of cladogenesis by speciation scaled against subsequent geological duration in the stasis of species so generated. Most notably (in terms of subsequent commentary), two papers of the mid-1980s (Newman, Cohen and Kipnis, 1985; and Lande, 1986) developed mathematical models to show that single populations could move rapidly (in the "ecological time" of a human career) from one adaptive peak to another in the absence of environmental change. (The major previous stumbling block had been set by problems in envisaging how a population could move *down* an adaptive peak, against any force of selection, to inhabit a valley, and therefore become subject to selection up an adjacent peak. The basic solution—that the descent must be rapid—allows sufficient impetus against selective forces, and also links the models to themes of speedy anagenesis.)

Lewin (1986) used these studies to write a news and views feature for *Science* entitled "Punctuated equilibrium is now old hat," while also recognizing that ecologically rapid anagenesis does not address the scale or level (not to mention the reality of *changing* environments in our actual world) of punctuated equilibrium's central concern. We welcome such plausible models of ecologically punctuated anagenesis as a contribution to understanding the panoply of causes that yield punctuational change at other levels. But this smaller-scale phenomenon, however fascinating and important, bears little relevance to the causes of stasis within species during geological time (or to the cladogenetic sources of geological punctuation as a slow branching event in ecological time).

We may order the major propositions for explaining stasis at the scale of punctuated equilibrium as an array running from conventional resolutions based on Darwinian organismic selection to more iconoclastic proposals invoking either higher levels of causation or less control by selection and adaptation. (Much of the genuine interest in the otherwise tedious and tendentious debate on the theoretical novelty of punctuated equilibrium lies in the legitimate weights that will eventually be assigned to the various proposals of this array.)

STABILIZING SELECTION. For most evolutionists who chose to see nothing new in punctuated equilibrium, the previously unacknowledged frequency of stasis (admitted, albeit sometimes begrudgingly, as an unexpected finding) could only indicate a stronger role than previously envisaged for the conventional mechanism of stabilizing selection. Although this putative explanation of stasis within paleospecies achieved an almost canonical status among evolutionists who tried to forge complete compatibility between punctuated equilibrium and the Modern Synthesis, and although we all acknowledge stabilizing selection as too important and pervasive a phenomenon to hold no relevance for this issue, a complete explanation of stasis in these conventional terms seems implausible both on empirical grounds, and also by the basic logic of proper scaling.

As often emphasized in this book, if stasis merely reflects excellent adaptation to environment, then why do we frequently observe such profound stasis during major climatic shifts like ice-age cycles (Cronin, 1985), or through the largest environmental change in a major interval of time (Prothero and Heaton, 1996)? More importantly, conventional arguments about stabilizing selection have been framed for discrete populations on adaptive peaks, not for the *totality* of a species—the proper scale of punctuated equilibrium—so often composed of numerous, and at least semi-independent, subpopulations. A form of stabilizing selection acting among rather than within subpopulations may offer more promise—as Williams (1992) has proposed (see discussion under point 6)—but such forms of supraorganismal selection fall into a domain of heterodoxies, not into this category of conventional explanations that would leave the Modern Synthesis entirely unaffected by the recognition of stasis as a paleontological norm.

DEVELOPMENTAL AND ECOLOGICAL PLASTICITY. If stabilizing selection holds that species don't change because they have achieved such excellence in current adaptation, this second proposal (of Wake, Roth and Wake, 1983) proposes that species don't change (in an evolutionary and genetic sense) because they can usually accommodate to environmental alteration by exploiting the plasticity (behavioral and developmental) permitted within their existing genetic and ontogenetic system—thus calling upon the physiologist's entirely different meaning of the term "adaptation" (improvement in functionality by exploiting possibilities within a norm of reaction, as in the enlarged lungs of people who inhabit the high Andes), rather than the usual evolutionary meaning in our profession.

(Although I have roughly ordered this list of proposed explanations for stasis from Darwinian conventionality towards more challenging proposals, I don't regard any item as excluding any other—indeed, I would be surprised if all cannot claim at least some measure of validity, for once again we deal with an issue of relative frequencies and differential circumstances—and I don't regard any pair as establishing a contradiction. In particular, these first two proposals, although different in implications about styles and reasons for limited change in species, remain primarily complementary in their common attribution of stasis to reasons based on satisfactory current status—the first on immediate optimality of overt features, the second on inherent plasticity within a current, and presumably adaptive, norm of reaction.)

Wake et al. (1983), for example, document how salamanders, artificially raised to encounter only fixed potential prey, will learn to eat immobile objects, thus contradicting "the widespread assumption that amphibians feed only on moving prey" (p. 216), and also permitting substantial "adaptation" (physiologist's sense again) to feeding regimes without disturbing the stasis of evolved form. In a thought-provoking conclusion, Wake et al. (1983, p. 219) site (and cite) stasis as one component of a more general attitude towards stability of systems and preference for non-change, with evolution conceptualized as a "default option" in the history of life—in contrast with the usual view of active and normative change embodied in the first explanation of stabilizing selection:

> Stasis is but the most rigid form of the stability that pervades living systems. Thus organisms have evolved as systems resistant to change, even genetic change. While changing environmental conditions may ultimately necessitate change in the system, until some critical point the system remains stable and compensating. The living system is sometimes envisioned metaphorically as a kind of puppet, with enormous numbers of strings, each controlled genetically, or as a blob of putty that can flow in any direction given sufficient force (selection). Our metaphor is the living system as a balloon, with the environment impinging as countless blunt probes. The system compensates environmental and genetic changes, and persists by evolving minimally.

DEVELOPMENTAL CONSTRAINT. This proposal veers more towards heterodoxy in ascribing stasis to an internally specified inability to change (thereby implying frequent suboptimality of adaptation), rather than to lack of adaptive impetus for change due to current optimality (explanation one) or flexibility within a current constitution (explanation two). (This notion of inability stands forth most clearly in the strict definition—too strict in my view (see *SET,* Chapter 10)—of constraint as absence of genetic variation for a particular and potentially useful alteration, as in the consensus concept of Maynard Smith et al., 1985.)

In our original paper on punctuated equilibrium, Eldredge and I (1972), basing our arguments partly on Mayr's (1954, 1963) concept of genetic revolutions in speciation of peripherally isolated populations, but more on Lerner's notions (1954) of ontogenetic or developmental, but especially of genetic, "homeostasis," proposed such constraint as the primary reason for stasis. We wrote (1972, pp. 114–115):

> If we view a species as a set of subpopulations, all ready and able to differentiate but held in check only by the rein of gene flow, then the stability of species is a tenuous thing indeed. But if that stability is an inherent property both of individual development and the genetic structure of populations, then its power is immeasurably enhanced, for the basic property of homeostatic systems, or steady states, is that they resist change by self-regulation. That local populations do not differentiate into species, even though no ex-

> ternal bar prevents it, stands as strong testimony to the inherent stability of species in time.

This proposal became one of the most widely controverted aspects of punctuated equilibrium, especially in linkage with other, largely independent concepts like the prevalence of neutral change (Kimura, 1968), and the exaptation of originally nonadaptive spandrels (Gould and Lewontin, 1979), also viewed as challenges to the more strictly adaptationist concept of Darwinian evolution then prevalent. I now believe that these criticisms, with respect to the issue of stasis in paleospecies through geological time, were largely justified—and that the theme of constraint, while not irrelevant to the causes of stasis in punctuated equilibrium, does not play the strong role that I initially advocated. (However—and perverse as this may seem to some detractors—my conviction about the general importance of constraint vs. adaptationism at other more appropriate scales has only intensified, particularly in the context of revolutionary findings in developmental genetics—see *SET*, Chapter 10.)

I have changed my initial view for two primary reasons. First, the arguments of Mayr and Lerner, the intellectual underpinnings of our initial proposals about constraint, have not held up well under further scrutiny, particularly in the privileging of small populations as especially, if not uniquely, endowed with properties that permit the breaking of stasis. Further modeling has led most evolutionists to deny that any major impediment for such change can be ascribed to large populations. Second, I now realize that my arguments for the channeling of potential direction and limitation of change apply primarily to levels above species—to aspects of the developmental *Baupläne* of anatomical designs that usually transcend species boundaries, rather than to resistance of populations against incorporating enough genetic change to yield reproductive isolation from sister populations.

THE ECOLOGY OF HABITAT TRACKING. This explanation for stasis, long favored by my colleague Niles Eldredge (1995, 1999), offers a first alternative (in this list) based on the structuring of species-individuals as ecological entities, rather than on adaptations or capacities of component organisms—thus taking explanation to a higher descriptive level of the evolutionary hierarchy. Otherwise, however, habitat tracking ranks as a conventional Darwinian explanation in calling upon stabilizing selection to confer stasis upon populations that react to environmental change in their geographic locale not by evolutionary alteration to new conditions, but rather by moving with their favored habitat to remain in an unchanged relationship with their environment of adaptation. Eldredge writes (1999, p. 142): "Paradoxically (and contrary to at least superficial Darwinian expectations) . . . stabilizing natural selection will be the norm even as environmental conditions change—so long, that is, as species are free to relocate and 'track' the familiar habitats to which they are already adapted. Rather than remaining in a single place and adapting to changing conditions, species move. And so they tend to remain more or less the same even if the environment keeps on changing."

I place this otherwise conventional explanation towards the heterodoxical end of my list because habitat tracking embodies the remarkably simple and obvious (in one sense), yet profound and unconventional view (in another sense) that evolutionary change represents a last resort, and not a norm for most times, in the response of populations to their environments. (The second explanation of plasticity also invokes this theme, but from the organism's, rather than the population's, perspective.) Habitat tracking also emphasizes the cohesion, and evolutionary reality, of supraorganismic individuals—an essential theme in the hierarchical reconstruction of Darwinian theory (see *SET*, Chapter 8). This subtly unconventional notion of change as a last resort or default option puts one's mind in a much more receptive state towards the reality of stasis as a genuine and fundamental phenomenon in evolutionary theory.

THE NATURE OF SUBDIVIDED POPULATIONS. With this fifth category, we finally enter the realm of truly—that is, causally—macroevolutionary explanations based on the reality of supraorganismal individuals as Darwinian agents in processes of selection. In a brilliant paper that may well become a breakthrough document on this perplexing subject, Lieberman and Dudgeon (1996) have explained stasis as an expected response to the action of natural selection upon species subdivided (as most probably are) into at least transiently semi-autonomous populations, each adapted (or randomly drifted) to a particular relationship with a habitat in a subsection of the entire species's geographic range.

Lieberman and Dudgeon derived their ideas (see also McKinney and Allmon, 1995, for interesting support) in the context of Lieberman's extensive multivariate morphometric analysis of two brachiopod species from the famous Devonian Hamilton fauna of New York State (see pp. 222–229). Lieberman noted profound stasis (with much morphological "jiggling" to and fro but no net change) over 6 million years (Lieberman, Brett, and

Eldredge, 1994, 1995); but he also studied samples of each species from each of several paleoenvironments through time. Paradoxically (at least at first glance), Lieberman documented several cases of measurable change in single discrete and continuous paleoenvironments through the section—but not for the entire species integrated over all paleoenvironments (an argument against habitat tracking, explanation 4 above, as a primary explanation for stasis). "It was found," Lieberman and Dudgeon write (1996, p. 231), "that more change occurred through time within a single paleoenvironment than across all paleoenvironments."

Interestingly, such a conclusion also builds a strong argument against the standard explanation of stabilizing selection (number one of this list) for stasis in paleospecies—because demes tracking single and stable environments through time should show no, or at least less, change than the species as a whole, not more. Lieberman and Dudgeon write (p. 231): "If stabilizing selection played a prominent role in maintaining stasis one would expect to find relatively little morphological change through time within a single environment." Williams (1992) has made a similar argument, at a lower scale, against stabilizing selection by emphasizing that the copiously, and lovingly, documented efficacy of natural selection in short-term situations of human observation—from beaks of Darwin's finches to industrial melanism in *Biston betularia*—makes stabilizing selection doubtful as a general explanation for such a pervasive phenomenon as stasis within paleospecies.

But when we consider this finding in supraorganismal terms, with demes as Darwinian individuals, an evident and sensible interpretation immediately emerges. A temporally coherent population may adapt gradually and continually while tracking one of several paleoenvironments inhabited by a species. But how can these anagenetic changes spread adaptively through an entire species composed of several other subpopulations, each adapted to (and tracking) its own paleoenvironment through time? No single morphology can represent a functional optimum for all habitats. In this common, and probably canonical, situation for species in nature, stability emerges as a form of "compromise" in most circumstances, a norm among "competing" minor changes that are, themselves, probably distributed more or less at random around a standard configuration, with each particular solution generally incapable, in any case, of spreading through all other demes of the species in the face of better locally adaptive configurations in most of these demes.

Of course, one can think of several obvious alternative structures where gradual change might be noted—lack of metapopulational division, with the entire species acting as a single deme, or some accessible and general biomechanical advantage that might be adaptive in all demes. But such circumstances may be uncommon—however important by cumulation in the overall history of life—in any general sample of species within a clade at any given time, thus accounting for the predominant relative frequency of stasis among all species, and for the relative rarity of anagenetic change within species as well.

Lieberman and Dudgeon summarize their proposed explanation by writing (1996, p. 231)

> Stasis may emerge from the way in which species are organized into reproductive groups occurring in separate environments. . . . The morphology of organisms within each of these demes may change through time due to local adaptation or drift, but the net sum of these independent changes will often cancel out, leading to overall net stasis . . . Only if all morphological changes across all environments were in the same direction in morphospace, or if morphological changes in a few environments were very dramatic and in the same direction, would there be significant net change in species morphology over time. . . . Thus, as long as a species occurs in several different environments one would predict on average it should be resistant to change.

The theoretical modeling of Allen, Schaffer and Rosko (1993) offers intriguing support in an implication not discussed by the authors. Allen et al. argue that the demic structure required for Lieberman's explanation of stasis strongly buffers species against extinction in chaotic ecological regimes. As an evident corollary, species selection must favor this architecture for species if such chaotic circumstances often prevail (or even just occur sporadically enough to impact a species' fate) over the geographic and temporal ranges of most species in nature. Thus, stasis would attain a predominant relative frequency among paleospecies because higher-level selection so strongly favors the persistence of species composed of multiple, semi-independent demes—the architecture that, as a consequence, engenders stasis by Lieberman's argument. Allen et al. (1993, p. 229) write:

> Even when chaos is associated with frequent rarity, its consequences to survival are necessarily deleterious only in the case of species composed of a single population. Of course, the majority of real world species . . . consist of multiple populations weakly coupled by migration, and in this circumstance

> chaos can actually reduce the probability of extinction . . . Although low densities lead to more frequent extinction at the local level, the decorrelating effect of chaotic oscillations reduces the degree of synchrony among populations and thus the likelihood that all are simultaneously extinguished.

NORMALIZING CLADE SELECTION. I cite Williams's (1992, p. 132) term for what most evolutionists would identify as a form of interdemic selection within species. (Williams uses "clade selection" as a general descriptor for all forms of selection among gene pools rather than among genes or gene combinations in organisms.) Williams also notes, as did Lieberman in a different context, the paradox of such strong empirical evidence for predominant stasis in the light of abundant data on substantial change within populations during the geological eyeblink of human careers in observation and experiment.

Williams therefore proposes, using Bell's work on stickleback fishes as a paradigm, that the environments of many demes within most species tend to be highly transient in geological terms, whereas one primary environment (often the original context of adaptation for the species) often tends to be highly persistent. (This phenomenon, however well recorded in sticklebacks, need not extend to a generality for species in nature, as Williams would readily admit in citing sticklebacks as a paradigm, not a claim for nature's normality. Sticklebacks exhibit this pattern because they generate successful, but also transient, freshwater demes from a persisting saltwater stock of lower population density.) Williams (1992, p. 134) therefore argues: "Clade selection acts against freshwater populations either because they cannot compete in mature freshwater faunas or because their habitats and ecological niches are ephemeral. The freshwater forms come and go in rapid succession, but the species complex endures in much the same form for long periods of time . . . [based on] the implied rapid extinction and intense clade selection against all but the conservative marine form . . . The appearance of stasis in the fossil record would result from an enormous variability in the persistence of ecological niches."

I am more attracted to Lieberman's suggestions, based on averaging among demes with no net change among persistent demes adapted to differing habitats, than to Williams's hypothesis, based on differential survival of one stable deme in a persistent habitat—if only because Lieberman has generated empirical evidence for longterm survival in several habitats within his two brachiopod species, whereas Williams's stickleback example may represent an unusual situation in the drastically different habitats (fresh vs. marine waters) of his transient vs. persistent demes. Still, I applaud these two suggestions for stasis based on the structuring of species-individuals as collections of deme-individuals, with differential selection acting upon demes in an irreducibly macroevolutionary mode. These proposals therefore occupy the heterodox end in a spectrum of proposed explanations for stasis—for they challenge the Darwinian orthodoxy of primacy or exclusivity for organismal selection. I especially appreciate Williams's openness towards explanations in this form, given his previous and highly influential preferences for formulating all evolutionary explanation, except when absolutely unavoidable, at the level of genic selection (in his famous book, Williams, 1966, as discussed in *SET,* pp. 550–554).

In summary, then, the assertion of predominant stasis in the geological history of most paleospecies—one of the two primary claims of punctuated equilibrium—has provoked an interesting debate in evolutionary theory, with implications for some of the most basic concepts and perspectives in our science. First, and if only as a comment about the contemporary sociology of science, the recognition of stasis as a norm of controlling relative frequency at the level of punctuated equilibrium (at least for conventional sexual species of Metazoa), has spurred general interest in phenomena of stability and non-change throughout other levels of evolutionary inquiry (see, for example, Maynard Smith, 1983). We do not yet know (see fuller discussion on pp. 236–240) whether or rather how much stasis across all scales might be attributed to structural similarity in nature's materials and processes—thus rendering this common pattern as an interesting parallelism (to use our evolutionary jargon) with genuinely homologous causal elements across scales, rather than a fortuitous convergence of similar overt patterns for disparate and merely analogous reasons. But at least we stand at the threshold of such an inquiry.

Second, and even more generally, the validation of predominant stasis as a norm would impel us to recast the basic problematic of evolution itself. If, following our conventional assumptions from Darwin to now, change represents the norm for a population through time, then our task, as evolutionary biologists, lies in specifying how this expected and universal phenomenon operates. But if, as punctuated equilibrium suggests, stasis represents the norm for most populations at most times; and if, moreover, stasis emerges as an active norm, not merely a passive consequence (as the modeling of Jackson and Cheetham, 1995, strongly suggests in documenting stasis at too high a relative frequency for models based on neutralism, directional selection, or any set of assumptions that do *not* include some active force promoting stasis directly)—then evolutionary change itself must be reconceptualized as the infrequent breaking of a conventional and expected state, rather than as an inherent and continually operating property of biological materials, ecologies and populations.

A phenomenon marking the disruption of normality holds a very different philosophical status than a phenomenon representing the ordinary architecture of biological space and time. Evolutionary change, regarded as an occasional disrupter of stasis, requires a different set of explanatory concepts and mechanisms—a different view of life, really—from evolutionary change, defined as an anagenetic expectation intrinsically operating in most populations most of the time. Punctuated equilibrium proposes that the macroevolutionary key to this new formulation lies in speciation, or the birth of new higher-level individuals at discrete geological moments (corresponding to long intervals at the scale of a human lifetime). Macroevolution, in this view, becomes an inquiry into modes and mechanisms for breaking the stasis of existing species, and generating new species, conceived and defined as discrete higher-level Darwinian individuals—and not a question about how species-individuals gradually change their parts and constitutions through time (as in conventional Darwinism). But even if this particular formulation at geological scales eventually yields more limited impact or utility than proponents of punctuated equilibrium suspect, the more general redefinition of evolution as a set of rare incidents in the breaking of stasis, rather than the pervasive movement of an expected and canonical flow, still poses an interesting challenge for rethinking a fundamental proposition about the nature and history of life.

PUNCTUATION, THE ORIGIN OF NEW MACROEVOLUTIONARY INDIVIDUALS, AND RESULTING IMPLICATIONS FOR EVOLUTIONARY THEORY

I have argued throughout this book that sets of related implications for expanding and reformulating the structure of Darwinian theory, particularly in applications to macroevolution, flow from each of the two major components of punctuated equilibrium—stasis as a norm for the duration of paleospecies, and punctuation (on geological scales) for their cladogenetic origin. The punctuational origin of species by cladogenesis provides our strongest rationale for regarding species as true evolutionary individuals in Darwin's causal world—rather than as arbitrarily delineated segments of transforming continua, and therefore not as genuine entities at all (a position maintained by both Darwin and Lamarck in some of their most forceful passages). If, following what I called the "grand analogy," species represent "items" or "atoms" of macroevolution in the same sense that organisms operate as fundamental interactors for natural selection in microevolution (see *SET*, pp. 714–744), then many features of the mechanics and patterning of macroevolution must be reformulated. For macroevolution then becomes a process irreducibly fueled by the differential birth and death of species (just as microevolution, under natural selection, is powered by the differential reproductive success of organisms)—and not, as Darwin and his successors have long held, a phenomenology ultimately built by, and extending causally from, the accumulating consequences of continuous organismic adaptation in transforming populations.

In this sense, punctuated equilibrium—by crowning the case for stable species as atoms of macroevolution—challenges all three legs of the essential Darwinian tripod: the first leg of organismal focus most directly, by establishing the higher-level species-individual as a potent causal agent of evolution as well; the second leg of functionalism more indirectly by affirming, as generators of macroevolutionary patterns, several modes of explanation that do not flow from organismal adaptation, or even rest upon an adaptational base at all; and, most comprehensively, the third leg of extrapolationism by validating a hierarchical view of pattern and causality, and by denying that the mechanisms of macroevolutionary change all flow from our uniformitarian understanding of how natural selection, working in the organismal mode, can alter populations on the scale of human observation in historical time.[1]

To illustrate the expansive and reformative potential of the species-organism as a causal agent in macroevolution, I will discuss the three major topics that punctuated equilibrium has helped to redefine during the past two decades:

Trends

In Chapter 8 of *SET*, I proposed that trends among species in clades may differ substantially from trends among organisms in populations as an "allometric" result of varying weights attached to the three major causal processes at disparate scales of organism-individuals and species-individuals—drives, and the two sources of sorting, drift, and selection (see *SET*, pp. 714–744 for full development of an argument only summarized here). At the conventional organismic level, drives from below assume little importance because the organism-individual so effectively suppresses the selective proliferation of lower-level individuals within its own body. In most circumstances, the sorting process of drift also contributes little to sustained trends because population sizes (of organisms in demes) usually exceed the small numbers required for maximal efficacy of such a stochastic force. Thus, of the three potential mechanisms, trends at the organismal level usually arise by selection.

But this understandable, and theoretically defendable, domination of selection at the focal level favored by traditional Darwinism does not extrapolate well to the higher causal level of species (as Darwinian individuals) in clades (as populations). When we shift our focus to this upper level, all three processes can claim significant potential weight in theory. (We cannot yet estimate the actual empirical weights due to paucity of research on a topic so recently defined—but see Wagner, 1996, for a breakthrough study based on quantitative and statistical discrimination of all three modes for various trends in the evolution of Paleozoic gastropods—see *SET*, pp. 733–735 for a summary of his particular conclusions.) Since the species-individual does not preferentially suppress its own transformation by directional alteration of subparts (organisms), macroevolutionary trends may often be propelled by drives from below. Such drives may arise either by the orthodox route of anagenetic transformation in populations *via* organismic natural selection ("ontogenetic drive" in my terminology of Table 8-1 in *SET*), or by the unorthodox process of directional speciation ("reproductive drive" in my terminology).

When Wright's Rule holds (see *SET*, pp. 731–735), and species arise at random with respect to the direction of a sustained trend in a clade, then we must invoke sorting processes among species. Sorting by drift can be highly effective at the species level because *N* tends to be small in relevant populations (species within clades), in contrast with the traditional Darwinian level (organisms within demes), where the magnitude of *N* usually precludes effective drift for major traits of organismal phenotypes.

A traditional Darwinian perspective might therefore lead us to denigrate the efficacy of the species level as a locus of causation for trends. If species do not marshall sufficient "strength" to stifle their own transformation by drives from below, or sufficient numbers to "prevent" the propagation of a cladal trend by random sorting, then species must pale as evolutionary agents before the strength of organisms (which manifest enough functional integrity to resist any differential proliferation of subparts, and also maintain sufficient population size to forestall random, and potentially nonadaptive, transformation of their collectivities).

I would, however, suggest that such an attitude stymies evolutionary theory as a restricting bias in the category that Francis Bacon called *idola theatri,* or idols of the theater, in his brilliant early 17th-century analysis of mental impediments to understanding the empirical world. Bacon defined idols of the theater as constraining mental habits bred by allegiance to conventionalized systems of thought. In the present case, we fall into the bad habit of reading susceptibility to drive and drift as signs of weakness in an evolutionary individual because the Darwinian agent that we understand best, and that we have previously deemed exclusive—the organism—happens to resist these modes of change as an active consequence of its inherent structure. But nature builds her scales with strong allometry, and not in a fractal manner with every higher level formed as an isometrically enlarged version of each lower level enfolded within (Gould and Lloyd, 1999).

I suggest that we challenge this idol of our traditional Darwinian conceptualizations, and at least open ourselves to the opposite view that the species-individual's capacity for change by drive and drift, as well as by selection, defines a potential source of strength for this hierarchical level as an exploiter of the full panoply of available causes for trends. Perhaps we should pity the poor organism for its self-imposed restrictions. Or perhaps, rather, we should praise the organism for managing to achieve so much with such a limited range of mechanisms! (Pardon my metaphorical lapses. I am, of course, suggesting that we view the different interplay of potential forces at various hierarchical levels as sources of distinctiveness and strength for each. We will gain a better understanding of evolutionary mechanics when we try to identify the particular capabilities of each level rather than attempting to establish a single "gatekeeper's" criterion for ranking levels in linear order by their quantity of a single enabling power analogous to such fictions as IQ.)

In any case, this allometric expansion of potential mechanisms for trending at the species level offers significant promise for fracturing by redefinition (rather than solving in conventional terms) one of the great conundrums of paleontology—an issue much fretted over, and bruited about, but usually (and finally) cast aside with vague statements of hopeful confidence that traditional explanations will suffice once we finally record enough details in any given case. At least in terms of dedicated pages in our professional literature,[2] trends represent the cardinal subject of macroevolution (with differential waxing and waning of diversity within and among clades, especially as influenced by episodes of mass extinction, as the second great theme of evolutionary discussion in paleontology).

Paleontology has long been trapped in the dilemma of recognizing only one conventional model for the explanation of trends, and then finding little credible evidence for the model's adequacy. By the expectations of all three central precepts in Darwinian logic, and by our habits of restricting explanations of sustained organismic trends to selectionist causes (given valid arguments for rejecting the alternatives of drift and drive at the organismic level, as discussed above), increasing adaptation of organisms must also propel macroevolutionary trends under extrapolationist premises.

(All Darwinians understand, of course, that natural selection only yields adaptation to immediate environments—a notion not conducive to sustained directional trends through geological time, given the effectively random fluctuation of most environmental configurations through substantial geological intervals. Consequently, most sustained trends have been interpreted as generalized biomechanical improvements conferring advantages across most or all experienced environments, and arising from Darwin's own preference for domination of biotic over nonbiotic competition in the history of life. See the discussion of Darwin's rationale for this defense of "progress" in evolution in *SET*, Chapter 6, pp. 467–479.)

Discourse about trends dominates the traditional literature of evolutionary paleontology, both at the most general level of universal phenomenology (Cope's law of increase of size, Dollo's law of irreversibility, Williston's law of reduction and specialization of modular segments, etc.), and as a dominating theme for the history of almost any individual clade. We all know the particular tales for textbook groups—increasing brain size in hominids; larger body size, fewer toes, and higher crowned cheek teeth in horses; increasing symmetry of the cup in Paleozoic crinoids, with eventual expulsion of the anal ray to the top of the calyx; complexification in ammonite suture lines; reduction in number of stipes in graptolite colonies. These summary themes for clades, all based on the concept of general biomechanical improvement through time, distill the essence of traditional paleontology.

I do not deny that generalized organismic advantage may explain some of these classic trends. I do not, for example, challenge the traditional notion that increasing perfection of radial symmetry may confer adaptive benefits in feeding upon sessile organisms like stalked crinoids (see Moore and Landon, 1943, for the classic statement). But I also note that other classic trends, apparently ripe for explanation in biomechanical terms, have stubbornly resisted any reasonable hypothesis ever proposed—most notably the complexifying ammonite suture, which does not clearly confer greater resistance to shell crushing, and does not evidently aid the growing animal by increasing surface area of tissues covering the septa.

But other trends, despite their prominence, have never generated even a plausible hypothesis of biomechanical advantage. Why should fewer-stiped graptolite colonies "do better" in any usual sense of organismic (or, in this case, astogenetic) advantage? In such circumstances, we need to expand our explanatory net by considering alternative causal resolutions based on differential success of species as Darwinian individuals engaged in processes of sorting. Instead of focusing upon the putative biomechanical virtues of fewer stipes, we should be asking how and why such a character might correlate with the propensities of species for branching or for resistance to extinction—the "birth" and "death" processes that regulate sorting at this higher level.

I am not even confident that we should preferentially attribute traits with more plausible organismic advantages—including the enlarging brain of hominids as an obvious example—to conventional microevolutionary explanations, without seriously considering unorthodox possibilities based on causal correlations of such traits with propensities for speciation, or on the sheer good fortune of nonadaptive hitchhiking due to fortuitous presence in the subclade growing to domination for other reasons. We should be paying more attention to interesting and plausible proposals like Sacher's (1966) on lifespan and developmental timing as the primary target of selection in hominid evolution, with large brains on a facilitating causal pathway to advantageous retardations of development (Gould, 1977).

In any case, and most generally, the need to describe trends—when punctuated equilibrium dominates the geometry of evolutionary change within a clade—as differential success of stable species, rather than as extrapolated anagenesis of populations, requires, in itself and as a "one liner" of extensive reformatory power, a radical reformulation (in the literal sense of reconstruction from the very *radices,* or roots, of the subject on up) of the primary topic in our macroevolutionary literature. Jackson and Cheetham (1999, p. 76) conclude: "Granted the prevalence of punctuated equilibria, macroevolutionary trends must arise through differential rates of origination and extinction, and not by adaptive evolution within single species. All of this is compatible with traditional neodarwinian evolutionary biology, but was unexpected before the theory of punctuated equilibria."

In summary, the efficacy of drifts and drives, in addition to selection, for generating trends at the hierarchical level of species as Darwinian individuals, suggests a rich, and virtually unexploited, domain of alternative explanations that might break through the disabling paradox of our current inability to resolve such a salient phenomenon in our preferred mode of adaptive advantages to organisms. Species-level explanations of trends in organismal phenotypes add at least two categories of potential resolution to our usual search for organismic benefit.

First, the trending character may be causally significant not for its phenotypic consequences to the organism, but for its role in influencing rates and directions of speciation in populations of organisms bearing the trait. If fossorial features of burrowing rodents (Gilinsky, 1986), or nonplanktonic lifestyles of marine molluscan larvae (Hansen, 1978, 1980), help to generate populational traits that enhance speciation rates, then a trend spreading such organismic features through a clade may arise by positive sorting of species rather than by general adaptive advantage of the phenotype itself. The organismal phenotypes may enjoy no general advantage at all, and may only produce adaptation to relatively ephemeral habitats within the clade's potential range. In fact, both fossorial rodent species (for reasons of small population size) and nonplanktonic molluscan species (for limited geographic ranges) may experience a reduced average geological longevity relative to surface dwelling (for rodents) and planktonic (for mollusks) clade members. But these impediments may be overbalanced by enhancement of speciation rates, thus driving the trend.

Second, both the low *N* for species in clades (relative to organisms in most populations), and the remarkably (and, for most people, counterintuitively) high frequency of fortuitous but significant correlations between pairs of traits in systems built by genealogical branching (Raup and Gould, 1974) virtually guarantee that trending by drift will be much more common in sorting of species-individuals than in conventional sorting of organism-individuals. After all, branching evolution imparts a set of autapomorphic traits (through a unique common ancestor) to any subclade of species—and we can scarcely believe that each of these traits establishes the basis of selective existence and success for each species in the entire monophyletic group. Therefore, any process that favors the relative proliferation of any subclade for any reason will automatically engender a positive trend for any included autapomorphic character, whatever the causal basis of the general trend.

When we combine this spur to drift by hitchhiking with the observation that many successful clades go through severe bottlenecks (often as single surviving species) during their geological existence, we obtain even more compelling reasons for considering drift as a major source of macroevolutionary trends, however much we may reject analogous processes as substantial generators of trends in phenotypic characters controlled by organismal selection. For example, ammonites endured two severe bottlenecks at two major mass extinction events—suffering reduction, perhaps to two surviving lineages in the Permo-Triassic debacle, and to a single lineage in the closing Triassic event. In this light, why should we regard explanations based on general biomechanical advantages for organisms as preferable to the obvious blessing of good fortune upon any trait belonging to the phenotype of single lineages that manage to squeak through such profound bottlenecks? Few other evolutionary processes can promote traits from partial representation to exclusivity within a population (of species-individuals within a clade in this case) so quickly and so decisively.

The paleontological literature has just begun to reconceptualize trends in speciational terms. Initial results offer much encouragement, both for revising traditional explanations of particular temporal sequences, and for posing new questions requiring tests by different kinds of data. New insights often emerge just by framing the subject in terms of numbers and longevities of taxa rather than gradual fluxes of form. In their study of trends in Mississippian crinoids, for example, Kammer, Baumiller, and Ausich (1997) reach a conclusion that surprised them only because such a reasonable idea had not previously been formulated in operational terms (p. 221): "Results of this study indicate that among Mississippian crinoids niche generalists had greater species longevity than niche specialists. Although logical, few data have previously been developed to test this relationship."

The complexity of the subject then becomes apparent when the authors discuss why the obvious implication—that a trend towards generalists should sweep through the clade—fails because *ceteris paribus* (all other things equal) does not hold at several levels and strengths of correlation. First of all, "niche generalists tend to have fewer species per clade than niche specialists" (Kammer et al., 1997, p. 221), thus illustrating the most pervasive and powerful macroevolutionary constraint recognized so far at the speciational level (see general discussion in *SET*, pp. 739–741): the forced and intrinsic negative correlation between speciation rate and longevity. Secondly, these patterns hold "only during times of background extinction when Darwinian natural selection prevails" (p. 221). Mass extinctions may then impact species at random, without preferential regard for their ecological status or prospects for longevity in normal times.

Similar questions may be asked at all scales, including general patterns for life itself. For example, many robust paleontological data sets show a general tendency for increased longevity in marine invertebrate species through geological time. Even a famously iconoclastic thinker like David Raup, who devoted so much of his career to exploring the power of random systems to render observed patterns of the fossil record, interpreted this result as our best case for a meaningful concept of "progress," defined as increasing adaptive excellence of organisms (and leading to greater resistance to extinction). But several authors (Valentine, 1990; Gilinsky, 1994; Jablonski, Lidgard and Taylor, 1997) have reread this result in terms of species sorting as a tendency "for high-turnover taxa to be replaced over geologic time by low-turnover taxa" (Jablonski, et al., 1997, p. 515).

As Gilinsky (1994) notes, such a pattern, thus reformulated, may bear little or no relationship to general adaptive excellence at the organismal level. If speciation and extinction rates generally operate in balance (as they do), then some clades may be designated more "volatile" (in Gilinsky's terminology) as marked by their high rates of speciation and extinction, and others as more stable for lower rates of both these defining processes at the species level. In an abstract and general sense, both "strategies" may be regarded as equal in yielding the same result of steady cladal persistence, but with volatile clades showing more variation around a stable mean. However, in our real world of fluctuating environments and, especially, mass extinctions, volatility may doom clades in the long run because any reduction to zero (however "temporary" and reversible in abstract modeling) extinguishes all futures in our actual world of material entities. Since volatile clades, on average, must cross the zero line more frequently than stable clades, a general trend to increasing species longevity may only arise as an indirect consequence of the higher vulnerability of volatile clades and the consequent accumulation of stable clades through geological time.

Several recent articles on bryozoan evolution illustrate the utility of a speciational approach to trends—if only as a method for setting base lines and making distinctions, all the better to document patterns *not* attributable to differential speciation. For example, Jablonski, Lidgard and Taylor (1997) found that "the generation of low-level novelties are effectively driven by speciation rates" (p. 514), whereas the origin of major apomorphies of larger groups do not correlate so clearly with numbers of speciation events, but rather with their magnitude (with such rare events favored in certain times and environments).

In documenting a speciational basis as a "null hypothesis" of sorts for identifying evolutionary patterns arising by other routes, McKinney et al. (1998) compared the differential successes of cheilostome and cyclostome bryozoan clades through time. In a fascinating discovery, they noted that, in times of joint decline for both clades (late Cretaceous and post-Paleocene after a Danian spike in diversification), "the relative skeletal mass of cyclostomes declined much more precipitously than did relative species richness" (p. 808). The authors could therefore identify an important trend by standardizing species numbers: "There is a long-term trend for the average cheilostome species to generate a progressively greater skeletal mass than the average cyclostome species. This could result from a gradual trend toward relatively larger colony sizes within cheilostomes, a greater number of colonies per cheilostome species, or both" (pp. 808–809).

In a similar vein, but at the larger scale of the phylum's initial Ordovician radiation, Anstey and Pachut (1995) found no relationship between number of speciation events and the establishment of defining apomorphies among major subgroups at the base of the clade. They write (1995, pp. 262–263): "The morphologies recognized as higher taxa of bryozoans were not built up through a gradual accumulation of species differences but appear to have diverged very rapidly in the initial radiation of the phylum . . . The processes producing the major branching events and familial apomorphies, therefore, apparently were not driven by speciation, and likewise could not have resulted from species selection or species sorting."

The Speciational Reformulation of Macroevolution

Beyond the immediate expansion of explanations for the signature phenomenon of trends, the recasting of macroevolution as a discourse about the differential fates of stable species (treated as Darwinian individuals) carries extensive implications for rethinking both the pageant of life's history and the causes of stability and change in geological time.

I am particularly grateful that Ernst Mayr (1992, p. 48)—the doyen of late-20th-century evolutionary biologists, and the inspirer of punctuated equilibrium through his views on peripatric speciation (though scarcely an avid supporter of our developed theory)—has identified the required speciational reformulation of macroevolution as the principal component of "what had not been recognized [in evolutionary theory] before" Eldredge and I codified punctuated equilibrium. Mayr continues, stressing the species as the macroevolutionary analog of the organism considered as the "atom" of microevolution:

> It was generally recognized that regular variational evolution in the Darwinian sense takes place at the level of the individual and population, but that a similar variational evolution occurs at the level of species was generally ignored. Transformational evolution of species (phyletic gradualism) is not nearly as important in evolution as the production of a rich diversity of species and the establishment of evolutionary advance by selection among these species. In other words, speciational evolution is Darwinian evolution at a higher hierarchical level. The importance of this insight can hardly be exaggerated.

As a most general statement, and extending Mayr's views from his specific words (cited just above) to his most characteristic philosophical observation, Darwinism's major impact upon western thinking transcends the replacement of a fixed and created universe by an evolutionary flux. As thoughtful evolutionists have always noted, and as Mayr has particularly stressed in our times by contrasting "essentialist" and "populational" ways of thinking, a fundamental revision in our concept of the essence of reality—from the Platonic archetype to the variable population—may represent Darwin's most pervasive and enduring contribution to human understanding. For what could be more profound or portentous than our switch from "fixed essences" to "sensibly united groups of varying items," as our explanation for the reality behind our names for categories in our parsings of the natural world?

Yet however successful we have been in executing this great philosophical shift at the level of microevolution—where we understand that no archetype for a seahorse, a sequoia or a human being exists; where an enterprise named "population genetics" stands at the core of an explanatory system; and where we have all been explicitly taught to view change as the conversion of intrapopulational variation into interpopulational differences—we have scarcely begun to execute an equally important reconceptualization for our descriptions and explanations of macroevolution. We still encapsulate the pageant of life's history largely as a set of stories about the trajectories of abstracted designs through time. Beetles and angiosperms flourish; trilobites and ammonites disappear. Horses build bigger bodies and fewer toes; humans evolve larger brains and smaller teeth. Meanwhile, in grand epitome, life itself experiences a general rise in mean complexity as a primary definition (in popular culture at least) of evolution itself.

We know, of course, that flourishing implies more species, while extinction marks an ultimate reduction to zero. But we still tend to visualize these patterns of changing diversity as consequences of the status of designs, with the Darwinian optimum standing for the old Platonic archetype as an ideal and guarantor. (And if we manage to construe such quintessentially populational phenomena as the waxing and waning of groups in this Platonic mode, then we will surely not be tempted to reformulate anatomical trends in average form within groups—a phenomenon far more congenial to our essentialist mythologies—in the Darwinian language of changing frequencies in variable populations.) In this way, the two major phenomena of macroevolution—phenotypic trends within groups and changes in relative diversities among groups—have stubbornly resisted the reformative power of Darwin's deepest insight. Moreover, this resistance arises for the most ironical of recursive reasons—namely, that other biases of the Darwinian tradition, particularly the reductionist and extrapolationist premises discussed as the essential Darwinian tripod, have forestalled an application of Darwin's deepest insight to nature's grandest scale!

We have conceptualized macroevolution as temporal fluxes of adaptive form, with either some notion of an average phenotype for a clade (corresponding to vernacular ideas of the "ordinary" or "normal"), or some extreme value (representing our view of the "promise" or "potential" of a collectivity) standing as a surrogate or summary for all variation, in both form and number of species, within the clade under consideration. This attitude has led us to embrace a set of patent absurdities that rank, nonetheless, as received wisdom about the history of life, and that we continue to support in a passive way because the fallacies can only become apparent when we reconceive macroevolution as an inquiry centered upon the changing sizes and variabilities of clades based on the fates of their component Darwinian individuals—their species, construed as "atoms" of macroevolution (at least for sexually reproducing organisms).

Thus, we usually summarize the history of life as a drama about generally increasing complexity of form (with bacteria, beetles and fishes left successively behind, despite their evident prosperity) when, at the very most, such a theme can only apply to a small group of species on the right tail of life's general distribution—and when, by any fair criterion generally employed by evolutionists, bacteria have always dominated the history of life from their origins in exclusivity more than 3.5 billion years ago to their current mastery of a much more diverse world. And we have chosen horses as our textbook and museum hall example of progressive evolution triumphant, when modern equids represent a pitiable remnant of past diversity, a small clade entirely extinguished in its original, and formerly speciose, New World home, and now surviving as only a half dozen or so species of horses, asses and zebras at the more hospitable termini of past migrations. Horses, moreover, represent failures within a failure—for the once proud order of Perissodactyla now persists as only three small clades of threatened species (tapirs, rhinoceroses and equids, with the last receiving an artificial boost for human purposes), while the once minor order of Artiodactyla now dominates the guild of large, hoofed herbivorous mammals as one of evolution's great success stories. But we will not grasp these evident patterns, truly generated by changing diversity of species, if we continue to dwell in a conceptual prison that frames the history of life as a flux from monad to man, and the phylogeny of horses as a stately race from little splay-footed eohippus to the one-toed nobility of Man O' War.

The key to a more expansive formulation—also a more accurate depiction in the language of probable causes based on genuine evolutionary agents—lies in recasting this discourse about fluxes of means or extreme values within clades as a history of the differential origins and fates of species, as organized by nature's genealogical system (that is, evolution itself) into the monophyletic groups of life's tree. If we can accomplish this speciational reformulation of macroevolution, we will understand that many classical "trends" emerge as passive consequences of temporal variation in numbers of species within a clade often enhanced when structural constraints channel the potential directions of such variation, and not as selectively driven vectors in the biomechanical form of "average" organisms within the clade. We should be asking questions about numbers of actual species, rather than rates of flux for anatomical archetypes or abstractions. We should be studying the dynamics of differential species success as the causal basis of macroevolutionary pattern, not placing our hopes for explanation upon undefinable optima for competitive triumph of organic designs.

We will finally recognize that causes for the evolution of form and the evolution of diversity do not interact in the conceptual opposition that defined Lamarck's original formulation of evolution (see *SET,* Chapter 3), and that persists today in common statements (see G. G. Simpson and J. S. Huxley as quoted in *SET,* pp. 562–563, or F. Ayala, 1982) that speciation (or cladogenesis) builds the luxury of iterated variation, while a different, and altogether more important, process of transformational anagenesis fashions the trends of form that culminate in such glories as the human brain and the dance language of bees. We will then grasp that many—though not all—phenomena in the evolution of form arise as noncausally correlated consequences of patterns in the changing diversity of species.

If heaven exists (and the management let Darwin in), he must be greeting this prospect with the same thought that the founders of America emblazoned on the new nation's Great Seal: *annuit coeptis* (he smiles on our beginnings). For Darwin tried (see *SET,* Chapters 2 and 3) to disassemble Lamarck's sterile dichotomy between fundamental, but probably unresolvable, causes of progressive evolution in form, and secondary, albeit testable, causes of lateral diversification—and to reformulate all evolution in terms of the previously trivialized tangent, while branding the supposed main line as illusory. I am proposing an analogous reform at the level of macroevolution—with the "diversity machine" of speciation, previously labelled as secondary and merely luxurious, also recognized as the generator of what we perceive as trends in form within clades. Again, the humble and testable factor, once relegated to a playground of triviality, becomes the cause of a supposedly higher process formerly judged orthogonal, if not oppositional. But this time, both the atom of agency and the cause of change reside at the higher level of macroevolution, and must therefore be accessed in the unfamiliar framework of deep time, rather than directly observed in human time. Our intellectual resources are not unequal to such a task, and we could not ask for a better leader than Darwin.

To illustrate how such a speciational reformulation might proceed, let us consider, at three levels of inquiry, the consequences of documenting macroevolution as expansion, contraction and changing form in the distribution of all species within clades through time. I regard this unfamiliar categorization as an empowering substitute for the usual tactic of summarizing the history of a clade by some measure of central tendency, or some salient extreme that catches our fancy, and then plotting the trajectory of this archetypal value through time.

LIFE ITSELF. In popular descriptions of evolution, from media to museum halls, but also in most technical sources, from textbook pedagogy to monographic research, we have presented the history of life as a sequence of increasing complexity, with an initial chapter on unicellular organisms and a final chapter on the evolution of hominids. I do not deny that such a device captures something about evolution. That is, the sequence of bacterium, jellyfish, trilobite, eurypterid, fish, dinosaur, mammoth, and human does, I suppose, express "the temporal history of the most complex creature" in a rough and rather anthropomorphic perspective. (Needless to say, such a sequence doesn't even come close to representing a system of direct phylogenetic filiation.)

But can such a sequence represent the history of life, or even stand as a surrogate either for the fundamental feature of that history, or for the central causal processes of evolutionary change? When we shift our focus to the full range of life's diversity, rather than the upper terminus alone, we immediately grasp the treacherous limitations imposed by misreading the history of an extreme value as the epitome of an entire system. The sequence of increasingly right-skewed distributions with a constant modal value firmly centered on bacteria throughout the history of life (Fig. 5-1) represents only a cartoon or an icon for an argument, not a quantification. The full vernacular understanding of complexity cannot be represented as a linear scale, although meaningful and operational surrogates for certain isolated aspects of the vernacular concept have been successfully designated for particular cases—see McShea, 1994.

I do not see how anyone could mistake the extreme value of a small tail in an increasingly skewed distribution through time for the evident essence, or even the most important feature, of the entire system. The error of construing this conventional trend of extremes as the essential feature of life's history becomes more apparent when we switch to the more adequate iconography of entire ranges of diversity through time. Consider just three implications of the full view, but rendered invisible when the sequential featuring of extremes falsely fronts for the history of the whole.

1. The salience of the bacterial mode. Although any designation of most salient features must reflect the interests of the observer, I challenge anyone with professional training in evolutionary theory to defend the extending tip of the right tail as more definitive or more portentous than the persistence in place, and constant growth in height, of the bacterial mode. The recorded history of life began with bacteria 3.5 billion years ago, continued as a tale of prokaryotic unicells alone for probably more than a billion years,

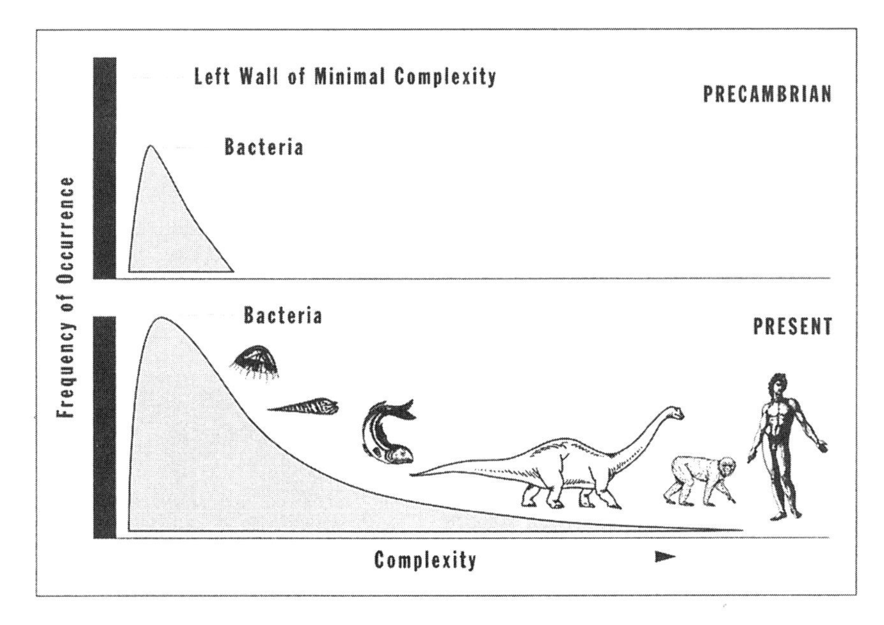

5-1. This cartoon of changing form and range in the histogram of complexity through life's phylogeny illustrates how we fall into error when we treat extreme values as surrogates or epitomes of entire systems. A view that emphasizes speciation and diversity might recognize the constancy of the bacterial mode as the outstanding feature of life's history. From Gould, 1996.

and has never experienced a shift in the modal position of complexity. We do not live in what older books called "the age of man" (1 species), or "the age of mammals" (4000 species among more than a million for the animal kingdom alone), or even in "the age of arthropods" (a proper designation if we restrict our focus to the Metazoa, but surely not appropriate if we include all life on earth). We live, if we must designate an exemplar at all, in a persisting "age of bacteria"—the organisms that were in the beginning, are now, and probably ever shall be (until the sun runs out of fuel) the dominant creatures on earth by any standard evolutionary criterion of biochemical diversity, range of habitats, resistance to extinction, and perhaps, if the "deep hot biosphere" (Gold, 1999) of bacteria within subsurface rocks matches the upper estimates for spread and abundance, even in biomass (see Gould, 1996, for a full development of this argument). I will only remind colleagues of Woese's "three domain" model for life's full genealogy (see Fig. 5-2), a previously surprising but now fully accepted, and genetically documented, scheme displaying the phylogenetic triviality of all multicellular existence (a different issue, I fully admit, from ecological importance). Life's tree is, effectively, a bacterial bush. Two of the three domains belong to prokaryotes alone, while the three kingdoms of multicellular eukaryotes (plants, animals, and fungi) appear as three twigs at the terminus of the third domain.

2. The cause of the bacterial mode. "Bacteria," as a general term for the grade of prokaryotic unicells lacking a complex internal architecture of organelles, represent an almost ineluctable starting point for a recognizable fossil record of preservable anatomy. As a consequence of the basic physics of self-organizing systems and the chemistry of living matter—and under any popular model for life's origins, from the old primordial soup of Haldane and Oparin, to Cairns-Smith's clay templates (1971), to preferences for deep-sea vents as a primary locale—life can hardly begin in any other morphological status than just adjacent to what I have called (see Fig. 5-1) the "left wall" of minimal conceivable preservable complexity, that is, effectively, as bacteria (at least in terms of entities that might be preserved as fossils). I can hardly imagine a scenario that could begin with the precipitation of a hippopotamus from the primordial soup.

Once life originates, by physico-chemical necessity, in a location adjacent to this left wall (see Kauffman, 1993), the subsequent history of right-

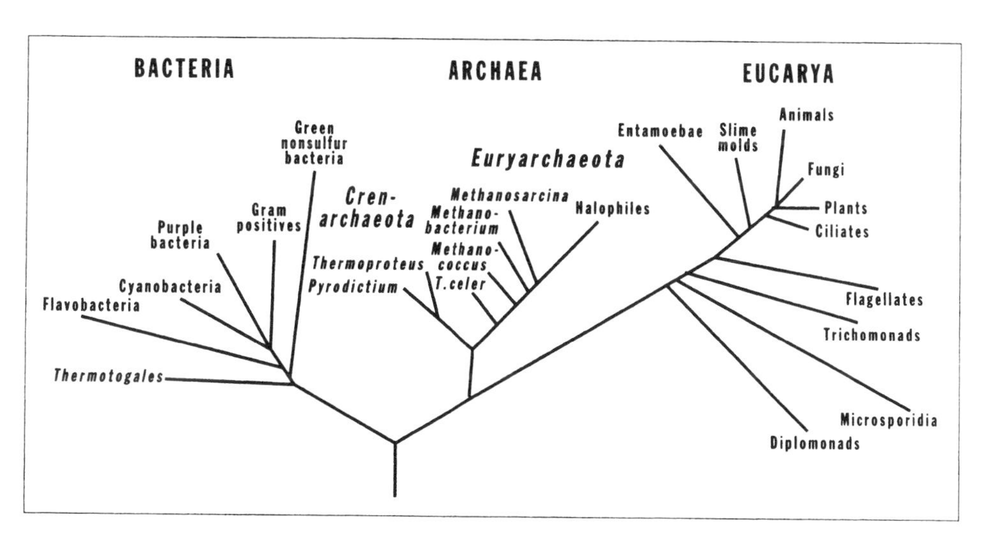

5-2. In life's full genealogy, all three multicellular kingdoms grow as twigs at the terminus of just one branch among the three great domains of life's history. The other two domains are entirely prokaryotic. From work of Woese and colleagues, as presented in Gould, 1996.

skewed expansion arises predictably as a fundamental geometric constraint of this initial condition combined with the principles of Darwinian evolution—that is, so long as the most genuine trend of life's history then prevails: "success" measured variationally, in true Darwinian fashion, as expansion in diversity and range through time.

If life continues to add taxa and habitats, then structural constraints of the system virtually guarantee that a right tail of complexity will develop and increase in skew through time as a geometric inevitability, and not necessarily for any overall advantage conferred by complexity. As noted above, life must begin, for physico-chemical reasons, next to the left wall of minimal complexity. Little or no "space" therefore exists between the initial bacterial mode and this natural lower limit; variation can expand only into the "open" domain of greater complexity. The vaunted trend to life's increasing complexity must be reconceived, therefore, as a drift of a small percentage of species from the constant mode of life's central tendency towards the only open direction for expansion. To be able to formulate this alternative view at all, we must reconceive the history of life as expansion and contraction of a full range of taxa under constraints of systems and environments, rather than as a flux of central tendencies, valued extremes, or salient features.

3. The right tail as predictable, but passively generated. A critic might respond that he accepts the reformulation but still wishes to assert a vector of progress as life's central feature in the following, admittedly downgraded, way: yes, the vector of progress must be construed as the expanding right tail of a distribution with a constant mode, not as a general thrust of the whole. But this expanding tail still arises as a predictable feature of the system, even if we must interpret its origin and intensification as the drift of a minority away from a constraining wall, rather than the active trending of a totality. The right tail had to expand so long as life grew in variety. This tail therefore originated and extended for a reason; and humans now reside at its present terminus. Such a formulation may not capture the full glory of Psalm 8 ("Thou hast made him a little lower than the angels"), but a dedicated anthropocentrist could still live with this version of human excellence and domination.

But the variational reformulation of life's system suggests a further implication that may not sit well with this expression of human vanity. Yes, the right tail arises predictably, but random systems generate predictable consequences for passive reasons—so the necessity of the right tail does not imply active construction based on overt Darwinian virtues of complexity. Of course the right tail might be driven by adaptive evolution, but the same configuration will also arise in a fully random system with a constraining boundary. The issue of proper explanations must be resolved empirically.

By "random" in this context, I only mean to assert the hypothesis of no overall preference for increasing complexity among items added to the distribution—that is, a system in which each speciation event has an equal probability of leading either to greater or to lesser complexity from the ancestral design. I do not deny, of course, that individual lineages in such systems may develop increasing complexity for conventional adaptive reasons, from the benefits of sharp claws to the virtues of human cognition. I only hold that the entire system (all of life, that is) need not display any overall bias—for just as many individual lineages may become less complex for equally adaptive reasons. In a world where so many parasitic species usually exhibit less complexity than their freeliving ancestors, and where no obvious argument exists for a contrary trend in any equally large guild, why should we target increasing complexity as a favored hypothesis for a general pattern in the history of life?

The location of an initial mode next to a constraining wall guarantees a temporal drift away from the wall in random systems of this kind. This situation corresponds to the standard paradigm of the "drunkard's walk" (Fig. 5-3), used by generations of statistics teachers to illustrate the canonical random process of coin tossing. A drunkard exits from a bar and staggers, entirely at random, along a line extending from the bar wall to the gutter

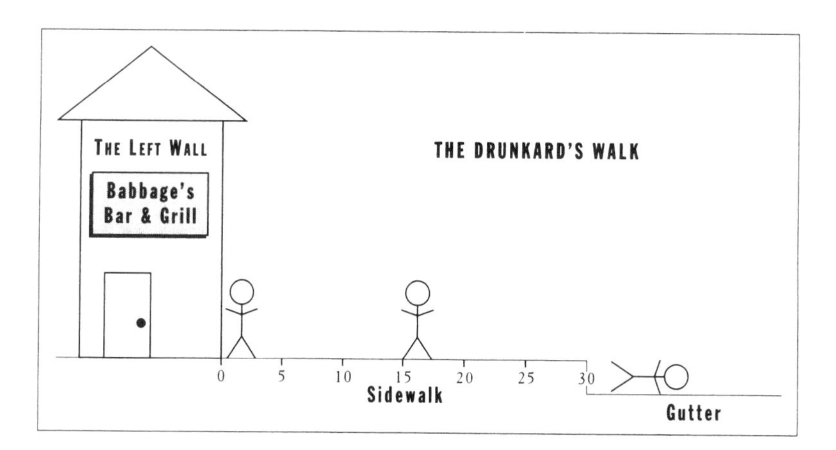

5-3. The standard statistical model of the "drunkard's walk" shows that even the expanding right tail of life's right-skewed histogram of complexity may arise within a random system with equal probabilities for the movement of any descendant towards either greater or lesser complexity. From Gould, 1996.

(where he passes out and ends the "experiment"). He winds up in the gutter on every iteration of this sequence (and with a predictable distribution of arrival times) simply because he cannot penetrate the bar wall and must eventually "reflect off" whenever he hits this boundary. (Of course, he will also end up in the gutter even if he moves preferentially towards the bar wall; in this case, the average time of arrival will be longer, but the result just as inevitable.)

The issue of active drive (a small bias in relative frequency fueled by the general Darwinian advantages of complexity) vs. passive drift (predictable movement in a random system based on the model of the drunkard's walk) for the expansion of the right tail must be resolved empirically. But the macroevolutionary reformulation of life's history in variational terms establishes a conceptual framework that permits this question to be asked, or even conceived at all, for the first time. Initial studies on mammalian vertebrae and teeth, foraminiferal sizes, and ammonite sutures have been summarized in Gould, 1996, based on pioneering studies of McShea, Boyajian, Arnold, and Gingerich. This initial research has found no departure from the random model, and no overall preference for increase in complexity in studies that tabulate *all* events of speciation.

GENERAL RULES. The older literature of paleobiology focused on the recognition and explanation of supposedly general "rules" or "laws" regulating the overt phenomenology of life's macroevolutionary pattern. As the modern synthesis developed its core of Darwinian explanation, several leading theorists (see especially Haldane, 1932, and Rensch, 1947, 1960) tried to render these laws as large-scale expressions of evolution's control by adaptive anagenesis in populations under Darwinian natural selection.

This subject fell out of favor for several reasons, but in large part because non-adaptationist explanations, deemed less interesting (and certainly less coordinating) than accounts based on natural selection, provided an adequate compass for most of these "laws." Thus, for example, Dollo's law of irreversibility (see Gould, 1970b) only restates the general principles of mathematical probability for the specific case of temporal changes based on large numbers of relatively independent components. And Williston's law of reduction and specialization in modular segments may only record a structural constraint in random systems, thus following the same principles as my previous argument about the expanding right tail of complexity for life's totality. Suppose that, in overall frequency within the arthropod clade, modular species (with large numbers of similar segments) and tagmatized species (with fewer fused and specialized groupings of former segments) always enjoy equal status in the sense that 50 percent of habitats favor one design, and 50 percent the other. (I am, of course, only presenting an abstract "thought experiment," not an operational possibility for research. Niches don't exist independent of species.) But suppose also that, for structural reasons, modular designs can evolve toward tagmatization, but tagmatized species cannot revert to their original modularity—an entirely reasonable assumption under Dollo's law (founded upon the basic statements of probability theory) and generalities of biological development. Then, even though tagmatization enjoys no general selective advantage over modularity, a powerful trend to tagmatization must pervade the clade's history, ultimately running to completion when the last modular species dies or transforms.

However, one of these older general rules has retained its hold upon evolutionary theory, probably for its putative resolvability in more conventional Darwinian terms of general organismic advantage: Cope's Law, or the claim that a substantial majority of lineages undergo phyletic size increase, thus imparting a strong bias of relative frequency to the genealogy of most clades—a vector of directionality that might establish an arrow of time for the history of life.

A century of literature on this subject had been dominated by proposed explanations in the conventional mode of organismic adaptation fueled by natural selection. Why, commentators asked almost exclusively, should larger size enjoy enough general advantage to prevail in a majority of lineages? Proposed explanations cited, for example, the putative benefits enjoyed by larger organisms in predatory ability, mating success, or capacity to resist extreme environmental fluctuations (Hallam, 1990; Brown and Maurer, 1986).

The speciational reformulation of macroevolution has impacted this subject perhaps more than any other, not because the theme exudes any special propensity for such rethinking—for I suspect that almost any conventional "truth" of macroevolution holds promise for substantial revision in this light—but because its salience as a "flagship," but annoyingly unresolved, issue inspired overt attention. Moreover, the conventional explanations in terms of organismal advantage had never seemed fully satisfactory to most paleontologists.

The rethinking has proceeded in two interesting stages. First, Stanley (1973), in a landmark paper, proposed that Cope's Law emerges as a passive consequence of Cope's other famous, and previously unrelated, "Law of the

Unspecialized"—the claim that most lineages spring from founding species with generalized anatomies, under the additional, and quite reasonable, assumption that the majority of generalized species also tend to be relatively small in body size within their clades.

These statements still suggest nothing new so long as we continue to frame Cope's Rule as anagenetic flux in an average value through time—that is, as a conventional "trend" under lingering Platonic approaches to macroevolution. But when we reformulate the problem in speciational terms—with the history of a Cope's Law clade depicted as the distribution of all its species at all times, and with novelty introduced by punctuational events of speciation rather than anagenetic flux—then a strikingly different hypothesis leaps forth, for we now can recognize a situation precisely analogous (at one fractal level down) to the previous construction of life's entire history: an evolving population of species (treated as stable individuals), in a system with a left wall of minimal size (for the given *Bauplan*), and a tendency for founding members to originate near this left wall (Fig. 5-4).

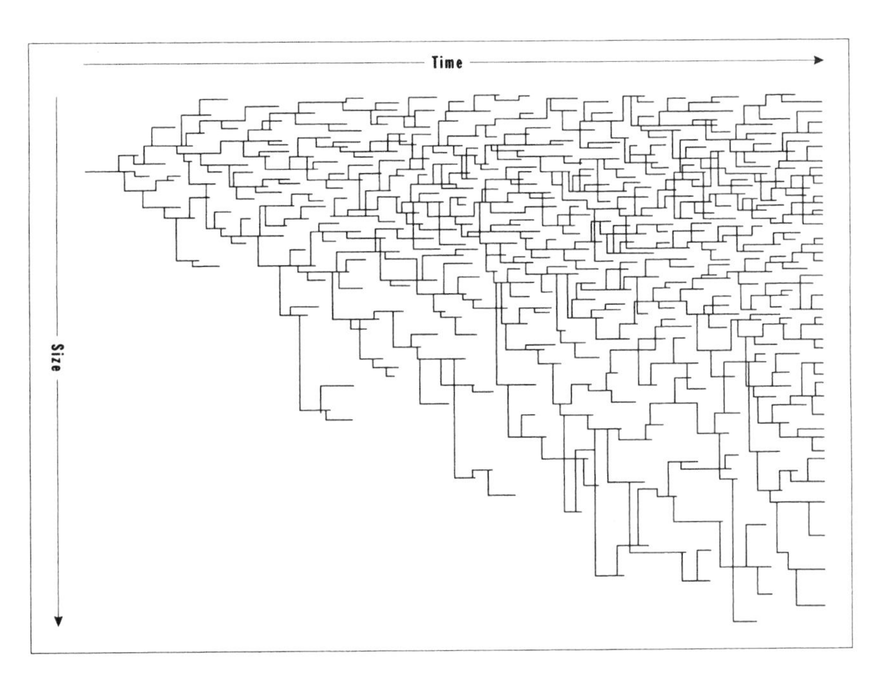

5-4. Cope's Law shown, under a speciational perspective, as a differential movement of speciation events towards larger size from a constraining boundary imparted by a small founding member of the lineage. Adapted from Stanley, 1973.

Therefore, just as for all of life in my previous example, if the clade prospers with an increasing number of species, and even if new species show no directional tendency for increasing size (with as many species arising smaller than, as larger than, their ancestors), then the *mean* size among species in the clade must drift to the right, even though the *mode* may not move from initial smallness, just because the space of possible change includes substantial room in the domain of larger size, and little or no space between the founding lineage and the left wall. Thus, as Stanley (1973) stated so incisively, Cope's Law receives a reversed interpretation as the structurally constrained and passive evolution (of an abstracted central tendency, I might add) *from small size,* rather than as active evolution *towards large size* based on the organismic advantages of greater bodily bulk under natural selection.

But we must then carry the revision one step further and ask an even more iconoclastic question: does Cope's Law hold at all? Could our impressions about its validity arise as a psychological artifact of our preferential focus upon lineages that grow larger, while we ignore those that remain in stasis or get smaller—just as we focus on fishes, then dinosaurs, then mammoths, then humans, all the while ignoring the bacteria that have always dominated the diversity of life from the pinnacle of their unchanging mode throughout geological time?

Again, we cannot even ask this question until we reformulate the entire issue in speciational terms. If we view a temporal vector of a single number as adequate support for Cope's Law, we will not be tempted to study *all* species in a monophyletic clade that includes signature lineages showing the documented increase in size. But when we know, via Stanley's argument, that Cope's Law can be generated as a summary statement about passively drifting central tendencies in random systems with constraining boundaries, then we must formulate our tests in terms of the fates of all species in monophyletic groups. Jablonski (1997) has published such a study for late Cretaceous mollusks of the Gulf and Atlantic coasts (a rich and well-studied fauna of 1086 species in Jablonski's tabulation) and has, indeed, determined that, for this prominent group at least, prior assertions of Cope's Law only represent an artifact of biased attention (see commentary of Gould, 1997a). Jablonski found that 27–30 percent of genera do increase in mean size through the sequence of strata. But the same percentage of genera (26–27 percent) also decrease in mean size—although no one, heretofore, had sought them out for equal examination and tabulation.

Moreover, and more notably for its capacity to lead us astray when we op-

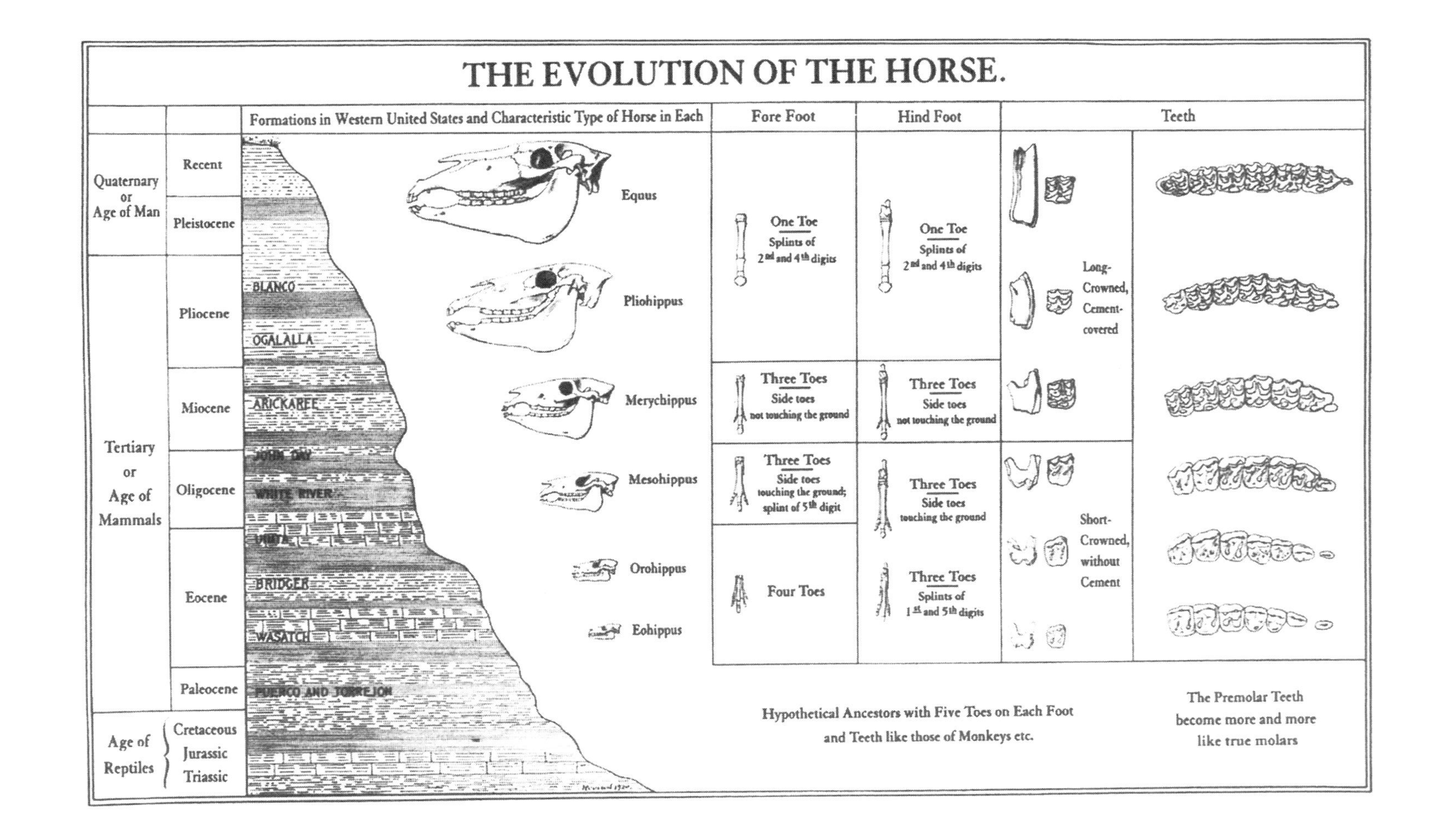
THE EVOLUTION OF THE HORSE.
Formations in Western United States and Characteristic Type of Horse in Each
Fore Foot
Hind Foot
Teeth
Quaternary or Age of Man
Recent
Pleistocene
Tertiary or Age of Mammals
Pliocene
Miocene
Oligocene
Eocene
Paleocene
Age of Reptiles
Cretaceous
Jurassic
Triassic
BLANCO
OGALALLA
ARICKAREE
JOHN DAY
WHITE RIVER
UINTA
BRIDGER
WASATCH
PUERCO AND TORREJON
Equus
Pliohippus
Merychippus
Mesohippus
Orohippus
Eohippus
One Toe
Splints of 2nd and 4th digits
Three Toes
Side toes not touching the ground
Three Toes
Side toes touching the ground; splint of 5th digit
Four Toes
One Toe
Splints of 2nd and 4th digits
Three Toes
Side toes not touching the ground
Three Toes
Side toes touching the ground
Three Toes
Splints of 1st and 5th digits
Long-Crowned, Cement-covered
Short-Crowned, without Cement
Hypothetical Ancestors with Five Toes on Each Foot and Teeth like those of Monkeys etc.
The Premolar Teeth become more and more like true molars

erate within a conceptual box defined by anagenetic flux rather than variation in numbers of taxa, an additional 25–28 percent of genera fall into a third category of generally and symmetrically increasing variation through the sequence—that is, the final range for all species within the genus includes species both smaller and larger than the extremes of the ancestral spread. I strongly suspect that a previous inclusion of these genera as affirmations of Cope's Law engendered the false result of dominant relative frequency for phyletic size increase. Older treatments of the topic usually considered extreme values only, and affirmed Cope's Law if *any* later species exceeded the common ancestor in size—thus repeating, in miniature, the same error generally committed for life's totality by ignoring the continuing domination of bacteria, and using the motley sequence of trilobite to dinosaur to human as evidence for a central and defining thrust. Obviously, from a variational and speciational perspective, successful genera with substantially increasing numbers of species through time will probably expand their range at both extremes of size, thus undergoing a speciational trend in variation, not an anagenetic march to larger sizes!

PARTICULAR CASES. This lowest macroevolutionary level of individual monophyletic clades has defined the soul of paleontological discourse through the centuries. Only the histories of particular groups can capture the details that all vivid story-telling requires; the "why" of horses and humans certainly elicits more passion than the explanation (or denial) of Cope's Law, or the pattern of increasing mean species longevity in marine invertebrates through time. Yet, even here on such familiar ground, our explanations remain so near and yet so far—for these "closer" stories of particular histories must also be reexamined in a speciational light. Consider just two "classics" and their potential revisions.

1. Horses as the Exemplar of "Life's Little Joke." As noted above (see *SET*, Fig. 7-3 and pp. 580–581), the line of horses, proceeding via three major trends of size, toes and teeth from dog-sized, many-toed, "eohippus" with low-crowned molars to one-toed *Equus* with high-crowned molars (see Fig. 5-5 for W. D. Matthew's classic icon, linearly ordered by stratigraphy) still marches through our textbooks and museums as the standard-bearer for adaptive trending towards bigger and better.

5-5. The classic icon, drawn by W. D. Matthew early in the 20th century, of the linear evolution of horses depicted as a ladder of progress towards larger size, fewer toes, and higher crowned teeth.

I do not deny that, even in a refomulated speciational context, several aspects of the traditional story continue to hold. MacFadden (1986), for example, has documented a clear cladal bias towards the punctuational origin of new species at larger sizes than their immediate ancestors, so both the iconic transition from ladders to bushes, and the recognition that several speciational events lead to smaller sizes, even to dwarfed species, throughout the range of the lineage, does not threaten (but rather reinterprets in interesting ways) the conventional conclusion that horses have generally increased in size through the Cenozoic. Moreover, I do not doubt the usual adaptational scenario that a transition from browsing in soft-turfed woodlands to grazing on newly-evolved grassy plains (grasses did not evolve until mid-Tertiary times) largely explains the adaptive context for both the general reduction in numbers of toes from splayed feet on soft ground to hoofs on harder substrates, and the increasing height of cheek teeth to prevent premature wear from eating grasses of high silica content.

Nonetheless, a speciational reformulation in terms of changing diversity as well as anatomical trending tells a strikingly different, and mostly opposite, story for the clade as a whole. Modern perissodactyls represent but a shade of their former glory. This clade once dominated the guild of large-bodied mammalian herbivores, with speciose and successful groups, especially the titanotheres, that soon became extinct, and with diversity in existing groups far exceeding modern levels. (The rhinoceros clade once included agile running forms, the hyrachiids, wallowing hippo-like species, and the indricotheres, the largest land mammals that ever lived.) Modern perissodactyls exist as three small clades of threatened species: horses, rhinos, and tapirs.

Horses have declined precipitously from their maximal mid-Tertiary abundance to modern marginality in both place and number. As O. C. Marsh proved to T. H. Huxley in a famous incident in the history of paleontology (see Gould, 1996, pp. 57–58), horses evolved in America only to peter out and disappear completely in their native land, surviving only as a few lines of Old World migrants. In North America alone, from 8 to 15 million years ago, an average of 16 species lived contemporaneously. As recently as 5 million years ago, at least six sympatric species of horses lived in Florida alone (MacFadden et al., 1999). Only 6–8 species of horses inhabit the entire earth today.

Moreover, if one wished to argue that a particular affection for *Equus* permits a restricted focus on this pathway through the bush alone, with the pruning of all other pathways regarded as irrelevant; and if one then claimed that this sole surviving pathway illustrates the predictable excellence of increasing adaptation by expressing a pervasive cladal trend (so that *any* surviving lineage would still tell the same adaptive tale for its own particular sequence); then the speciational view must debunk this last potential version of the old triumphalist attitude as well. If all paths through the equid bush led to the same Rome of modern *Equus*, or even if all major and prosperous paths—as measured by species range, diversity, or any conventional attribute of phyletic success—moved in the same general direction, then one might separate issues of species numbers, where the decline of equids through time cannot be denied, from the question of cladal direction, where the classic trend could still be asserted as predictable and progressive, even while the clade declined in diversity.

But reasonable, and very different, alternative scenarios for cladal direction remained viable until the recent restriction of the clade to the single genus *Equus*. MacFadden's (1988) study of all identifiable ancestral-descendant pairs showed 5 of 24, or more than 20 percent, leading towards decreased size. The general bias remains clear, but alternative scenarios also remain numerous and entirely reasonable (albeit, obviously, unrealized in the unpredictable history of life).

For example, the dwarfed genus *Nannippus* lived as a highly successful equid subclade for 8 million years (far longer than *Equus* so far) at substantial species diversity (at least four named taxa) in North America, becoming extinct only 2 million years ago. Suppose that, in the reduction of the clade to one genus, *Nannippus* had survived and the forebears of *Equus* died? What, with this small and plausible alteration by "counterfactual history," would we then make of the vaunted and canonical trend in the evolution of horses? *Nannippus* did not exceed ancestral eohippus in size, and still grew three toes on each foot. *Nannippus* molars were as relatively high-crowned as those of modern *Equus*, but I doubt that any biologist would challenge my noncynical scenario based on psychological salience: If *Equus* had died, and *Nannippus* survived as the only genus of modern horses, this "insignificant" clade would now be passing beneath our pedagogical notice as just one more "unsuccessful" mammalian line—like aardvarks, pangolins, hyraxes, dugongs, and several others—equally reduced to a shadow of former success.

The situation becomes even more paradoxical, and also entirely general, when we take an honest look at our iconographic prejudices in the light of speciational reformulations for macroevolution. Consider the true "success stories" of mammalian evolution—the luxurious clades of rodents, chiropterans (bats), and, among large-bodied forms, the antelopes of the artiodactyl clade that expanded so vigorously as perissodactyls constantly declined within the guild of large-bodied hoofed herbivores. Has anyone ever seen, either in a textbook or a museum hall, a chart or a picture of these truly dominating clades among modern mammals? We don't depict these stories because we don't know how to draw them under our restrictive anagenetic conventions.

If we define evolution as anagenetic trending to a "better" place, how can we depict a successful group with copious modern branches extending in all directions within the cladal morphospace? Instead, and entirely unconsciously of course, for we would laugh at ourselves if we recognized the fallacy, our conventions lead us to search out the histories of highly unsuccessful clades—those now reduced to a *single* surviving lineage—as exemplars of triumphant evolution. We take this only extant and labyrinthine path through the phyletic bush, use the steamroller of our preconceptions to linearize such a tortuous route as a main highway, and then depict this straggling last gasp as the progressive thrust of a pervasive trend. I refer to this pattern of "life's *little* joke" (see Gould, 1996).

2. Rethinking Human Evolution. Ever since Protagoras proclaimed that "man [meaning all of us] is the measure of all things," Western intellectual traditions, bolstering our often unconscious emotional needs, have invariably applied the general biases of our analytic procedures, with special energy and focused intensity, to our own particular history. As a cardinal example, concepts of human evolution long labored under the restrictive purview (now known to be empirically false) of the so-called "single species hypothesis" (see Brace, 1977)—the explicit claim that a maximal niche breadth, implied by the origin of hominid consciousness, made the coexistence of more than one species impossible, since no two species can share the same exact habitat under the "competitive exclusion" principle, and since consciousness must have expanded the effective habitat of hominid species into the full ecological range of conceivable living space. Human evolution could therefore be viewed as a single progressive series gradually trending towards our current pinnacle.

Needless to say, such a scheme precluded any speciational account of human history because only one taxon could exist at any one time, and trends had to record the anagenetic transformation of the only existing entity. (I do not think that such a relentlessly limiting scenario has ever been presented *a priori*, or so stoutly defended in principle even by a special name of its own, for any other lineage in the history of life.) This theoretical stricture enforced several episodes of special pleading for apparently contrary data. For example, early evidence for two distinct and contemporaneous australopithecine lineages (the so-called gracile and robust forms, now universally regarded as separate taxa) inspired dubious proposals about sexual dimorphism. The same theoretical constraint led many researchers to regard European neanderthals as necessarily transitional between *Homo erectus* and modern humans, even though the empirical record indicated a punctuational replacement in Europe about 40,000 years ago, with no evidence for any anatomical intermediacy.

I freely confess my partisan attitude, but I do think that reforms of the past two decades have centered upon a rethinking of this phylogeny in speciational terms, with punctuated equilibrium acting both as a spur for reformulation and a hypothesis with growing empirical support. Claims for stasis have been advanced, and much debated, for the only two species with appreciable longevity of a million years or more—*Australopithecus afarensis*, aka "Lucy," and *Homo erectus* (Rightmire, 1981, 1986, for support; Wolpoff, 1984, for denial). Lucy's earlier case seems well founded, while claims for our immediate ancestor, *Homo erectus*, have inspired more controversy, with arguments for gradual trending towards *Homo sapiens*, at least for Asian populations, generating substantial support within the profession. (But if the reported date—see Swisher et al., 1996—of 30,000 to 50,000 years for the Solo specimens holds, then the geologically youngest Asian *H. erectus* does not stand at the apex of a supposed trend.)

More importantly, hominid "bushiness" has sprouted on all major rungs of the previous ladder implied by the single-species hypothesis. The hominid tree may grow fewer branches than the comparable bush of horses, but multiple events of speciation now seem to operate as the primary drivers of human phylogeny (see Leakey et al., 2001, for a striking extension to the base of the known hominid bush in the fossil record), while humans also share with horses the interesting feature of present restriction to a single surviving lineage, albeit temporarily successful, of a once more copious array. (I would also suggest, on the theme of "life's little joke," that the contingent happenstance of this current restriction has skewed our thinking about human phylogeny away from more productive scenarios based on differential speciation.)

Speciation has replaced linearity as the dominant theme for all three major phases of hominid evolution (see Tattersall and Schwartz, 2000; and Johanson and Edgar, 1996, for booklength retellings of hominid history centered upon this revisionary theme of bushiness vs. linearity). First, before the origin of the genus *Homo,* the australopithecine clade differentiated into several, often contemporaneous, species, including, at a minimum, three taxa of "robust" appearance (*A. boisei, A. robustus,* and *A. ethiopicus*), and at least two of more "gracile" form (*A. afarensis* and *A. africanus*). Some of these species, including at least two of "robust" form, survived as contemporaries of *Homo.* Historically speaking, the death of the single species hypothesis may be traced to Richard Leakey's discovery, in the mid 1970s, of two undeniably different species in the same strata: the most "hyper-robust" australopithecine and the most "advanced" *Homo* of the time (the African form of *Homo erectus,* often given separate status on cladistic criteria as *Homo ergaster*). (Needless to say, no true consensus exists in this most contentious of all scientific professions—an almost inevitable situation, given the high stakes of scientific importance and several well known propensities of human nature, in a field that features more minds at work than bones to study. Nonetheless, despite endless bickering about details, I don't think that any leading expert would now deny the theme of extensive hominid speciation as a central phenomenon of our phylogeny—see Johanson and Edgar, 1996.)

Second, the crucial period of 2–3 million years ago, spanning the origin of initial diversification of the genus *Homo,* also represents the time of maximal bushiness for the hominid clade, then living exclusively in Africa. The correspondence of a time of maximal speciation with anatomical change of greatest pith and moment in our eyes—that is, the origin of our own genus *Homo,* with an extensive expansion of cranial capacity—probably records a causal process of central importance in our evolution, not just an accident of coincidental correlation. Vrba's turnover-pulse hypothesis, an extension of punctuated equilibrium (see pp. 224–229), represents just one causal proposal for this linkage. As many as six hominid species may have coexisted in Africa during this interval, including three members of the genus *Homo.*

Third, the central theme of bushiness persisted far longer than previous conceptions of human evolution had ever allowed, right to the dawn of historical consciousness. Under earlier anagenetic views, European neanderthals marked a transitional stage in a global passage. But under speciational reformulations, and acknowledging extensive anatomical distance between neanderthals and moderns, *Homo neandertalensis* must be construed as a European offshoot of local *Homo erectus* populations, with *Homo sapiens* evolving in a separate episode of speciation, probably in Africa on strong genetic and more tenuous paleontological grounds, and then replacing neanderthals by migration. Similarly, Asian *Homo erectus* populations may not have passed anagenetically into *Homo sapiens,* but may also have been replaced by migrating stocks of *Homo sapiens,* originally from Africa. If the redating of the Solo *Homo erectus* specimens can be confirmed (as mentioned above), then this Asian replacement occurred at about the same time as the death of neanderthals in Europe.

This information implies the astonishing conclusion, at least with respect to previous certainties, that three human species still inhabited the globe as recently as 40,000 years ago—*Homo neanderthalensis* as the descendant of *Homo erectus* in Europe, persisting *Homo erectus* in Asia, and modern *Homo sapiens,* continuing its relentless spread across the habitable world. This contemporaneity of three species does not match the richness of an entirely African bush with some half a dozen species about 2 million years ago, but such recent coexistence of three human species does require a major reassessment of conventional thinking. The current status of our clade as a single species represents an oddity, not a generality. Only one human species now inhabits this planet, but most of hominid history featured a multiplicity, not a unity—and such multiplicities constitute the raw material of macroevolution.

This recasting of human evolution in speciational terms documents the extent of proffered revision, but the scope of reform gains even greater clarity when we recognize the pervasive nature of the speciational theme as a guide for resolving paradoxes, understanding puzzles of popular misconception, and offering new formulations to break impasses in almost every nook and cranny of discussion about the evolutionary history of our own lineage. When common claims seem askew or confused, I would venture to suggest that the first and best strategy for breakthrough will usually lie in a speciational reformulation for any puzzling issue. Consider just three bugbears of popular confusion in serious newspapers and magazines, and in books on general science for lay audiences, all finding a potentially simple and elegant resolution in speciational terms.

1. The ordinariness of "out of Africa." During the 1990s, popular articles on human evolution, at least in American media, focused upon one issue above all others: the supposed dichotomy between two models for the origin

of *Homo sapiens.* The press labelled these alternatives as, first, the "multiregional" hypothesis (also dubbed the "candelabra" or the "menorah" scheme to honor our religious pluralism in metaphorical choice) representing the claim that *Homo erectus* migrated from Africa about 1.5 to 2.0 million years ago and established populations on the three continents of Africa, Europe, and Asia. All three subunits then evolved in parallel (with enough gene flow to maintain cohesion) towards more gracile and substantially bigger-brained *Homo sapiens.* The second position, usually called "out of Africa" or "Noah's ark," holds that *Homo sapiens* emerged in one coherent place (presumably Africa from genetic evidence of relative similarities among modern humans) as a small, speciating population geographically isolated from the ancestral *Homo erectus* stock. This new species, probably arising less than 200,000 years ago, then migrated out of Africa to spread throughout the habitable world, displacing, or perhaps partly amalgamating with, any surviving stocks of *Homo erectus* encountered along the way.

The dichotomous division, as presented by the media, may have been a bit stark and unsubtly formulated. But I can raise no major objection either to the basic categorization or to most press reports about details of explanation and evidence. Still, I became more and more puzzled, and eventually amused in a quizzical or sardonic way, by a remarkable fallacy in basic interpretation that pervaded virtually every article on the subject. Almost invariably, popular presentations labelled the multiregional view as the conventional expectation of evolutionary theory, and out-of-Africa as astonishingly iconoclastic, if not revolutionary.

But all professionals must recognize that the exactly opposite situation prevails. Out-of-Africa presents a particular account, "customized" for human evolution, of the most ordinary of all macroevolutionary events—the origin of a new species from an isolated, geographically restricted population branching off from an ancestral range, and then, if successful, spreading to other suitable and accessible regions of the globe. By contrast, multiregionalism should be labelled as iconoclastic, if not a bit bizarre. How could a new species evolve in lockstep parallelism from three ancestral populations spread over more than half the globe? Three groups, each moving in the same direction, and all still able to interbreed and constitute a single species after more than a million years of change? (I know that multiregionalists posit limited gene flow to circumvent this problem, but can such a claim represent more than necessary special pleading in the face of a disabling theoretical difficulty?) Do we advocate such a scenario for the evolution of any other global species? Do we ever suspect that rats evolved on several continents, with each subgroup moving in the same manner towards greater ratitude? Do pigeons trend globally towards increased pigeonosity? When we restate the thesis in terms of non-human species, the absurdity becomes apparent. Why, then, did our media not grasp the singular oddity of multiregionalism and recognize out-of-Africa—especially given the cascade of supporting evidence in its favor—as the most ordinary of evolutionary propositions?

I can only conclude that popular views of evolution conceptualize the process, in general to be sure but particularly as applied to humans, as a linear and gradual transformation of single entities. Most consumers would be willing to entertain the speciational alternative for lineages (like rats and pigeons) that impose no emotional weight upon our psyches. But when we ask the great Biblical question—"what is man that thou art mindful of him?" (Psalm 8)—we particularly yearn for explanations based on anticipated global progress, rather than contingent origins of small and isolated populations in limited local regions. We want to regard our origin as the necessary, or at least predictable, crest of a planetary flux, not as the chancy outcome of a single event unfolding in a unique time and place.

This example, I believe, best illustrates the deep-seated nature of prejudices that must be overcome if we wish to grasp the truly Darwinian character of macroevolution as change wrought by differential success of favored individuals (species) within variable populations (clades)—thus finally breaking the Platonic chain of defining evolution as improvement of an archetypal form. Eldredge (1979) has advocated this transition by contrasting "taxic" approaches to evolution with the older "transformational" view. In the particular context of human evolution (Gould, 1998), I have labelled multiregionalism as a "tendency theory," and out-of-Africa as an "entity theory." However much we may yearn to regard ourselves as the apotheosis of an inherent tendency in the unfolding of evolution, we must someday come to terms with our actual status as a discrete and singular item in the contingent and unpredictable flow of history. If we could bring ourselves to view this prospect as exhilarating rather than frightening, we might attain the psychological prerequisite for intellectual reform.

2. The unsurprising stability of *Homo sapiens* over tens of thousands of years. In a second example of a subject generally reported by the press with factual accuracy accompanied by ludicrously backwards explanation, the supposedly astonishing stability of human bodily form from Cro-Magnon

cave painters to now has provided notable fodder for science journalists during the past decade. Consider, for example, the lead article in the prestigious "Science Times" section of the *New York Times* for March 14, 1995, entitled: "Evolution of Humans May at Last Be Faltering." The opening sentence reads: "Natural evolutionary forces are losing much of their power to shape the human species, scientists say, and the realization is raising tantalizing questions about where humanity will go from here." (Professionals should always become suspicious when we read the universal and anonymous justification, "scientists say.")

These stories begin from the same foundational fallacy and then proceed in an identically erroneous way. They start with the most dangerous of mental traps: a hidden assumption, depicted as self-evident, if recognized at all—namely, a basic definition of evolution as continuous flux. Under this premise, the correct observation that *Homo sapiens* has experienced no directional trending for at least 40,000 years seems outstandingly anomalous. Reporters therefore assume that this unique feature of human evolutionary history requires a special explanation rooted in mental features that we share with no other species. They then generally assume, to complete the cycle of false argument, that human culture, by permitting the survival of marginal people who would perish in the unforgiving world of raw Darwinian competition, has so relaxed the power of natural selection that evolutionary change (popularly defined as "improvement") can no longer occur in *Homo sapiens.* This situation supposedly raises a forest of ethical questions about double-edged swords in the cure of diseases arising from genetic predisposition, the spread of genes for poor vision in a world of cheap eyeglasses, *et cetera ad infinitum.* (Pardon my cynicism based on some knowledge of the history of such arguments, but the neo-eugenical implications of these claims, however unintended in modern versions, cannot be ignored or regarded as just benignly foolish.)

This entire line of fallacious reasoning, with all its burgeoning implications, immediately collapses under a speciational reformulation. Once people understand *Homo sapiens* as a biological species, not a transitory point of passage in the continuous evolutionary progression of nature's finest achievement, the apparent paradox disappears by conceptual transformation into an expectation of conventional theory. Most species—especially those with large, successful, highly mobile, globally spread, environmentally diverse, and effective panmictic populations—remain stable throughout their history, at least following their origin and initial spread, and especially under the model of punctuated equilibrium that seems to apply to most hominid taxa. Change occurs by punctuational speciation of isolated subgroups, not by geologically slow anagenetic transformation of an entirety.

So if speciation usually requires isolated populations, how (barring science fiction scenarios about small groups of people spending generations in space ships hurtling towards distant stars) can a global species like *Homo sapiens*, endowed with both maximal mobility and an apparently unbreakable propensity for interbreeding wherever its members travel, ever expect to generate substantial and directional biological change in its current state? Most species should evoke predictions of stability, but *Homo sapiens* must lie at an extreme end of confidence for such an expectation. Thus, there is no solution to the supposed paradox of human stability because there is no problem. *Homo sapiens* has been stable for tens of thousands of years, and any proper understanding of macroevolution as a speciational process must yield this very expectation.

The same resolution applies to the extensive, and almost preciously silly, literature on human biological futures. (I do not speak of the real issues surrounding genetic engineering as an interaction of culture and nature, but of the fallacious and conjectural scenarios that treat presumptive human futures under a continuing regime of natural selection.) We have all seen reconstructions of improved future humans with bigger brains and disappearing little toes (perhaps balanced by the calloused butts of perennial couch potatoes). We also note the same features in reconstructions of advanced extraterrestrial aliens like ET, and in conjectural restorations of the hyperbrainy bipedal dinosaurs that might now rule the world if a bolide hadn't struck the earth 65 million years ago.

Again, this entire theme is moonshine. The only sensible biological prediction about human futures envisions continued stability into any time close enough to warrant any meaningful speculation. In any case, cultural change, in its explosive Lamarckian mode, has now so trumped biological evolution, that any directional trend in any allelic frequency can only rank as risibly insignificant in the general scheme of things. For example, during the past ten thousand years, any distinctive alleles in the population of native Australians must have declined sharply in global frequency as relative numbers of this subgroup continue to shrink within the human population as a whole. At the same time, cultural change has brought most of us through hunting and gathering, past the explosive new world triggered by agriculture, and into the age of atomic weaponry, air transportation and the electronic revolution, not to mention our prospects for genetic engineering and our capacity for environmental destruction on a global scale.

We have done all this, for better or for worse, with a brain of unaltered structure and capacity—the same brain that enabled some of us to paint the caves of Chauvet and the ceiling of the Sistine Chapel. What could purely biological and Darwinian change accomplish, even at a maximal rate (a mere thought experiment in any case, since we can only predict future stability for the short times that can justify any reasonable claims for insight), in the face of this explosive cultural transformation that our unchanging brains have unleashed and accomplished?

3. The conventional rate (and unconventional mode) of supposedly rapid trends traditionally cited as testaments to our uniqueness. When we recognize that human evolution occurred largely by differential success and replacement among species within a phyletic bush—and not by anagenetic transformation in measures of central tendency for a single, coherent entity in constant flux—then almost every standard claim about the tempo and mode of this process must be reformulated, and often substantially revised. My first two examples treated broad issues of maximal public attention. I now close this section with a smaller, but stubbornly persistent, error in order to clarify and exemplify the fractal reach of this speciational reformulation into the smallest nooks and crannies of conventional wisdom.

Throughout my professional life, I have read, over and over again, the almost catechistic claim that the increase in cranial capacity from *Homo erectus* to *Homo sapiens* (variously specified as a 50 percent growth from about 1000 cc to 1500 cc as a lower estimate, to a doubling from 750 to 1500 as an upper bound) represents a stunning example of evolution at maximal rate, something so unusual and unprecedented that we must seek a cause in the particular adaptive value, and potential for feedback, of human consciousness. (Again, I suspect that our mental predisposition to commit such errors arises from an overwhelming desire to find something unique not only in the *result*—a defensible proposition—but also in the biological *mechanism* of human consciousness.) This claim for a maximal rate presumes anagenetic transformation, with the high value then calculated by spreading the total increment over the short amount of available time (from 100,000 years as a lower bound to a million years as an upper limit).

Two major errors, one obvious and almost ludicrous, the other more subtle and speciational, promote this "urban legend" that would disappear immediately from the professional literature if people only stopped to think before they copied canonical lines into their textbook manuscripts. First, the claim isn't even true within its own assumption of anagenetic transformation. Such a rate should not be designated as rapid at all when we recognize the proper scaling between our estimates of selection's strength in ecological time and its effect in geological time. Williams (1992, p. 132), for example, cites a standard claim and then presents some calculations:

> Even some widely recognized examples of rapid evolution are really extremely slow. Data on Pleistocene human evolution are interpretable in various ways, but it is possible that the cerebrum doubled in size in as little as 100,000 years, or perhaps 3000 generations (Rightmire, 1985). This, according to Whiten and Byrne (1988) is "a unique and staggering acceleration in brain size." How rapid a change was it really? Even with conservative assumptions on coefficient of variation (e.g. 10%) and heritability (30%) in this character, it would take only rather weak selection ($s = 0.03$) to give a 1% change in a generation. This would permit a doubling in 70 generations. An early hominid brain could have increased to the modern size, and back again, about 21 times while the actual evolution took place. Indeed, it is plausible that a random walk of 1% increases and decreases could double a quantitative character in less than 3000 generations.

If this first error of scaling has been identified before, the second and deeper fallacy of false assumptions about mode has generally escaped the notice of critics. The anagenetic assumption that trends represent the flux of a central tendency in a species's global transition to a better form must be replaced, in most cases, with a speciational account of trends as differential success of certain species within a clade. When we add the additional observation that punctuated equilibrium describes the history of most species, the absurdity of the conventional claim becomes immediately apparent.

The change from *Homo erectus* to *Homo sapiens* did not occur in a gradual and global flux throughout the range of *Homo erectus*, but as an event of speciation, geologically rapid under punctuated equilibrium and local in geography (probably in Africa). By Williams's argument above, the incorporation of the entire increase into the geological moment of speciation represents no surprise whatsoever. Why, then, do we misconstrue the rate as maximal over a much longer anagenetic transformation? We fall into this trap as a simple artifact of the happenstance that this particular event of speciation occurred very recently—100,000 to 250,000 years ago by most current estimates. The full transition probably occurred during the geological moment of speciation, but we erroneously interpret the change as progressing in gradualist fashion, and at constant rate, from this time of origin until the present day. Since the moment of origin happened to occur only a short time ago, the false anagenetic rate becomes very high (for we integrate the full change over a small interval). But if the identical punctuational event had occurred, say, 2 million years ago—an entirely plausible, but actually unrealized, situation—then the same episode of speciation would be read as anagenetically slow because the identical amount of change would now be falsely spread over a much longer interval. In other words, the claim for a maximal anagenetic rate in a trait marking the apotheosis of human success only records a false interpretation of global anagenesis for the recent cladogenetic origin of this trait in a punctuational event of speciation.

Ecological and Higher-Level Extensions

The basic logic and formulation of punctuated equilibrium does not proceed beyond the level of species treated as independent individuals, or "atoms" of macroevolution. All biologists recognize, of course, that extensive ecological interactions bind each "atom" to others in complex ways. The "bare bones" structure of punctuated equilibrium does not include specific claims about ecological aggregations made of species as component parts. In this sense, punctuated equilibrium operates as a "null hypothesis" of sorts, by regarding each species as making its own way through geological time, with interactions treated as important components within the set of background conditions needed to explain the particulars of any history. In this sense, punctuated equilibrium treats time homogeneously, and species as independent agents; the theory therefore includes no inherent, or logically enjoined, predictions about the nature of temporal clumping in the ecological interactions among species.

But we know, as a basic fact and preeminent source of perennial controversy about the fossil record, that clumping occurs as a major pattern in the history of life—from minimal (and obvious) expression in joint disappearances during mass extinctions, to various theories that stress more pervasive and tighter clumping at several lower levels, all coordinated by the broad notion that ecosystems (or "communities," or other terminological alternatives, each with different theoretical implications) operate coherently during normal times as well, keeping species together in "organic" ways that falsify the null hypothesis of independent status.

As I write this at the end of the 1990s, the implication of punctuated equilibrium for higher-level theories about the pulsing and clumping of species in putatively stable communities through considerable stretches of geological time has become the most controversial and widely discussed "outreach" from the basic theory that Eldredge and I formulated in 1972. Whatever my own opinions on the major alternatives now under debate—and I confess to a somewhat cautious, if not downright conservative, stance retaining a maximal role for the null hypothesis of species as independent Darwinian individuals making their own way through geological time—I take pride in the role that punctuated equilibrium has played in building an intellectual context for making such a debate possible in the first place.

One can't even pose meaningful questions about higher-level aggregations of species unless species themselves can be construed as stable and effective ecological or evolutionary agents—a status best conferred by punctuated equilibrium's recognition of species as true Darwinian individuals (see *SET,* pp. 604–608). Under Darwin's personal view of species as largely arbitrary names for transitory segments of lineages in continuous anagenetic flux, such questions make no sense, and the entire potential subject remains undefined. (Some tradition exists for paleontological study of clumping in the distribution of named species through time, but this literature has never achieved prominence because researchers could not shake an apologetic feeling that they had based their studies on chimerical abstractions. For reasons well beyond accident or non-causal correlation, the beginning of serious and extensive research on this subject has coincided with the development and acceptance of punctuated equilibrium, a theory that recognizes these species as genuine evolutionary individuals.)

Precedents for studies of coordination above the species level can be found in such formulations as Boucot's (1983) twelve EEU's (ecological evolutionary units) for the entire Phanerozoic, or Sepkoski's (1988, 1991) three successive EF's (evolutionary faunas) for the same interval. But these works address the different subject of how major environmental shifts, including (but not restricted to) the substrates of mass extinctions, impact biotas defined at the family level and above. The subject of temporal interactions among species as basic macroevolutionary units raises a different set of questions about the nature of the "glue" that binds such sets of Darwinian individuals together at time scales matching their average durations (as contrasted with global geological changes that may coordinate—but probably not actively "bind"—biotas for intervals greatly exceeding the average lifespan of species). Some pioneering studies—most notably Olson's (1952) remarkable work on "chronofaunas" of late Paleozoic terrestrial vertebrates—have offered intriguing suggestions about the potential "glue" that may bind species into evolutionary ecosystems. However, as stated above, the subject could not be readily conceptualized until punctuated equilibrium provided a theoretical rationale for viewing species as legitimate Darwinian individuals.

The intense discussion of punctuational patterns at the level of species aggregations or ecosystems (extensive stability of species composition in regional faunas, followed by geologically rapid overturns and replacements of large percentages of these species) has centered upon two theoretical schemes and their proposed exemplars in the fossil record (see extensive symposium of 18 articles, edited by Ivany and Schopf, 1996, and entitled "New perspectives on faunal stability in the fossil record"). Working with the famous and maximally documented Hamilton faunas (Devonian) of New York State, and then extending their work up and down the stratigraphic record for a 70 million year interval of Paleozoic time in the Appalachian Basin, Brett and Baird (1995) documented 13 successive faunas, each including 50 to 335 invertebrate species, and each showing considerable stability both for the history of any species throughout its range (the predicted stasis of punctuated equilibrium; see Lieberman, Brett, and Eldredge, 1995, for quantitative evidence), and, more importantly in this context, in the virtually constant composition of species throughout the fauna's range.

Each fauna persists for 5 to 7 million years until replaced, with geological rapidity, by another strikingly different fauna including only 20 percent or fewer carryovers from the preceding unit. As a defining attribute, 70 to 85 percent of species in the fauna persist from the earliest strata to the very end, remaining in apparently stable ecological associations (with characteristic numerical dominances of taxa) to forge a pattern that Brett and Baird call "coordinated stasis." Eldredge (1999, p. 159) writes of coordinated stasis: "It is a true, repeated pattern, the most compelling and at the same time underappreciated pattern in the annals of biological evolutionary history."

Vrba (1985a,b) found a similar pattern in the maximally different ecosystem of vertebrates in Pliocene terrestrial environments of southern and eastern Africa. The geologically rapid faunal replacement, following an extensive period of previous stability, occurred in conjunction with a 10–15 degree drop in global temperatures that lasted some 200,000 to 300,000 years, and began about 2.7 to 2.8 million years ago. In generalizing this pattern as the "turnover-pulse hypothesis," Vrba emphasizes the role of environmental disruption in prompting the transition and, especially, the coordinated effects of both extinction and speciation as consequences of disruption—extinction by rapid change and removal of habitats favored by species of the foregoing fauna, and origination by fragmentation of habitats and resulting opportunities for speciation by geographic isolation of allopatric populations. As an example of the role granted to increased propensities for speciation, as promoted by the same environmental events that decimated the previous faunas, Vrba links the origin and initially rapid speciation (at least three taxa) of the genus *Homo* to this Pliocene turnover-pulse, a proposition that has generated substantial interest and debate.

The two propositions—Brett and Baird's coordinated stasis, and Vrba's turnover-pulse—identify similar patterns in prolonged stasis and punctuational replacement for linked groups of species. However, the two formulations differ in proposed explanations for this common pattern. Brett and Baird identify rapid environmental turnover as the trigger for collapse of incumbent faunas, but tend to view the prolonged stability of each fauna as a consequence, at least in large part, of internal ecological dynamics. New faunas come together largely by migration of separate elements from other regions (rather than originating primarily by local speciation *in situ,* as in Vrba's model), but then maintain stability by ecological interactions. Vrba, on the other hand, tends to attribute both fundamental aspects of the pattern—prolonged stasis and abrupt replacement—to vicissitudes and stabilities of the physical environment. As noted above, she also attributes the construction of new faunas to local speciation following fragmentation of habitats induced by environmental change, whereas Brett and Baird stress migration for the aggregation of new faunal associations. In Vrba's view, rapid physical changes induce the turnover by (at least local) extinction, and then also engender the subsequent stability as a propagated effect—because interspecific interactions play little role in regulating faunal stability, which must then arise as a basic expectation from punctuated equilibrium about the independent behavior of individual species. Ivany (1996, p. 4) accurately describes this aspect of Vrba's model: "stasis intervals between are in essence side consequences requiring no additional explanation beyond that required to explain stasis in individual lineages."

The developing debate in the paleontological literature has focused upon two issues of markedly different status. First, does the pattern actually exist—with sufficiently crisp and operational definition in any single case, and with sufficiently frequent occurrence among all cases—to warrant an assertion of evolutionary generality (see numerous examples and discussion in the Ivany and Schopf symposium, 1996, cited above)? I remain entirely optimistic on this point, if only because the "type" example of the Hamilton Fauna seems well and extensively documented.

However, and to confess a personal bias, my feelings of caution about unmitigated endorsement also arise from a substantial worry under this heading. If a capacity for individuation establishes the basis, or even just ranks as an important criterion, for status as an evolutionary agent in a Darwinian world, then our logical inability to render the faunas of coordinated stasis or turnover-pulse as coherent individuals does cause me concern. I do recognize that higher units of ecological hierarchies (see Eldredge, 1989) generally lack the coherence of individuals defining most levels of the standard genealogical hierarchy—because ecological associations cannot "hold" their component members as tightly as genealogical individuals enfold their subparts. Species, at a high level of the genealogical hierarchy, function as excellent Darwinian individuals because their subparts (organisms) remain tightly bounded by potential for interbreeding within, and prevention without. But ecological units like "faunas" must be constructed in a more "leaky" manner, for I cannot imagine a force that could hold taxonomically disparate forms together by ecological interaction with anything like the strength that species can muster to "glue" component organisms into a higher individual. However, and in response to my own doubt, the demic level of the genealogical hierarchy manifests a similar intrinsic leakiness because no strong "glue" exists, in principle, to prevent the passage of component parts (organisms) from one deme to another. Still, and after much debate, the efficacy of interdemic selection now seems well established, at least in certain important evolutionary settings (see *SET,* pp. 648–652, and Sober and Wilson, 1998).

Second—and more importantly in raising a theoretical issue at the heart of evolutionary studies—if the pattern of coordinated stasis and turnover-pulse does exist with sufficient clarity and frequency, then what forces hold faunas together at such intensity and for such long intervals, especially in the light of intrinsic capacity for "leakiness," as mentioned above? (Theoretical debate on this issue has rightly centered upon the putative causes of coordination in faunal stability, not on the rapidity of overturn. All formulations agree in ascribing quick transitions between faunas to direct effects of environmental perturbation.) Roughly speaking, two proposals of strikingly different import have dominated this debate. Some authors—in what we may call the "conservative" view, not for any intrinsic stodginess, but for envisioning no new or unconventional explanatory principles—hold that faunal stasis requires no active coordinating force at all, but arises as a side consequence of the environmentally triggered overturns themselves. (Vrba's formulation, as noted above, tends to this interpretation.) All active control then falls to the extrinsic causes of rapid overturn, with the coordination in between merely recording the predicted behavior, under punctuated equilibrium, of species acting as independent entities. In other words, we see temporal "packages" of coordinated stasis because external forces impose coincident endings and beginnings.

But other authors (see Morris, 1996; Morris et al., 1995) advocate active causal mechanisms, at the level of interaction among species, for holding the components of ecosystems together during periods of stasis—a notion generally called "ecological locking," and envisaging an explicit and intrinsic "glue" to build and then to hold the coordination of coordinated stasis. Morris, for example, cites the work of O'Neill et al. (1986) on mathematical theories of ecological hierarchies, in advancing a "claim that ecosystems in frequently disturbed settings become hierarchically organized such that the effects of large, low-frequency disturbances do not propagate through the system and cause disruption" (Ivany, 1996, p. 7). Other proposals for "intrinsic" mechanisms of coordination have invoked the general concept of "incumbency," and tried to designate theoretical reasons why established associations of species, even if non-optimal and only contingently or adventitiously built, may resist displacement by active mechanisms rooted in the behavior and construction of such aggregations.

These admittedly somewhat fuzzy and operationally ill-defined proposals address, nonetheless, the core of a vitally important issue within the developing hierarchical extension of Darwinian theory: how far "up" a hierarchy of levels do active causal forces of evolutionary change and stability extend? Do such causes generally weaken, or become restricted to peripheral impacts, at these higher levels? If so, can we attribute such diminution to increasingly limited opportunities for devising "glues" that might bind components into coherent individuals at these higher levels? Can "glues" for higher units in ecological hierarchies be strong enough, even in theory, to achieve the sufficient bounding (and bonding) that higher levels of the genealogical hierarchy (like species) can and do attain?

I do not know how this debate will develop, and how, or even whether, these questions can be operationally defined and activated. We remain stymied, at the moment, because so little thought, and so little empirical work, has been devoted to operational criteria for distinguishing alternatives—particularly for defining the different expectations of coordination as a passive consequence of joint endings vs. an active result of ecological locking during intervals of stasis. Perhaps such distinctions can be defined and recognized in the statistics of faunal associations (varying strengths and numbers of paired correlations, similarities in joint ranges and relative abundances of groups of species: what numbers of taxa, and what intensities of coordination, imply active locking beyond the power of passive response among independent items to accomplish?). Given the notorious imperfections of the geological record, and the daunting problems of consensus in taxonomic definition, I recognize the extreme difficulty of such questions. But the issues raised are neither untestable nor non-operational, and the concepts involved could not be more central to evolutionary theory. Whatever the future direction of this debate, punctuated equilibrium has proven its mettle in prompting important extensions beyond its original purview, and in proposing a fruitful strategy of research, based on a new way of viewing the fossil record, that broke some longstanding impasses in paleontological practice. At the very least, punctuated equilibrium has raised some interesting and testable questions that could not be framed under previous assumptions about evolutionary mechanisms and the patterns of life's history.

As a final note and postscript, either extreme alternative for the explanation of faunal stasis—passive consequence or active ecological locking—bears an interesting implication for the significance of punctuated equilibrium. (Of course, I would be shocked if either extreme eventually prevailed, or if a future consensus simply melded aspects of both proposals into harmony. I suspect that the reasons behind coordinated stasis are complex, multifarious, and informed by other modes and styles of explanation as yet unconceived.) If coordination arises as a passive consequence, then our original version of punctuated equilibrium, proposed to explain the pattern of individual species, also suffices to render this analogous pattern at the higher level of faunas as well—thus increasing the range and strength of our mechanism. But if coordination must be forged by higher-level mechanisms of active ecological locking, then punctuated equilibrium provided the basis, both logically and historically, for regarding species as evolutionary individuals, the conceptual prerequisite for Darwinian theories of causation at the level of aggregations of species.

Punctuation All the Way Up and Down? The Generalization and Broader Utility of Punctuated Equilibrium (in More Than a Metaphorical Sense) at Other Levels of Evolution, and for Other Disciplines In and Outside the Natural Sciences

GENERAL MODELS FOR PUNCTUATED EQUILIBRIUM

If the distinctive style of change described by punctuated equilibrium at the level of speciation—concentration in discrete periods of extremely short duration relative to prolonged stasis as the normal and actively maintained state of systems—can be identified in a meaningful way at other levels (that is, with sufficient similarity in form to merit the same description, and with enough common causality to warrant the application in more than a metaphorical manner), then general mathematical models for change in systems with the same fundamental properties as species might also be expected to generate a pattern of punctuated equilibrium under assumptions and conditions broad enough to include nature's own. In this case, we might learn something important about the general status and range of application of such a pattern—thus proceeding beyond the particular constraints and idiosyncrasies of any biological system known to generate this result at high relative frequency.

Many scholarly sources in the humanities and social sciences, with Thomas Kuhn's theory of scientific revolutions as the most overt and influential, have combined with many realities of late-20th-century life (from the juggernaut of the internet's spread to the surprising, almost sudden collapse of communism in the Soviet Union, largely from within) to raise the general critique of gradualism, and the comprehensive acceptability of punctuational change, to a high level of awareness, if not quite to orthodoxy. But the greatest spur to converting this former heresy into a commonplace, at least within science, has surely arisen from a series of mathematical approaches, some leading to little utility despite an initial flurry of interest, but others of apparently enduring worth and broad applicability. These efforts share a common intent to formalize the pattern of small and continuous inputs, long resisted or accommodated by minimal alteration, but eventually engendering rapid breaks, flips, splits or excursions in systems under study: in other words, a punctuational style of change. These proposals have included René Thom's catastrophe theory, Ilya Prigogine's bifurcations, several aspects of Benoit Mandelbrot's fractal geometry, and the chief themes behind a suite of fruitful ideas united under such notions as chaos theory, nonlinear dynamics, and complexity theory.

Empirical science has also contributed to this developing general movement by providing models and factual confirmations at several levels of analysis and for several kinds of systems, with catastrophic mass extinction theory first seriously proposed in 1980 and then strongly promoted by increasingly firm evidence for bolide triggering of the late Cretaceous event, as an obvious input from paleontology. I take pride in the role that punctuated equilibrium has played as a spur for this larger intellectual transformation—for our 1972 proposal, formulated at one level of biological change, provided some general guidelines, definitions and terminology, and also provoked a good deal of interest in the application of this general style of change to other fields of study and other levels of causality. This extension has proceeded so far that some scientists and scholars from other disciplines (see Gersick, 1991; Mokyr, 1990; Den Tex, 1990; Rubinstein, 1995, for example) now use punctuated equilibrium as the general term for this style of change (while we would prefer that punctuated equilibrium retain its more specific meaning for the level of speciation, with punctuational change or punctuationalism used for the generality).

Some recent mathematical work has explicitly tried to model punctuational change at the level and phenomenology of our original theory. Rand and Wilson (1993, p. 137), for example, following Rand et al.'s (1993) "general mathematical" model for "Darwinian evolution in ecosystems," applied their basic apparatus to the problem of speciation in individual taxa within ecosystems, primarily to test whether or not the pattern of punctuational equilibrium would emerge. "We do not mean," they write (1993, p. 137), "a large multispecies extinction event but rather the sudden disappearance of an evolutionary stable state causing a species to undergo very rapid evolution to a different state."

Under basic trade-off "rules" of bioenergetics and ecological interaction, which they call "constraints" (correlations, for example, between a prey's increase in population size and the exposure of individuals to a predator's attention), punctuated equilibrium emerges as a general pattern. Gradual change may prevail in systems without such constraints, but as the authors state, and to say the very least about nature's evident complexity(!), "the absence of such constraints is biologically unrealistic" (1993, p. 138). Moreover, they argue, reasonable features of the model's internal operation, and recognized properties of natural ecosystems, suggest a general status for punctuational change at all levels: "In this note," Rand and Wilson write (1993, p. 137), "we wish to address the important issue of gradualism against punctualism in evolutionary theory. We discuss this in terms of a simple illustrative example, but emphasize that . . . our results apply quite generally and are ubiquitous and wide ranging."

Explicit modeling of other levels has also yielded punctuational change as an expectation and generality under realistic assumptions. Elsewhere, I discuss models for punctuational anagenesis within populations in ecological time (see p. 175), erroneously interpreted by some critics as a demonstration that punctuated equilibrium emerges from ordinary microevolutionary dynamics and therefore embodies nothing original—although such studies should be interpreted as illustrating the potential generality of punctuational change by rendering the same pattern as an anticipated result of different processes (transformation of a deme rather than origin of a new species by branching) at a lower hierarchical level (intra rather than interspecific change).

Punctuational change has been modelled more frequently at the most evident level above punctuated equilibrium—coordinated and rapid change in several species (or the analogs of separate taxa in modelled systems) within communities or faunas. Per Bak's "sandpile" model of self-organized criticality (see Bak and Sneppen, 1993; Sneppen et al., 1994; and commentary of Maddux, 1994) have generated both particular interest and legitimate criticism. Maddux (1994, p. 197), noting the "minor avalanche of articles on the theme," began his commentary by writing: "That physicists are itching to take over biology is now well attested . . . But surely only a brave physicist would take on Darwin on his home ground, the theory of evolution, let alone Gould and Eldredge on punctuated equilibrium."

Bak's models operate by analogy to metastable sandpiles, where grains may accumulate for long periods without forcing major readjustment (the analog of community stability), whereas, at a critical point, just one or a few added grains will trigger an avalanche, forcing the entire pile to a new and more stable configuration (the analog of mass extinction and establishment of new faunas, not to mention the straw that broke the camel's back). In his basic model, Bak assigns random fitnesses, chooses the "species" with the smallest fitness, and then reassigns another random number both to this item and to the two neighboring species of its line (to stimulate interactions among taxa in communities). He also randomly selects a certain number of other points for similar reassignment (to acknowledge that interconnections among taxa need not link only the most obviously related or adjacent forms).

This procedure often generates waves of rapidly cascading readjustments, propagated when some species receive small numbers in the reassignment of fitnesses, and then must change, taking their neighbors and also some distant forms with them, by the rules of this particular game—a play that admittedly cannot mimic nature closely (if only because the model includes no analogs of extinction or branching), but that may give us insight into expected rates and patterns of change within simple systems of partly, and largely stochastically, linked entities. In any case, Bak and his colleagues have formalized the general notion that small inputs (random reassignment of fitness to just one entity of lowest value) to simple systems of limited connectivity among parts (changes induced in a few other entities by this initial input) can lead to punctuational reformation of the entire system.

In a similar spirit, the substantial research program known as Artificial Life (AL to aficionados) takes an empirical, if only virtual, approach to such questions by generating and tracking evolving systems operating under simple rules in cyberspace. I regard such work as of great potential value, but often philosophically confused because researchers have not always been clear about which of two fundamentally different intentions they espouse: (1) to build systems that mimic life with enough fidelity to state something useful about actual evolution on earth; or (2) to construct alternative virtual worlds so explicitly unlike actual life in their minimality that we can ferret out some abstract properties, applicable to any genealogical system, by using models that permit perfect tracking of results and also operate with sufficient simplicity to identify the role of any single component.

Ray, a pioneer in these studies of "evolution in a bottle" or "synthetic life in a computer" (1992, p. 372), started his "Tierra" system by designing a block of RAM memory as "a 'soup' which can be inoculated with creatures" (1992, p. 374), and then beginning with a "prototype creature [that] consists of 80 machine instructions," with "the 'genome' of the creatures consisting of the sequence of machine instructions that make up the creature's self-replicating algorithm."

"When the simulator is run over long periods of time, hundreds of millions or billions of instructions, various patterns emerge" (1992, p. 387). Obviously, the results depend crucially on the human mental protoplasm that sets the particular rules and idiosyncrasies of the virtual system. Ray found, for example (and unsurprisingly), that "under selection for small sizes, there is a proliferation of small parasites and a rather interesting ecology." Similarly, "selection for large creatures has usually led to continuous incrementally increasing sizes . . . This evolutionary pattern might be described as phyletic gradualism" (p. 387).

But under the much more "reality mimicking" condition of no consistent directional selection for size, Ray found "a pattern which could be described as periods of stasis punctuated by periods of rapid evolutionary change, which appears to parallel the pattern of punctuated equilibrium described by Eldredge and Gould" (p. 387). Ray then describes his frequently replicated and longest running results in more detail (pp. 387–390):

> Initially these communities are dominated by creatures with genome sizes in the 80's. This represents a period of relative stasis, which has lasted from 178 million to 1.44 billion instructions . . . The system then very abruptly (in a span of 1 or 2 million instructions) evolves into communities dominated by sizes ranging from about 400 to about 800. These communities have not yet been seen to evolve into communities dominated by either smaller or substantially larger size ranges. The communities of creatures in the 400 to 800 size range also show a long-term pattern of punctuated equilibrium. These communities regularly come to be dominated by one or two size classes, and remain in that condition for long periods of time. However, they inevitably break out of that stasis and enter a period where no size class dominates . . . Eventually the system will settle down to another period of stasis dominated by one or a few size classes which breed true.

All models previously discussed have generated punctuational patterns at explicit and particular levels of evolutionary change (anagenetically within demes, for the origin of species by branching, and in coordinated behavior of groups of species within entire faunas and ecosystems). This range of success suggests that the apparent ubiquity of punctuational patterns at substantial, if not dominant, relative frequencies may be telling us something about general properties of change itself, and about the nature of systems built of interacting components that propagate themselves through history. Some preliminary work has attempted to formalize these regularities, or even just to identify them through a glass darkly (see, for example, Chau, 1994, on Bak's models).

Bak has tried to specify two "signatures of punctuated equilibrium" in very general properties of systems: "a power-law distribution of event sizes where there is no characteristic size for events, but the number of events of a certain size is inversely proportional to some power of that size"; and a property that Bak calls 1/f noise, "where events are distributed over all time-scales, but the power or size of events is inversely proportional to some power of their frequency" (Shalizi, 1998, p. 9). Since we can document such inverse relationships between magnitude and frequency in many natural systems—indeed, R. A. Fisher (1930) began his classic defense of Darwinism with a denial of efficacy for macromutations based on their extreme rarity under such a regularity—punctuational change may emerge as predictably general across all scales if Bak's conditions hold.

The intellectual movement dedicated to the study of complex dynamical systems and their putative tendencies to generate spontaneous order from initial randomness—a prominent fad of the 1990s, centered at the Santa Fe Institute and replete, as all fashions must be, with cascades of nonsense, but also imbued with vital, perhaps revolutionary, insights—has identified punctuated equilibrium as a central subject of inquiry. A defining workshop, held in Santa Fe in 1990, specified three primary illustrations or consequences of this discipline's central principle, "the tendency of complex dynamical systems to fall into an ordered state without any selection pressure whatsoever": the origin of life, the "self-regulation of the genome to produce well defined cell types"; and "the postulated sudden waves of evolutionary change known as 'punctuated equilibrium.'"

Stuart Kauffman, the leading biological theorist and mathematical modeler of this movement (see *SET,* Chapter 11, pp. 1208–1214 for a discussion of his work on structuralist approaches to adaptive systems), stressed the generality of punctuational change by beginning with simple models of co-evolution and then obtaining punctuational change at all levels as a consequence. *Science* magazine's report of this 1990 meeting linked Kauffman's multilevel work to the ubiquitous emergence of punctuated equilibrium from models of highly disparate systems and processes—all suggesting a generality and an intrinsic character transcending any particular scale or phenomenology: "This pattern of change and stasis itself evolves," says Kauffman. "In the subtly shifting network of competition and cooperation, predator and prey, a fast-evolving species might suddenly freeze and cease to evolve for a time, while a formerly stable species might suddenly be forced to transform itself into something new. The fossil record of the latter process would then resemble 'punctuated equilibrium': a pattern of stasis interrupted by sudden change, which some paleontologists now believe to be the norm in real evolution. . . . This same pattern of stasis punctuated by sudden change also showed up in a number of other ecosystem models presented at the workshop, even when those models seemed superficially quite different. Does this mean some more general mechanism is at work, some theory that could account for the behavior of these models—and perhaps real life—no matter how they are structured?"

A false and counterproductive argument has enveloped this work during the past few years. Bak, in particular, has noted that punctuations at the highest level, corresponding to simultaneous extinction of a high percentage of components in a system, can be generated from internal dynamics alone, and require no environmental trigger of corresponding (or even of any) magnitude. He and others then draw the overextended inference that because such large scale punctuations can arise endogenously, the actual mass extinctions of the fossil record therefore need no exogenous trigger of environmental catastrophe, or any other external prod. This claim, emanating from a theoretical physicist with little knowledge of the empirical archives of geology and paleontology, and emerging just as persuasive evidence seems to have sealed the case for bolide impact as a trigger of at least one actual mass extinction (the end Cretaceous event 65 million years ago), could hardly fail to raise the hackles of observationally minded scientists who, for reasons both understandable and lamentable, already bear considerable animus towards any pure theoretician's claim that success in modeling logically entails reification in nature.

The obvious solution—if human emotions matched human logic in clarity, or the empirical world in complexity—would welcome the mathematical validation of potential endogenous triggers (often of small initial extent) for punctuational change as a partner with well-documented exogenous triggers (of great extent in one well documented case, but perhaps also of potentially small magnitude as well). Instead of waging a false battle for preference or exclusivity of one alternative between two plausible arguments, we should recognize instead the complementary and general theme behind both proposals—their common role as sources for punctuational change (which then achieves higher status as a truly general pattern in nature), and in their mutual reinforcement for revising and expanding the

Darwinian paradigm on all three supporting legs of its essential tripod. For the punctuational style of change—disfavored by Darwin, who recognized the necessary status of gradualism within the logic of his world view—now emerges as a primary consequence of repairs and reinforcements upon all legs of the tripod: the expansion beyond small uniformitarian inputs for the external triggers and causes of leg three (thus granting environment an even greater role than Darwin himself, who so brilliantly introduced the concept to defeat previous internalist theories of change, had envisioned); and the recognition that constraints of systems (leg two)—not only overt natural selection—acting at all levels of a causal and genealogical hierarchy (leg one) can also generate punctuations from within.

PUNCTUATIONAL CHANGE AT OTHER LEVELS AND SCALES OF EVOLUTION

A Preliminary Note on Homology and Analogy in the Conceptual Realm

The simple documentation of punctuational patterns at scales other than the speciational status of punctuated equilibrium (and, therefore, presumably attributable to different causes as well) gives us little insight into the key question of whether or not punctuated equilibrium, in either its observed phenomenology or its proposed mechanics, can lay claim to meaningful generality in evolutionary studies. Rather, the overt similarity in pattern must be promoted to importance through an additional claim, akin in the world of ideas to the weight that an assertion of homology would carry in assessing the value of taxonomic characters. What, then, would make an example of punctuational change from another scale (where the immediate speciational cause of punctuated equilibrium could not apply) effectively "homologous" to punctuated equilibrium—that is, sufficiently similar by reason of phenomenological "kinship" that the similar pattern across disparate scales may be read as revealing the shared components of a common explanation?

We rank some similarities across scales as capricious enough to be deemed accidental, and therefore devoid of causal meaning. The appearance of a "face" on a large mesa on the surface of Mars—an actual case by the way, often invoked by fringe enthusiasts of extraterrestrial intelligence—bears no such conceptual homology to faces of animals on earth. We label the similarity in pattern as accidentally analogous—even though the perceived likeness can teach us something about innate preferences in our neural wiring for reading all simple patterns in this configuration (a line below two adjacent circles) as faces. (An actual face and the accidental set of holes on the mesa top may stimulate the same pathway in our brain, but the two patterns cannot be deemed causally similar in their own generation—that is, *as faces.*)

Identity of specific cause will rarely be available to provide a basis for asserting meaningful homology, rather than misleading analogy, between common patterns at disparate scales. Punctuated equilibrium, for example, gains power and testability in proposing a particular scale-bound reason for an observed phenomenon—the expression of ordinary speciation in geological time, in this case. Since most theories win strength by such specificity, conceptual homologies across scales must seek other definitions and rationales. A punctuational pattern below the scale of punctuated equilibrium (change within a single deme for example), or above (temporal clumping in the origin or extinction of many species within a fauna), could not, in principle, be explained by the specific causes of punctuated equilibrium itself.

Therefore, meaningful "homology" in this conceptual sense must generally be sought in properties that are genuinely held in common across systems and scales, and that operate to channel the different causes of these various scales into the same recognizable and distinctive pattern. Moreover, such homology becomes all the more interesting if the particular efficient causes at different scales—the actual pushers and movers of immediate change in each case—remain evidently disparate, thus implying, by elimination, that the observed commonality of pattern may arise from constraining channels of similar structural properties across scales. If all roads lead to Rome, then the eternal city ranks as a dominant and ineluctable attractor!

In the present case of punctuational patterns across markedly different scales of time and component entities, claims for conceptual homology rest upon an overarching hypothesis that punctuation records something quite general about the nature of change itself, and that the differing causes of punctuational change at each level—the waiting time between favorable mutations in bacterial anagenesis, the scaling of ordinary speciation as a geological moment in punctuated equilibrium, or the simultaneity of species deaths in mass extinction—must run in a common structural channel that sets and constrains the episodic nature of alteration.

If punctuated equilibrium gains this generality by conceptual homology, then both components of its name should achieve such transfer across nature's numerous scales of size and time. (The general mathematical models discussed presume such meaningful transfer as a primary rationale for their relevance.) The equilibrial component wins potential generality if active resistance to change can be validated as an important structural property of systems discrete and stable enough to be named and recognized as entities at any scale of nature (whatever the causes of stability, whether internal to self-integration or imposed from without upon an intrinsically less coherent structure—a fascinating question that should become an object of research, not the subject of prior definition). This property of *active* maintenance underlies our primary claim about stasis in punctuated equilibrium, and our insistence that stasis must be conceptualized and defined as a positive phenomenon, not as a disappointing or uninteresting absence of anticipated change. (Throughout this book, I have provided evidence, primarily in observed relative frequencies—far too high to originate either passively or randomly in a world of natural selection and genetic drift—for interpreting stasis as an actively generated property of systems, embodied in species at the scale of punctuated equilibrium, but necessarily recognized in structures of different status at scales both below and above species.)

The punctuational component, operationally measured by its short duration relative to periods of stasis within definitive structures of the same scale, would then achieve homological generality as the obverse to proposed reasons for stasis: the reinterpretation of change—at least in its usual, if not canonical, expression—as a rare and rapid event experienced by systems only when their previous stabilities have been stretched beyond any capacity for equilibrial return, and when they must therefore undertake a rapid excursion to a new position of stability under changed conditions.

Obviously, these "brave" statements about conceptual homology across disparate scales and immediate causalities must remain empty and meaningless without operational criteria for distinguishing—if I may again use the conventional evolutionary jargon in this wider context—meaningful similarity of genesis (homology) from misleading superficiality of appearance (analogy). As a first rule and guideline, we might look to the same basic precept of probability that regulates our general procedures in the study of overt similarity among separate phenomena: co-occurrence of substantial numbers of potentially independent parts as a sign of meaningful genetic (and conceptual) connection vs. resemblance based upon single or simple features, however visually striking, as far more likely to be unconnected, separately built, and perhaps not even meaningfully alike in any causal or functional sense (the complex and identical topology of arm bones in a bat's wing and a horse's foreleg as meaningfully homological vs. my face and the same disposition of holes on the Martian mesa as meaninglessly analogical). Thus, and in a practical sense, I focus much of the following discussion upon a search for what I will call "conjoints," or sets of independent features whose joint occurrence predisposes us to consider meaningful conceptual homology in punctuational patterns of change produced by different immediate causes at disparate scales of size and time. (I have used the same form of argument frequently throughout this book—as in emphasizing the usual conjunction of openness to saltational change, belief in the importance of internal channeling, and suspicion about adaptationist explanation in defining the biological worldview of structuralist thinkers—see *SET,* Chapters 4 and 5).

When such broad "homologies" of common structural constraint have been established across several realms of size and time, then we can most fruitfully ask some second-order questions about systematic, or "allometric," differences (see Gould and Lloyd, 1999) in the expression of common patterns along continua ordered by increasing magnitude among scales under consideration. For example, do internal forces of cohesion among subparts set the primary basis for active stasis, or does the "fit" of a structure into a balanced and well-buffered environment, made of numerous interacting entities, prevent change in a system otherwise fully capable of continuous alteration in the absence of such externalities? Does the balance between these internalist and externalist explanations change as we mount through scales of magnitude? Is the change systematic (and therefore "allometric" in the usual sense), or capricious with respect to scale itself, having no correlation with magnitude?

The important principle that meaningful similarity may reside in homology of structural constraint across scales, while particular causes that "push" phenomena through these constrained channels may vary greatly, has rarely been stated or exemplified with proper care, and has therefore usually been ignored by commentators on the role of events at one scale in the interpretation of others. Most regretfully, a frequent misunderstanding has then led to dismissal of meaningful commonality in pattern because a critic notes a strong difference in immediate causes for a pattern at two scales and then rejects, on this erroneous basis, any notion of an informative or integrative status for the similarity. Or, even worse, the critic may become intrigued by a cause just elucidated at one scale and then assume that the significance of such a discovery can only lie in extrapolating this particular and strongly scale-bound cause to debunk a different mechanism previously proposed to explain the same pattern at another level—rather than exploring the more fruitful and integrating hypothesis that a genuine basis for meaningful similarity in pattern might reside in homologous structural constraints that channel different causes to similar outcomes at the two distinct scales.

For example, an excellent science reporter for the *New York Times* erroneously argued that punctuations caused by long waiting times between rapidly sweeping, favorable mutations in bacterial anagenesis on a scale of months should lead us to reinterpret the speciational breaks of punctuated equilibrium (at geological scales) as similarly caused by quick and simple genetic changes! "The finding that all it takes is a few mutations and a little natural selection to generate punctuated evolution comes as a surprise. Researchers say numerous theories that are considerably more complex have been put forth to explain what might produce the punctuation seen in the fossil record. If bacteria are any indication, the rapid evolution documented in the fossil record might be the product of a very few simple, if quick, genetic changes" (Yoon, 1996).

But R. E. Lenski, the chief scientist in the bacterial study (Elena, Cooper, and Lenski, 1996), properly sought commonality with punctuated equilibrium in the domain of homologous reasons for punctuational patterns. Recognizing the disparities in scale, and the different causes thus implied, they rightly declined to apply the term punctuated equilibrium to their findings. Instead, they invoked the general term for the pattern itself as the title for their paper (Elena, Cooper, and Lenski, 1996): "Punctuated evolution caused by selection of rare beneficial mutations."

Punctuation below the Species Level

I have, at several points in this and the preceding chapter, discussed various empirical and theoretical studies that validate the pattern of substantial stability followed by rapid peak shifts in the anagenetic transformation of single populations during the microevolutionary time of potential human observation (see p. 175). I have also urged (to reiterate the theme of the preceding section) that such an important conclusion should not be read as an argument that punctuated equilibrium holds no interest for evolutionary theory because ordinary population genetics can produce patterns of stasis and punctuation—a common but erroneous claim rooted in the misinterpretation of punctuated equilibrium as a saltational theory in ecological time. Rather, this small-scale anagenetic conclusion for another domain of size and time should be read as welcome confirmation—based on causes different from the generators of punctuated equilibrium at a larger scale—for the broader claim that punctuational patterns may be common and robust across several spatial and temporal realms in nature.

But the most impressive affirmations of punctuational patterns at scales below punctuated equilibrium have emerged, in recent years, from a domain unparalleled (and unmatchable) for richness of empirical data on evolution over a sufficient number of generations to claim potential linkage with scales of substantial evolutionary change in nature: well-controlled experimental studies of bacterial lineages.

This field is now developing so rapidly that any particular study, as discussed here, will, no doubt, seem quite rudimentary by the time this book reaches the presses. But, as I write in 1999, an impressive case may be taken as indicative of possibilities and directions. By using strains of *E. coli* that pass through six generations in a single day, Elena, Cooper, and Lenski (1996; see also Lenski and Travisano, 1994) were able to study evolution in cell size for 10,000 continuous generations. By imposing constancy of environment (to limits of experimental perceptibility of course), and using a strain lacking any mechanism for genetic exchange (Elena et al., 1996), mutation becomes the sole, and experimentally well isolated, source of genetic variations.

The experimenters have followed 12 replicate populations, each founded by a single cell from an asexual clone, and each grown under the same regimen (of daily serial transfer, with growth for 24 hours in 10 ml of a glucose-limited minimal salts medium that can support *ca.* 5×10^7 cells per ml). At an average of 6.6 bacterial generations per cycle, the population undergoes a daily transition from lag phase following transfer, to sustained increase, to depletion of limiting glucose and subsequent starvation. At each serial transfer, a 1:100 dilution begins the next daily cycle with a minimal bacterial population of *ca.* 5×10^6 cells. Samples of the common ancestral population, and of selected stages in the history of each population, were stored at −80° C, and can be revived for competition experiments with the continually evolving populations—a situation that can only fill a paleontologist with envy, and with thoughts of beautiful and utterly undoable experiments from life's multicellular history (neanderthals or australopithecines released in New York City; tyrannosaurs revived to compete against lions in a field of zebras, etc.).

In each of the 12 populations, both fitness and cell volume increased in a punctuational manner through the 10,000 generations of the experiment. (The experimenters measured cell volume by displacement (Lenski and Travisano, 1994, p. 6809), and mean fitness of populations by the Malthusian parameter of realized rate of increase in competition against resuscitated populations of the common ancestor.) The general path of increase followed the same trajectory in all populations, but with fascinating differences of both form and genetics in each case—a remarkable commentary, at such a small and well-controlled scale, of the roles of detailed contingency and broad predictability in evolution (see the explicit discussion of Lenski and Travisano, 1994, on this point).[3]

Punctuational patterns occur at the two different scales of overall trajectory and detailed dynamics, even within the limited scope of this study. In the 1994 paper, Lenski and Travisano sampled each of the 12 populations once every 500 generations. They noted a rapid increase in mean cell size, well fit by a hyperbolic model (see Fig. 5-6), for the first 2000 generations in each population, followed by several thousand subsequent generations of little or no further increase—a pattern that they described as punctuational in one of the major conclusions of their paper (1994, p. 6809): "For *ca.* 2000 generations after its introduction into the experimental environment, cell size increased quite rapidly. But after the environment was unchanged for several thousand generations more, any further evolution in cell size was

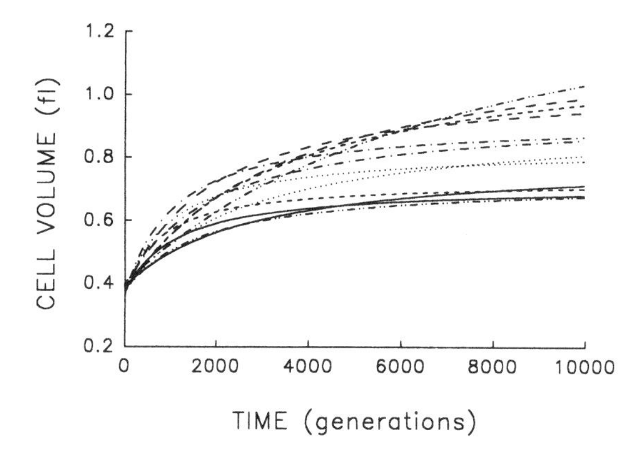

5-6. Punctuation in clonal evolution below the species level—from Lenski and Travisano, 1994. In 12 replicate populations of *E. coli* during ten thousand generations, mean cell size increases rapidly during the first two thousand generations, and then very little during subsequent evolution.

imperceptible . . . These data therefore indicate a rapid bout of morphological evolution after the population was placed in the experimental environment, followed by evolutionary stasis (or near stasis)."

But, as reported in a later paper (Elena, Cooper and Lenski, 1996), they then sampled each population at a much finer scale—every 100 generations for the first 3000 generations of the experiment. Now they found clear evidence of a punctuational "step pattern" (see Fig. 3-4, discussed previously) within the initial phase of rapid increase that they had previously fit with a simple hyperbolic model. The authors noted: "When mean fitness was measured every 100 generations over the period of faster change . . . a step function model, in which periods of stasis were interrupted by episodes of rapid change, gave a better fit to the data than did the hyperbolic model. Evidently, it was necessary to make measurements at sufficiently high frequency to resolve the punctuated dynamics."

This gain of insight by finer sampling raises the important methodological theme that proper choice for a scale of inquiry depends crucially upon the resolution needed to identify and characterize the underlying causal process of the observed pattern—in particular, to specify the *natural unit* or entity of change in the given system. (Such studies often face the paradox that, whereas the recognition of this principle requires no act of genius, empirical adequacy often founders upon a conceptual dilemma: We can specify a proper scale if we know the causal basis beforehand. But, more often than not, we undertake such research in order to discover an unresolved causal basis—thus bringing upon ourselves the classical problem of a single equation with two unknowns: the causal basis and the scale required for its identification, to complete the analogy.)

At the macroevolutionary scale of punctuated equilibrium *sensu stricto,* events of speciation represent the natural unit, and geological resolution must be sufficient to identify the occurrence and timing (relative to stasis, or any other pattern, in the species's subsequent geological history) of origination for these macroevolutionary "atoms." Several published studies have been fatally marred by the basic flaw of using a scale so coarse that a trend generated by multiple events of staircase speciation could not be distinguished in principle from the same result achieved by smooth anagenesis in an unbranched lineage. In the most widely discussed fallacy thus engendered, Cronin et al. (1980) claimed gradualism (explicitly against punctuated equilibrium) for major trends in hominid evolution because a temporal sequence of mean values moved in the same direction. But the successive points were so widely separated in time and morphology that the authors could not determine whether they had measured mean values of successive species during their periods of stasis, or had sampled points in an anagenetic continuum. (Punctuated equilibrium, after all, was proposed as an alternative explanation for phyletic trends of this kind, not as a denial of their existence!) The scale of measurement used by Cronin et al. may be compared with Lenski's first procedure of sampling every 500 generations. Both schemes are too coarse to "catch" the underlying causal unit of change—speciation, if punctuated equilibrium holds, for the macroevolutionary case of hominids; the infrequent origin and sweep of favorable mutations in bacterial anagenesis.

In a fascinating study, extending (to an utterly different realm of inquiry) the generality of this important point about appropriate scale of measurement for the recognition of punctuations, Lampl et al. (1992) note that human growth in body length has generally, and for centuries of study, been regarded as smooth (albeit highly variable in rate at different states of ontogeny) because "individuals have been traditionally measured at quarterly intervals during infancy, and annually or biannually during childhood and adolescence. Physiological data are mathematically smoothed and growth is represented as a continuous curve" (1992, p. 801). But by measuring a sample of 31 "clinically normal" Caucasian children at intervals ranging from daily to weekly between the ages of 3 days and 21 months, Lampl et al., using the language and concepts of punctuated equilibrium, found that "90 to 95 percent of normal development during infancy is growth-free and length accretion is a distinctly saltatory process of incremental bursts punctuating background stasis" (p. 801). In fact, Lampl et al. did not detect the pattern of quickness in change and prevalence of stasis until they measured subjects at their finest daily scales (for even the semi-weekly and weekly measurements smoothed out punctuations over intervals of stasis). They conclude (Lampl et al., 1992, p. 802): "Human length growth during the first two years occurs during short (less than 24 hours) intervals that punctuate a background of stasis. Contrary to the previous assumption that the absence of growth in developing organisms is necessarily pathological, we postulate that stasis may be part of the normal temporal structure of growth and development."

Lenski's bacterial populations generate large numbers of mutations (some 10^6 every day in each population, by the estimate of Elena et al., 1996). But the step-dynamics revealed by the finer scheme of sampling—a pattern "predicted . . . by a simple model in which successive beneficial mutations sweep through an evolving population by natural selection" (Elena et al., 1996)—presumably occur for two reasons: first, the well-known exponential principle, however intuitively paradoxical for most people untrained in science, that "many generations are required for the beneficial allele to reach a frequency at which it has an appreciable effect on mean fitness, but then relatively few generations are required for that allele to become numerically dominant" (op. cit.); second, and probably more important, the extreme rarity of favorable variants amidst the daily plethora of mutations, leading to "a substantial waiting period before a beneficial mutation even occurs" (op. cit.). Thus, at the proper scale (for resolving the causal mechanism) of sampling every 100 generations, the plateaux of stasis mark the waiting times between favorable sweeps, while punctuations express the sweep itself.

Finally—and to place some substantial empirical weight upon the keystone of my argument for the potential generality of punctuational change—Lenski and his colleagues have greatly increased our understanding of evolution by developing an artificial (in the best and fully positive sense of the word) experimental system purposely reduced to an absolute "bare bones" of Darwinian causal minimalism. With an identical genetic starting point for each replicate, an asexual clonal system that permits no genetic exchange among cells, and an unchanged environment (the regimen of daily transfer by controlled dilutions into a constant growth medium), this experiment leaves only two factors free to work and vary as potential agents of evolutionary change: the paired and essential Darwinian components of new mutations as raw material, and shaping by natural selection among cells that vary as a consequence of these mutations. The fact that punctuational dynamics prevail in this first truly adequate experiment in pure Darwinian minimalism must at least evoke a suspicion—even among biologists who, by custom and *faute de mieux*, have never questioned gradualism—that this episodic mode might be expressing something important about the general nature of change itself across the varying scales of nature's evolutionary construction.

Punctuation above the Species Level

Punctuated equilibrium stands on an "isthmus of a middle state" (to quote Alexander Pope out of specific context, but in proper structural analogy, see *SET*, p. 680)—a speciational bridge linking the microevolutionary history of discrete populations with the macroevolutionary waxing and waning of clades through geological time. I believe that the prevalence of punctuational change on the bridge itself (punctuated equilibrium *sensu stricto*), combined with a strong case for punctuational dynamics in Darwinian processes stripped to a lean and clean minimality in microevolution (see preceding section), behooves us to consider a generalization across all scales, by suggesting an examination of larger realms beyond the bridge of speciation. I shall therefore discuss potentially instructive examples (not comprehensive statistical generalities, a worthy goal not nearly in current reach) at three expanding levels: (1) consequences of accumulated events of ordinary speciation within the history of individual clades; (2) the origin of phenotypically complex and extensive evolutionary novelties; and (3) the history of biotas in ecological and evolutionary time.

STASIS ANALOGS: TRENDING AND NON-TRENDING IN THE GEOLOGICAL HISTORY OF CLADES. Do we find cladal patterns, generated by different causal mechanisms, that might be sufficiently "homologous" (see pp. 236–240) with punctuated equilibrium to warrant comparison based on a common deep structure? (I am, of course, not considering or reiterating here (see full discussion, pp. 186–194) the most important and direct impact of punctuated equilibrium upon cladal histories—its ability to explain trends as the differential success of species rather than the extrapolated result of adaptive anagenesis. This section treats *other* scales and causes of change with meaningful structural parallels to punctuated equilibrium at the species level.) Possible parallels for the punctuational aspect are treated in the later chapters of *The Structure of Evolutionary Theory*—rapid origin of extensive evolutionary novelties for cladal beginnings (*SET*, Chapter 10), and patterns of mass extinction for endings (*SET*, Chapter 12)—but we should also consider the almost entirely neglected analogs of stasis at the cladal level.

An apt comparison for clades may be made to philosophical and sociological reasons for the previous failure of evolutionary biology to study, or even to acknowledge, the phenomenon of stasis as the predominant feature of phyletic history in a large majority of species. Stasis, construed as absence of evolution, once designated a negative result unworthy of a category, or even a name. In a similar way (and I cast no stones from a sinless state, for I have followed this tradition myself throughout my career), the study of trends has consumed nearly all research on the history of clades. And why not? Trends tell stories, and evolution is a narrative science. Western tradition, if not universal human nature itself, has always favored directional tales of conquest and valor (with Darwinian analogs of competition and adaptation), while experiencing great discomfort with the aimless cyclicity of Ecclesiastes, however much we may admire the literary power: "The thing that hath been, it is that which shall be; and that which is done is that which shall be done; and there is no new thing under the sun" (Ecclesiastes 1:9).

But the undeniable salience of trends—a psychological comment about our focus of attention—bears no necessary relationship to the relative frequency or causal weight of this phenomenon in the natural history of clades. How many monophyletic clades feature sustained and substantial trends in major characters of functional importance—and what percentage of characters participates in trends that do exist in such clades? Indeed, we have no idea whatever, for no neutral compilations exist. No one has ever tabulated the number or percentage of non-trending clades within larger monophyletic groups. The concept of a non-trending clade—the higher level analog of a species in stasis—has never been explicitly formulated at all. If only one percent of clades exhibited sustained trends, we would still focus our attention upon this tiny minority in telling our favored version of the story of life's history.

(Ironically, stable lineages become salient enough to catch our attention only at the extreme that we call "living fossils"—species or lineages supposedly unchanged during such long stretches of geological time that their stability becomes a paradox in a world of Darwinian evolutionary flux and continuity. As a double irony—see pages 102–105 for a full discussion in the light of punctuated equilibrium's different reading—we have also thoroughly misinterpreted this phenomenon under the same gradualistic bias that inspired our notice in the first place! The classical "living fossils" (the inarticulate brachiopod *Lingula,* the horseshoe crab *Limulus polyphemus,* the extant coelacanth) are not long-lived as species [*Limulus polyphemus,* for example, has no fossil record at all], but rather belong to clades with such a low speciation rate that little raw material for cladal trending has been generated over the ages.)

I suspect that most clades, while waxing and waning in species diversity through time, show no outstanding overall directionality. But we do not know because the literature has never recognized, or attempted to tabulate, the frequency of such "Ecclesiastical" clades that change all the time, but "go" nowhere in particular during their evolutionary peregrinations. Paleontologists achieved no sense of the strength of punctuated equilibrium, even though Eldredge and I had formulated the theoretical apparatus, until researchers studied the relative frequencies of stasis and punctuation in entire faunas, or entire clades, by full sampling and with no predisposing bias to favor one kind of result—see discussion on pages 147–171 for this extensive and growing literature. Similarly, we will not know the general fate of clades until we ratchet this methodology one notch higher, and sample sets of clades not identified by our prior sense of their evolutionary "interest." Stasis is data at the species level. Non-trending is data at the clade level.

Budd and Coates (1992) broke conceptual ground in devoting an entire paper to such Ecclesiastical non-directionality in the actively evolving and speciating clade of montastraeid corals during 80 million years of Cretaceous time. As a rationale for their study, the authors state an analogy to the lower level phenomenon of punctuated equilibrium: "Just as the study of stasis within species has facilitated understanding of morphologic changes associated with speciation, we show that study of nonprogressive evolution offers valuable insight into how the causes of trends interact and thereby produce complex evolutionary patterns within clades, regardless of their overall direction."

The central theme of non-trending, identified by Budd and Coates for this large clade of massive, reef-building corals, stands as an empirical pattern in any case, but the (admittedly somewhat speculative) explanation proposed by the authors also builds an interesting framework for regarding such a signal as predictable and unsurprising, rather than anomalous. Their proposal also integrates the two principal Darwinian critiques of this book by attributing a causal pattern generated at the species level (the hierarchical expansion of my first theme) to the effects of architectural or developmental constraint (the structuralist or internalist perspective of my second theme) in channeling the possibilities and directions of natural selection.

Budd and Coates (1992) propose that monstastraeid species vary within a range set by minimal and maximal size of individual corallites on these large colonies. Such a notion does not debar classical trending, for the clade could originate in one small portion of the permitted range, and then strongly trend towards the other domain. But Budd and Coates argue that Cretaceous montestraeids already inhabited the full range, and that each end represented an adaptive configuration continually available and exploited throughout the clade's duration. Therefore, phenotypic evolution fluctuated between the two realized potentials of a fully populated domain of workable solutions.

The authors argue that "large-corallite" species (3.5 to 8.0 mm in diameter) maximize efficiency in removal of sediment, and tend to dominate in turbid waters; while "small-corallite" species (2.0 to 3.5 mm in diameter) prevail in clearer waters of the reef crest. Moreover, large-corallite species derive most nutrition by direct carnivory, whereas small-corallite species tend to feed upon their own symbiotic zooxanthellae. Budd and Coates then advance the claim—the more speculative aspect of their scenario—that montastraeid species remain constrained with this range by limitations at either end: an inability of still smaller corallites to develop and function adequately, and a restriction in septal number and strength that would not grant sufficient biomechanical support to still larger corallites.

These two arguments may validate and explain the basis of active non-trending in a persistently vigorous and successful clade. For if such constraints limit the range of corallite size, and if each end enjoys advantages in different environments continuously available in major parts of the habitat, then evolution might oscillate back and forth, with no persistent directional component, throughout cladal history.

Budd and Coates document such a directionless oscillation within the clade's developmental and adaptive boundaries during four successive divisions of Cretaceous time. The transition from interval 1 to interval 2 featured a differential production of small-corallite species from large-corallite ancestors, as well as a southward expansion of the clade's geographic range. Limited and directionless speciation, accompanied by predominant stasis within established species prevailed during intervals 2 and 3. Between intervals 3 and 4, large-corallite species radiated from small-corallite ancestors, and the geographic range of the clade became more restricted. In other words, the general pattern at the end of interval 4 differed little from the initial spread of morphology and geographic range at the outset of interval one, the inception of the study itself.

But the montastraeids remained a vigorous, successful, and evolutionarily active clade throughout Cretaceous times. Who are we to proclaim their pattern "boring" or unworthy of study because the evolutionary history of these corals does not resonate well with human preferences about "interesting" or "instructive" stories? Perhaps we should force ourselves to learn that patterns traditionally shunned for such quirky reasons of human appetite may hold unusual interest and capacity to teach—precisely because we have never sought messages that might challenge our complacencies. The predominant pattern of life's history cannot fail to be instructive—and such non-trending may well mark a norm of this magnitude, even if heretofore hidden in plain sight because we also see with our minds, and conventional concepts can be more blinding than mere ocular failure.

PUNCTUATIONAL ANALOGS IN LINEAGES: THE PACE OF MORPHOLOGICAL INNOVATION. I do not wish to resuscitate one of the oldest canards, and least fruitful themes of evolutionary debate: the claim for truly saltational, or macromutational—that is, effectively one-generational, or ecologically "overnight"—origin of new species or morphotypes. This ultimate extreme in punctuational change has never been supported by punctuated equilibrium, or by any sensible modern account of punctuational change in any form. Even if older evolutionists did advocate this mode of change (see my discussions of de Vries in *SET,* pp. 415–451 and of Goldschmidt in *SET,* pp. 451–466), they granted no exclusivity to its operation, and they also defended more continuationist, and more structurally plausible, accounts of rapid origin for morphological novelties—as in the developmentally rooted and theoretically sensible concept, based on mutations in "rate genes," embodied in Goldschmidt's unwisely named "hopeful monster," in contrast with the speculative scenario that he built upon his later concept of "systemic mutations"—see Schwartz (1999) for an interesting modern retelling and defense of this notion.

Just as punctuated equilibrium scales the geologically abrupt (but ecologically slow and continuous) process of speciation against the long duration of most species in subsequent stasis, punctuational hypotheses at higher levels regard the pacing of substantial phenotypic change in the origin of novel morphotypes as similarly episodic, with origins concentrated in very short episodes relative to periods of stability in basic design during the normal waxing and waning of clades—and perhaps with the ecologies, or the structural and developmental bases, of such episodes belonging to a distinctly different class of circumstances from those that regulate the ordinary pace of flux and speciation during the long-term geological history of most clades and morphotypes. In other words, a timekeeper with a metronome beating at an appropriate frequency for discerning the units and causes of evolution at each scale of nature's hierarchy might recognize an episodically (and rarely) pulsating, rather than an equably flowing, tempo as the dominant signal of change in all realms.

For example, Erdtmann (1986, p. 139) proposed that the active cladogenesis of early Ordovician planktic graptolites (an extinct subphylum of colonial organisms close to the chordate lineage) "operated on two levels:

gradualistic change involving species-level and intergeneric clades, and punctualistic (anagenetic) changes operating on supergeneric levels." He linked the rapid and extensive morphogenetic innovations of the punctuational mode, involving such basic features of colonial form as loss of bithecae and reduction in number of stipes, to major environmental changes marked by rapid eustatic shifts in sea level. Moreover, these punctuational innovations arise by an astogenetic mode (a term for the ontogeny of colonies) different from the developmental basis of most smaller changes that mark the flux of speciation during "normal" geological intervals. The punctuational innovations that produce new developmental patterns begin at the proximal end of the colony—that is, they affect the early ontogeny of the initial organisms of the developing aggregate, thereby pervading the life cycles of both the organism and the colony, and strongly affecting the global phenotype of the entire structure. The lower-level changes (smaller variations within existing developmental themes), on the other hand, tend to begin at the colony's distal end—that is, they arise late in the ontogeny of older organisms in the colony (for new organisms of the colony arise proximally, pushing older organisms to progressively more distal positions), and then proceed to earlier phylogenetic expression in both the colony's astogeny and the individual organism's ontogenies. These lower-level changes therefore affect only a small, and astogenetically late, portion of the colony's form—hence their much more limited capacity for yielding major morphological change.

This case provides an interesting astogenetic analog to the common claim that heterochronic changes in the early ontogeny of organisms gain a distinctive status among evolutionary mechanisms in their potential for rendering substantial phenotypic change (at a punctuational tempo) with minimal genetic alteration. In fact, the lability inherent in early ontogenetic changes of rate and regulation undergirds most theorizing about qualitatively different categories of evolutionary outcomes based on similar underlying magnitudes of raw genetic alteration—the most promising basis for a dominant punctuational tempo in the history of morphological innovation in evolution.

By linking constraints of preferred developmental channels with a punctuational tempo that precludes accumulative incremental selection as the sole cause of extensive evolutionary change, this familiar argument unites the two central themes of this book—hierarchical models of evolutionary mechanics with structuralist accounts of evolutionary stasis and directionality. Several recent volumes have explored the growing power and prestige of this argument, as provided by breakthroughs in unraveling the genetics of development (see *SET,* Chapter 10), combined with classical data of allometry and heterochrony (Raff, 1996; Schwartz, 1999; McNamara, 1997; McKinney and McNamara, 1991; McKinney, 1988).

Wray (1995) has recently summarized an emerging generality that integrates all components of the argument across a wide variety of organisms. His chosen title—"Punctuated evolution of embryos"—underscores the putative generality of change in this mode, with punctuated equilibrium as its major expression at the level of ordinary speciation, and his proposed linkage of development and ecology as its hypothesized primary source for the rarer, but highly consequential, phenomenon of the origin of morphotypic novelty.

Nearly two centuries of tradition proclaim the conservatism of early larval and embryonic phases of the life cycles—from von Baer's enunciation of his celebrated laws (1828, see discussion in Gould, 1977) to the standard evolutionary rationale that formative stages of early ontogeny become virtually impervious to change because cascading consequences, even of apparently minor alterations, would discombobulate the subtle complexities of development. Recent discoveries of "deep" genetic homologies and developmental pathways among animal phyla separated for more than 500 million years (see *SET,* Chapter 10) have tended to highlight this conventional view.

But Wray (1995) summarizes several recent studies of broad taxonomic scope—with best examples from the sea urchin *Heliocidaris,* the frogs *Eleutherodactylus* and *Gastrotheca,* and the tunicate *Mogula*—all showing that "similar species have . . . modifications in a variety of crucial developmental processes . . . that have traditionally been viewed as invariant within particular classes or phyla" (Wray, 1995, p. 1115). These substantial changes in the development of closely related forms exhibit three common properties: (1) They usually affect traits of timing and regulation in early development, including specification of cell fates and movement of cells during gastrulation. (2) They yield substantial changes in larval forms and modes of life, but often leave the adult phenotype largely unaltered. (3) They are associated with major changes in the ecology and life history strategies of larval or early developmental forms, and involve such major changes as loss of larval feeding ability (the echinoderm and frog examples) or capacity to disperse (tunicates).

Wray presents evidence that alterations in larval ecology "drive changes in development," not vice versa. Moreover, and most importantly, comparison of molecular and phenotypic modification shows that these "functionally profound changes in developmental mechanics can evolve quite rapidly" (ibid., p. 1116). For species of *Heliocidaris* with strikingly different developmental mechanisms, fewer than 10 million years have elapsed since divergence from common ancestry. Wray draws a general and punctuational conclusion from this evidence for ecologically driven change in mechanisms of early development—events that can occur very rapidly and do not compromise the conserved life styles of later development due to greater dissociation than traditional views would allow between successive phases of ontogeny: "Long periods of little net change, with functionally minor modifications in developmental mechanisms and larvae, seem to be the normal mode of evolution. This near stasis is interrupted on occasion by rapid, extensive, and mechanically significant changes that coincide with switches in life history strategy . . . Rapid modifications can arise in developmental mechanisms that have been conserved for hundreds of millions of years."

Note the striking similarity of language (with analogs of stasis and punctuation)—and of evolutionary style in tempo and mode—between punctuated equilibrium and Wray's description of phenotypic and ecological shift at these much larger scales of morphological change and temporal extent (with the analog of stasis persisting for hundreds of millions of years). These similarities in style and import seem to mark a genuine conceptual "homology" based on similar structural principles regulating the nature of change in complex systems.

As a final example of the fruitfulness and detailed testability of punctuational models for the origin of morphogenetic novelty, Blackburn's (1995) remarkable study of "saltationist and punctuated equilibrium models for the evolution of viviparity and placentation" deserves special notice for the author's clarity in specification of hypotheses, and for his richness and rigor in attendant documentation. Blackburn treats the multiple evolution of viviparity in squamate reptiles (lizards and snakes)—a much better case for studying this phenomenon than the group that most of us emphasize for parochial reasons (the Mammalia), for extant squamate species include many examples in all stages of the process. This taxonomic richness permits clear distinction of gradualistic vs. punctuational alternatives, and also provides sufficient data to distinguish between punctuated equilibrium and saltation as the dominant punctuational style. Moreover, and largely because the subject has been embraced as a workable surrogate for unresolvable questions about mammalian evolution, the origin of viviparity in squamate reptiles has become a classical case, and has therefore generated an extensive literature to illustrate (however unintentionally) some major biases of evolutionary argumentation. The power of Blackburn's study resides in three interrelated themes:

1. Gradualistic scenarios have dominated the classical literature in ways that authors rarely defend, or even recognize. In particular, previous workers have assumed that three apparent stages in the "perfection" of live bearing must represent an actual historical sequence gradually and incrementally evolved by all lineages that reach the "last stage"—viviparity or live birth, placentation for gas exchange and water intake, and placentotrophy for embryonic nutrition. Blackburn writes (1995, pp. 199–201): "Viviparity and placentation in squamates have stood for over half of a century as examples of gradual evolution . . . Even recent reviewers have not considered the applicability of alternative evolutionary models."

The supposed evidence for such gradualism consists largely of inferences drawn from structural series of extant forms, without affirmation, or even consideration, of an explicit phylogenetic hypothesis that the successive stages represent a cladistic sequence. (Even worse, a phylogenetic inference has often been based only upon the series itself—a flagrantly circular argument that validates the conclusion by the hypothesis supposedly under test.) For example, extant oviparous species do vary substantially in the stage of development at which the eggs are laid—and researchers have generally assumed that a linear ordering of such a series must represent an evolutionary continuum "on the way" to viviparity: "The inferred continuum of developmental stages at oviposition among squamates commonly is interpreted as evidence for a gradual increase in the proportion of development occurring in the female reproductive tract" (ibid., p. 201).

2. Blackburn marshalls an impressive array of data from a broad range of fields—taxonomy, development and geology, in particular—to affirm an alternative punctuational scenario for the evolution of live birth, with simple viviparity, placentation and placentotrophy as three distinct modes, not three way stations in a progressive sequence. (I suspect that our gradualistic biases have been particularly intrusive in this case because we unconsciously read the squamate story in a mammalian perspective that makes placentotrophy the "obvious" goal of any trend to live bearing.)

In taxonomy, viviparity has originated more than 100 times among squamate reptiles (ibid., p. 202). But cladistic data have provided not a single case of correspondence between branching order and the four structural stages of the hypothetical trend: ovipary, vivipary (live birth without placentation of embryos), placentation, and placentotrophy. Blackburn writes (1995, pp. 201–202): "Clines of phenotypic variation that can be invoked to support gradualistic evolution of viviparity and placentotrophy tend to be composites of unrelated species representing multiple lineages . . . Despite the documentation of over 100 evolutionary origins of viviparity in squamates . . . available evidence has not yet permitted construction of a single, complete phenocline of parity modes and embryonic nutritional patterns out of representatives of a single clade."

In structure and development, Blackburn coordinates several lines of evidence to argue that intermediary forms between any two stages in the hypothetical trend either cannot be found, or exist only rarely and in a tenuous state (because such transitional phenotypes would experience either architectural problems in construction or adaptive insufficiencies in function). For example, if viviparity evolved by progressive delay of oviposition, then we might expect, among extant species, "a full continuum of developmental stages . . . representing steps in the parallel evolutionary transformations that have occurred independently (and perhaps to different degrees) in various lineages" (p. 202). Instead (see Fig. 5-7), the distribution of developmental stages at oviposition shows marked bimodality, with species either depositing eggs containing embryos in the pharygula/limb bud stages (with near normality or minor left skewing for this lower mode, and no right skew in the direction of the putative trend) or else retaining the eggs to term and then giving birth to live young.

Moreover, the supposed development of placentation, and then of placentotrophy, only after the origin of live birth also derives no support from documented intermediary stages. In the traditional view, the shell membrane between fetal and maternal tissues must thin gradually, permitting an initial function of placental organs in uptake of water and exchange of gases. Placentotrophy then evolves later "as the placental supply of nutrients first supplements and then supplants provision by the yolk" (p. 208). But evidence from at least 19 independent clades of viviparous squamates indicates that all "have anatomically recognizable placentae derived from both the chorioallantois and the yolk sac" (p. 208). Thus Blackburn concludes, "the existence of a truly non-placental viviparous squamate has not been documented in over a century of investigation . . . The universal occurrence of

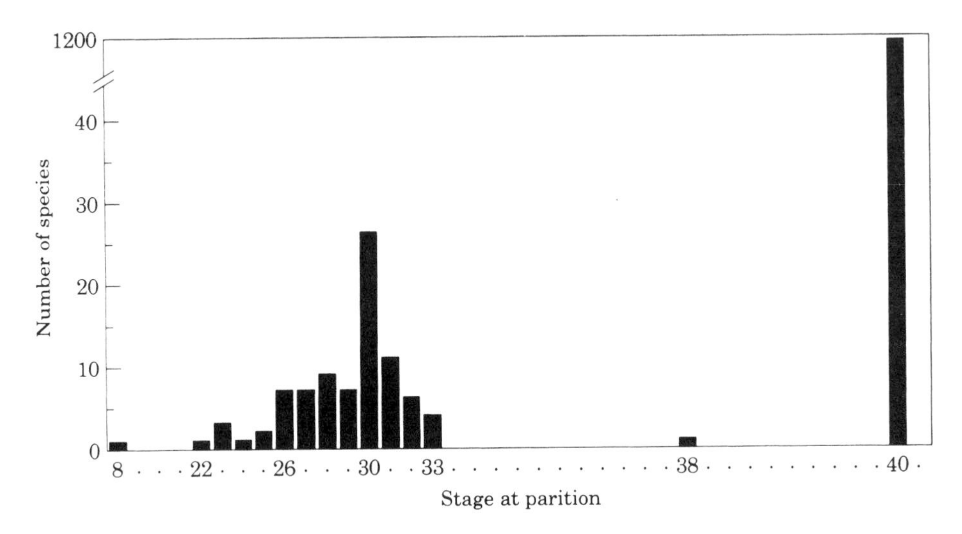

5-7. Punctuational change in the morphological evolution of lineages in squamate reptiles, Ovoviviparity does not evolve by progressive and gradualistic delay of oviposition, but rather shows marked bimodality with females either depositing eggs with embryos in their early limb bud stages or else retaining the eggs within their body to term, and then giving birth to live young (the right mode). The existence of a few intermediary species shows that the full sequence proceeds by punctuational steps and not by full saltation.

placentae in viviparous squamates is most consistent with the view that placental organs that accomplish gas exchange and water uptake evolve simultaneously with viviparity" (p. 209).

Similarly, no purely lecithotrophic (yolk feeding) placental squamates have been discovered, and all viviparous forms derive at least some nutrition through the placental organs. Thus, "available data are most consistent with the hypothesis that incipient placentotrophy is a necessary correlate of viviparity." The three "logical" steps of the hypothetically gradual trend become telescoped into a single structural transition, with an evident implication of punctuational origin.

Blackburn supplements this empirical evidence for a punctuational divide with both structural and functional rationales for the inviability of putative intermediary stages. Viviparity without placentation may be structurally unattainable because live birth requires that the eggshell become sufficiently thin "to permit gas exchange in the hypoxic uterine environment" (p. 211)—while such reduction may entail gas and water exchange (that is, incipient placentation) as a virtually automatic consequence. Thus, Blackburn argues (p. 210), "placentation is best viewed as a necessary correlate of viviparity, not as a 'reproductive strategy' *per se.*"

Intermediacy may be equally unlikely in functional and adaptationist terms as well. Both endpoints entail costs as well as putative benefits—oviparity in dangers and energetic requirements of nesting behavior, and in maternal loss of calcium in making eggshells (p. 211); viviparity in decreased maternal mobility, fecundity, or clutch size. A hypothetical intermediate that incurs both sets of costs—for example, the calcium drain from internal shells of hypothetical stage one (still too thick for placentation), combined with a heightened susceptibility to predators caused by compromised mobility—without winning greater compensation in combined benefits, could not compete against either end member of the supposed trend, and probably would not survive even if patterns of development permitted evolutionary access to this putatively transitional design.

Finally, *in geology,* the recent origin of most viviparous lines strongly supports a punctuational inference. A few origins of squamate viviparity may date to late Mesozoic or early Cenozoic times (p. 207), but most represent Pliocene or Pleistocene events. Moreover, taxonomic distribution fully supports the rapidity of full transition to placentotrophy. More than 60 percent of origins for viviparity "have occurred at subgeneric levels, and virtually all have arisen at subfamilial levels." Several origins can be traced to populations of a single species (with other populations remaining oviparous)—for example, a Pleistocene event within *Lacerta vivipara,* and an origin within the past 11 to 25 thousand years within the *Sceloporus aeneus* complex (p. 207). The extent of structural differences between oviparous and viviparous populations of these minimally distant forms (both temporally and phylogenetically) fully matches the phenotypic separation noted for the same features in cladistically distant lineages.

3. As an indication that data of natural history can provide combined criteria to permit fine and testable distinctions, Blackburn has been able to reject saltation and defend punctuated equilibrium as the probable cause and temporal basis of this well-documented punctuational pattern. Blackburn notes that "under the punctuated equilibrium model, typical oviparity and viviparity could represent regions of stasis, with prolonged oviparous egg-retention being a transitory, intermediate stage between them" (p. 206). The structurally well integrated and functionally well-adapted end members "would contrast with the instability of the evolutionary intermediate, and

prolonged oviparous egg-retention would be a relatively short-lived (and hence scarce) pattern" (p. 206). By contrast, of course, saltational models predict the structural unattainability and adaptive inviability of intermediates, and therefore envisage a one-step transition (with substantial opportunity, perhaps, for later adaptive fine-tuning).

In favor of punctuational origins (rapid transition between domains, based on structural properties of endpoints as well coordinated states that actively resist change, and with intermediary forms as unstable, and "driven" towards one of the endpoints) versus saltational events (truly sudden transition, scaled to the magnitude of the unit of change and enforced by the absolute structural inaccessibility of intermediary states), Blackburn cites several features of squamate viviparity. The putative intermediary stage of "prolonged oviparous egg retention" (p. 206), while empirically rare, structurally unstable, and adaptively compromised, does exist as the characteristic form of a few species in nature, not just as a facultative or transient state of a population in transformation. As the bimodality of Fig. 5-7 shows, prolonged oviparous egg retention does represent an attainable intermediary state between two endpoints. But few forms occupy this uncertain ground between advantageous configurations.

In a second indication of punctuational rather than saltational change, Blackburn notes that a viable first step to intermediacy does exist in nature as a strategy that can be activated under certain environmental conditions: "facultative egg-retention with continued intra-oviductal development" (p. 206). Several squamate species exhibit this phenotypic flexibility in development. "Such facultative retention could provide raw material for natural selection, making more likely the evolution of a pattern in which prolonged egg-retention was obligative."

As a third confirmed prediction, favoring punctuation over saltation when joined with the two previous arguments (but unable to make the distinction otherwise, while confuting gradualism in any case), the phenotypic similarity of oviparous and viviparous congeners affirms the relative ease and accessibility of transition: "As an evolutionary unstable pattern, prolonged egg-retention might lead to viviparity, or might revert to typical oviparity; thus the less genotypic change involved, the more probable the origin of viviparity would be" (p. 206).

Blackburn's final paragraph (p. 212) serves as an apt reminder about the restrictive nature of gradualistic bias, and of the power and inherent probability of punctuational alternatives in a world that may favor this mode of change as a general structural property of material organization at all scales: "For over 60 years, research on amniotes has assumed that gradualistic change is the sole mechanism by which viviparity, placentation, and placentotrophy could have evolved. Future empirical and theoretical analyses of reproduction in squamates and other vertebrates should not overlook the potential of non-gradualistic models as explanations for evolutionary change and the biological patterns it has produced."

PUNCTUATIONAL ANALOGS IN FAUNAS AND ECOSYSTEMS. If stable locations, reached by rapid movement though "perilous" intermediary terrain, sets the structural basis of punctuational change for the "internalities" of evolution in complex organic phenotypes, then a similarly episodic, rather than evenly flowing, mode of change might characterize the "externalities" of environments that regulate any coordinated evolutionary tempo among components of biotas and ecosystems. I have already considered the scale of environmental punctuation most immediately relevant to punctuated equilibrium—the geological history of regional biotas (see pp. 222–229 on coordinated stasis, the turnover-pulse hypothesis, and other notions of faunal transition by coincidence of punctuational extinction and origination in a high percentage of species within a previously stable biota). But the generality of such punctuational tempos in external controls might also extend to levels both below and above the direct mechanics of speciation itself.

In a general argument strikingly similar to Blackburn's for the evolution of squamate viviparity, Smith (1994) holds that gradualistic assumptions have stymied our understanding of evolutionary processes at the small scale of ecological immediacy in deep-sea faunas. No other environment has been so conventionally associated with plodding, incremental change through substantial periods of time. Smith begins his article by noting the "the deep-sea floor is traditionally perceived as a habitat where low food flux and sluggish bottom currents force life to proceed at slow, steady rates. In this view, benthic community structure is controlled by equilibrium processes, such as extreme levels of habitat partitioning, made possible by remarkable ecosystem stability" (Smith, 1994, p. 3).

As indicated by the title of his article—"Tempo and mode in deep-sea benthic ecology: punctuated equilibrium revisited"—Smith holds that we must revise this traditional view, and reconceive the deep-sea as a punctuational domain where "endogenous disturbances may be relatively frequent, and where pulses of food reach the seafloor from the upper ocean" (p. 3). In what he labels as a "parallel argument" to our punctuated equilibrium from a much lower scale of analysis—in other words, as a claim for conceptual homology of constraining structural principles (in the language of this section)—Smith discusses three examples of "pulsed events that 'punctuate' the apparent 'equilibrium' of the deep-sea floor" (p. 3), and that "may substantially influence processes of modern and past ecological significance including (1) maintenance of macrofaunal diversity and population structure, (2) deposit-feeder–microbe interactions and associated trace production, and (3) dispersal and biography of chemosynthetic communities at the deep-sea floor" (p. 3).

First, Smith documents the importance of "pulsed physical disturbance" in benthic faunas of the Nova Scotian Rise (4750–4950 m)—particularly of erosional "storms" that scour and redeposit sediments "to depths of millimeters-centimeters over areas encompassing at least tens of square kilometers" (p. 7), and that strongly influence both the composition and successional stage of local faunas.

In a second microbiotal example, Smith documents the importance of "phytodetrital pulses" in nutrition for the deep-sea macrofauna. The "slow and steady 'drizzle'" usually regarded as the gradual (and meager) planktonic contribution to sustaining deep-sea life "can be punctuated by downpours of 'phylodetritus' (i.e., detrital material composed primarily of relatively fresh phytoplanktonic remains), during which the flux of labile particulate organic carbon to the seafloor temporarily exceeds biological demand, yielding a carpet of 'food'" (p. 7).

Finally, and to add a third punctuational source of maximally different character from the physical and microfloral cases discussed above, Smith argues (p. 10) that "whale falls" produce occasional and (obviously) "huge local pulses" of organic matter that may decay to produce distinctive "chemosynthetic habitats" supporting faunal associations much like those documented at deep-sea vents. For example, in 1987, his team discovered a 21-meter whale skeleton at a depth of 1240 m: "The bones were covered with mats of sulfur bacteria and clusters of small mussels and limpets; nearby sediments harbored large vesicomyid clams" (p. 10)—for a total of 42 macrofaunal species, only nine of which also inhabited surrounding sediments. Smith concludes that "sunken whales may provide dispersal stepping stones for at least some of the species dependent on sulfide-based chemosynthesis."

Strong circumstantial evidence indicates considerable temporal and spatial influence for this source that most of us would surely have regarded as dubious, if not risible, at apparent face value of relative importance. A fossilized chemosynthetic community has been reported from a 35 million year old whale fall on the Northeast Pacific ocean floor (p. 10). "Whale skeletons," Smith concludes (p. 10), "may be the dominant source of chemosynthetic habitats over the vast sediment plains constituting most of the ocean floor."

At the opposite end of a hierarchy in spatial and temporal scales, punctuational models continue to gain in strength and acceptability for events that impact entire biotas at regional or even planetary scales—with catastrophic mass extinction as a "flagship" notion, spurred by nearly conclusive evidence for bolide impact as the trigger of the K-T global dying 65 million years ago (see *SET,* Chapter 12 for full treatment). An expansion of research away from the extinctions themselves, and towards the subsequent recovery phases as well, has strongly accentuated the episodic and punctuational character of this most comprehensive signal in the history of life.

Even after the Alvarez's impact hypothesis forced paleontologists to acknowledge the potentially catastrophic nature of at least some mass extinctions, students of fossils usually assumed that subsequent recoveries of global faunas must have been tolerably gradual. This expectation has not been fulfilled, and episodes of recovery from maximum decimation at the extinction to full reestablishment of previous levels of diversity occur more quickly, and in a much shorter percentage of the "normal" time (until the next mass extinction), than previously suspected. (Of course, no one expects that recoveries which require successive events of branching can be nearly as rapid as truly catastrophic extinctions, which can feature truly simultaneous killings—so the complete record of an extinction-recovery cycle will surely remain asymmetric. But the recoveries now seem to occur rapidly enough, in most cases, to invoke the central concept of punctuational change: origin in a tiny fraction of later existence in stasis.)

For example, Kerr (1994) begins his report on Peter Sheehan's work (in a commentary entitled "Between extinctions, evolutionary stasis") by writing (Kerr, 1994, p. 29): "More and more, paleontologists are learning that the full measure of a mass extinction can't be found in its immediate toll. Just as important is the wholesale reorganization of living communities that takes place afterward. And those brief recovery periods, lasting just a few million years, are all the more important because during the tens of or hundreds of millions of years that follow, until the next mass extinction, not much may happen."

Sheehan divides the last 640 million years into six major faunal packages that he calls EEU's, or Ecologic Evolutionary Units. Each lasts for 35 to 147 million years, and each ends at a mass extinction. The subsequent recovery periods for the new units occupy only 3 to 8 million years.

This recent affirmation of a strongly punctuational character for change (primarily extinction) at the highest level has led to a tendency, probably overextended—and I blame myself, in part, for propagating the theme, see Gould, 1985—for ascribing a dualistic character to the pulse of evolution, with punctuations of mass extinction alternating with a more stately flow in "normal" times between these macropulses. But this view may prove to be overly simplistic, although not wrong. When we assess each level of change by its own appropriate measuring rod (scaled to emphasize the relevant unit or units), all may be punctuational. We must dismiss as irrelevant and misleading the fact that punctuations at a small scale may "smooth out" to more gradual and continuous trends when inappropriately measured at too large a scale to reveal the causal mechanics, or even to identify the relevant unit, of change—a theme that I have emphasized throughout this book, in such examples as punctuated bacterial anagenesis, viewed as gradual when sampled too infrequently to note the steps of mutational sweeps; and cladal trends, viewed as anagenetic when sampled too broadly to discern the speciational jumps of punctuated equilibrium.

In a provocative work, Raup (1992) played devil's advocate by asking if all extinctions at all levels, from single local populations to global faunas, might be catastrophic—for he could not reject the "null hypothesis" of his "field of bullets" model (random and catastrophic removal, triggered by "bolides" of various sizes randomly shot towards the earth at frequencies inversely proportional to their size and effect) in favor of the traditional Darwinian model of gradual declines mediated by competitive inferiority in biotic interactions. I do not believe that such extreme punctuationalism could rule so completely (see full discussion of this argument in *SET,* Chapter 12, pp. 1323–1326). But finer analysis of the most famous cases of supposedly gradual, and biotically controlled, events may well require such a punctuational reinterpretation. Most outstandingly, perhaps the two most widely discussed and most generally accepted examples of geologically slow global diversification—the Ordovician spread of the great Paleozoic marine invertebrate fauna, and the Mesozoic "modernization" of invertebrate predators and prey (the classic example of a supposed biotic and gradualistic "arms race")—now seem to occur far more abruptly in each separate geographic region, with the previous impression of gradual construction based on a blurring of the different times of transition in each region (Jablonski, 1999, p. 2114).

In an important paper, Miller (1998) has generalized this claim by summarizing the increasing evidence for punctuational tempos in faunal change (both locally and regionally, and for both extinctions and the necessarily slower rediversifications)—with our conventional notions of gradual flux, particularly for build-ups, arising as an artifact of summation over displaced timings for rapid pulses in several regions. Miller first states the general observation and emerging principles (1998, pp. 1158–1159): "In recent years, local and regional studies of marine faunal patterns have converged on a similar theme—that biotic turnover occurred episodically through investigated stratigraphic intervals. There were comparatively broad intervals with little net turnover, punctuated by narrower intervals in which many taxa either emigrated or became extinct and were replaced by a roster of taxa that either originated in the area or immigrated into it. . . . Episodicity appears to be a general feature of regional stratigraphic packages." He then uses this finding to correct what may be a substantial error in traditional views (1998, p. 1159): "Thus, major faunal transitions in global-scale compilations, which seem to have transpired over protracted intervals of geological time, took place far more rapidly and episodically when evaluated regionally or locally. The transitions only appear gradual on a global scale because of variations in their timing from venue to venue."

Finally, Miller asserts a general "fractal" conclusion about punctuational change (ibid., p. 1159): "The processes that produced major mass extinctions simply represented the largest and most globally extensive of a spectrum of perturbations that produced episodic biotic transitions."

As a closing note in this context, Miller also offers a similar punctuational reinterpretation for the putatively best documented and most widely accepted case of global, geologically gradual, and broadly progressive change in life's history—the pattern that Vermeij (1987) has called "escalation" (largely, and with good reason, to avoid false implications and arguments in the traditional notion of "progress"), based on relayed "arms race" between predators and their prey, and on other kinds of similarly reciprocal biotic interaction through extensive time. This entirely sensible concept of escalation seemed to provide the best available argument for two deeply rooted and strongly held themes of traditional Darwinian extrapolationism: the predominant power of biotic interactions to shape patterns in the history of life, and trends towards the slow accumulation of biomechanically improved designs in major lineages.

The general argument sounds so reasonable, but when we rethink macro-evolution as a process based upon geologically rapid production of higher-level individuals by punctuational speciation as the primary units of change, then the mechanics of this usual interpretation of escalation become elusive. The pattern certainly exists—especially for Vermeij's (1977) classic case of increasing strength and efficiency in crab claws matched by growing intricacy and sophistication of adaptive defenses in molluscan shells. But how can such an arms race operate if the full trend proceeds by stepping stones of punctuational speciation for any increment, and not in the style of tit-for-tat anagenetic escalation, based on immediate organismal competition and more familiar to us through human models of "anything you can do, I can do better"—a point that Vermeij himself recognizes and finds puzzling (1987)?

Miller, on the other hand, affirms the gradual trend to escalation in biomechanical improvement—and I don't think that any party to this debate denies the reality of the pattern (for we have been arguing about mechanisms)—but finds the same unconventional (and punctuational), finer-scale theme upon "dissecting" the full result into component causal units. Again, each step in escalation seems episodic in each region, with the full trend thus rendered as a summation of punctuational events. Miller writes (p. 1159): "Although the case for these kinds of transitions over the sweep of the Phanerozoic is difficult to deny, the manner in which they transpired over shorter intervals is less certain. There is little evidence of gradual escalation through stratigraphic intervals at local or regional levels. The introduction of escalated forms appears to have occurred episodically, in concert with the broader class of changes in taxonomic composition discussed earlier, which suggests a role for physical mediation."

Two general points provide a fitting close for this section:

1. The probable generality of punctuation and stasis as a powerful—if not predominant—style of change across all scales must lead us to reassess our previous convictions about "important" and "interesting" phenomena in evolutionary theory and the history of life. Kerr admits the potential generality, in reporting punctuation at lower levels to complement Sheehan's similar claim for the broadest scale. But he closes his report by writing (1994, p. 29): "Sheehan sees these intervals as analogous to his longer, global

EEU's reaffirming that stability—as boring as it may be—is the evolutionary norm." But, to restate my mantra, and to emphasize its implications for understanding the history of earthly life and the psychology of human discovery, stasis is data—and data of such high generality, such unanticipated occurrence, and such theoretical interest simply cannot be boring.

2. The ubiquity—and the possibly canonical character—of punctuational change at all scales, from the shortest trends of bacterial anagenesis in single clonal lineages over weeks to months, to the broadest patterns of global waxing and waning of biotas through the history of life in deep time, can only recall the familiar tale, by now a cliché, of the Eastern sage who revealed the nature of the cosmos to his disciple: the globe of the earth rests on the back of an elephant who stands, in turn, on the back of a turtle. When asked by the disciple what one might find under the turtle, from its feet to the ultimate source of being, the sage simply replies: "it's turtles all the way down." I suspect that it is also punctuational change all the way down, from Permian extinctions to mutational sweeps through little laboratory populations of *E. coli.*

PUNCTUATIONAL MODELS IN OTHER DISCIPLINES: TOWARDS A GENERAL THEORY OF CHANGE

Principles for a Choice of Examples

In their symposium for the American Association for the Advancement of Science, and in their subsequent book, Somit and Peterson (1992) explored the wider role of punctuated equilibrium in suggesting similar modes of change in other disciplines. (Their edited book bears the title: *The Dynamics of Evolution: The Punctuated Equilibrium Debate in the Natural and Social Sciences.*) In discussing the "manner in which punctuated equilibrium theory renders its greatest contribution to the behavioral sciences" (1992, p. 12), they suggested (loc. cit.): "By providing a different metaphor for explaining social phenomena, the theory may assist us in better understanding human behavior in all of its manifestations."

I don't question either the widespread invocation or the extensive utility of the metaphorical linkage, and I list elsewhere (pp. 297–300) a range of such invocations across disciplines from economics to cartooning to guidelines for the self-help movement. But in discussing the application of punctuated equilibrium to other disciplines, I am more interested in exploring ways in which the theory might supply truly causal insights about other scales and styles of change, based on conceptual and structural "homologies" (as defined and discussed on pp. 236–240), rather than broader metaphors that can surely nudge the mind into productive channels, but that make no explicit claim for causal continuity or unification. Thus, in discussing the influence of punctuated equilibrium upon other disciplines, I will focus upon two kinds of potentially homological proposals.

First, where authors proceed beyond simple claims for broad similarities in jerky tempos of change to identify additional and explicit overlaps in the set of collateral principles that I called "conjoints" (see p. 239) in defining conceptual homology vs. analogy—including, for example, (1) claims that link punctuations to the origin of discretely individuated units arising by branching (a conceptual homolog of speciation), (2) discussions of the difference between punctuational and saltational modes, and (3) proposals about active causes for the maintenance of stasis. And second, where authors use the similarities between punctuated equilibrium and punctuational tempos in their own discipline to advance more than vaguely metaphorical suggestions for general theories about the nature of change in systems that may be said to "evolve," and to display historical continuity.

Examples From the History of Human Artifacts and Cultures

I presented arguments for punctuational models of human biological evolution in a previous section (pp. 212–222). But I have also been struck by the frequency of punctuational explanations advanced for patterns in the development of human artifacts and cultural history, processes that must "evolve" under causes and mechanisms quite different from the genetic variation and natural selection that regulate our Darwinian biology. Moreover, the Lamarckian character of human cultural change—the inheritance by teaching of useful innovations acquired during the life of an inventor—provides an entirely plausible mechanism for a more accumulative, progressive and gradual style of change in this realm than the Darwinian character of physical evolution (and the explicit denial of Lamarckian effects) should permit for the history of our anatomical changes. Thus, the discovery of punctuational patterns in cultural change might be viewed as even more surprising than the application of punctuated equilibrium to our morphological evolution.

For example, although more gradual and accumulative change may prevail in the history of tools following the origin of *Homo sapiens* (and, one presumes, a markedly increased capacity for cultural transmission), many scholars have noted, usually with surprise, a remarkable lack of change in the *Homo erectus* tool kit during more than a million years. For example, Mazur (1992, p. 229; see also Johanson and Edey, 1981, and Roe, 1980) states: "the early tool cultures were remarkably stable over long periods of time. The constancy of the Acheulean hand-ax tradition has been especially noted, for hand axes have been found at sites widely separated in distance and across a million years of *Homo erectus*'s existence look very similar to one another, their uniformity more striking than regional differences." Such collateral data support a view of *Homo erectus* as an individuated biological species, an entity rather than an arbitrarily defined segment of a continuity in anagenetic advance.

The history of scholarly research on European Paleolithic cave art provides an especially interesting example of how belief in progressivistic and gradual anagenesis can operate as a limiting preconception, and how punctuated equilibrium can play a salutary role as a potential corrective, or at least as a source of novel hypotheses for consideration. No aspect of the prehistory of artifacts has stunned or moved modern humans more than the parietal (wall) art of the great caves of Lascaux, Altamira, and many others, with their subtle and beautiful animal paintings that establish an immediate visceral link of aesthetic equality between the anonymous prehistoric artists and a Leonardo or Picasso. At least by standards of human history, these caves span a considerable range of time, from Chauvet at greater than 30,000 years BP (by radiocarbon evidence) to several at about 10,000 years of age.

Unsurprisingly, all great scholars of cave art have wanted to learn if any "evolution" can be discerned in the temporal sequence of these images. Two preeminent scientists built sequential schools of thought that virtually define the intellectual history of this subject in the 20th century. These men, the Abbé Henri Breuil and André Leroi-Gourhan, shared a firm belief that gradualistic evolution through a series of progressive stages provides a primary organizing theme for the history of parietal art (see Gould, 1998, for an epitome of their beliefs), even though, in other philosophical respects, their worldviews could not have been more different. (In fairness, no techniques of absolute dating were available to these scientists, so they used the traditional methods of art history in attempting to establish chronology by a series of developmental stages. But for human art in historical times, we can back up such aesthetic theories with known dates of composition, so the argument does not become intrinsically circular.)

The Abbé Breuil viewed the paintings functionally as a form of hunting magic (if you draw them properly, they will come). He accepted the linearly progressivist view of evolution that his late-19th-century education had inculcated, and that his religious convictions about human perfectability also favored. He therefore conceived the chronology of cave art as a series in styles of improvement, leading to rigidification and a final "senile" decline. In an early article of 1906, he wrote: "Paleolithic art, after an almost infantile beginning, rapidly developed a lively way of depicting animal forms, but didn't perfect its painting techniques until an advanced stage."

Leroi-Gourhan, a devoted follower of Lévi-Strauss and his functionalist school, embraced the opposite concept that cave art embodies timeless and integrative themes of human consciousness, based on dichotomous divisions that define our innate mental style of ordering the complex world around us. Thus, we separate nature from culture (the raw vs. the cooked in Lévi-Strauss's famous metaphor), light from darkness, and, above all, male from female. Leroi-Gourhan therefore read the caves as sanctuaries where the numbers and positions of animals (with, for example, horses as male symbols, and bisons as female) reflected our unchanging sense of natural order based on a primary sexual dichotomy—with an appended set of symbolic, and similarly dichotomized, attributes, including the conventional and sexist active vs. passive, and rational vs. emotional.

Given his view of cave art as representing the unchanging structure of human mentality, Leroi-Gourhan might have emphasized an implication of stasis for the duration of this form of expression. In fact, matching Breuil in commitment to the notion of gradualistic progress, Leroi-Gourhan contrasted a stability in conceptual intent with continual improvement in fidelity of artistic rendering for images of unchanged significance—that is, a gradual progression in overt phenotypes (the only aspect of change that an "evolutionist" might note and measure) contrasted with a constancy in symbolic meaning. Leroi-Gourhan wrote in 1967: "The theory . . . is logical and rational: art apparently began with simple outlines, then developed more elaborate forms to achieve modeling, and then developed a polychrome or bichromate painting before it eventually fell into decadence."

This scenario sounds eminently reasonable until one subjects the argument to further scrutiny, with an explicit effort to identify and question gradualistic biases. After all, we are not examining a lineage of enormous geological extent spanning a range of phenotypic complexity from amoeba to mammal, or even from one species to another. We are tracing about

20,000 years in the history of a single species, *Homo sapiens*, that remained anatomically stable throughout this time. Of course, cultural achievements can accumulate progressively while Darwinian biology remains unaltered. And, of course again, we assume that the first person who ever took ochre to wall could not render a mammoth with all the subtlety developed by later artists; some substantial learning and development must have occurred. But then, the earliest known cave paintings do not record these initial steps, for our oldest data probably represent a tradition already in full flower—so that we observe, in the total range now available to us, something analogous to the history of Western art from Phidias to Picasso (with much change in style, but not directional progress), not the full sweep from the first hominid who ever pierced a hole in a bear tooth and then strung the object around his neck, to the Desmoiselles of Avignon. Why then should we ever have anticipated a linear sequence of change in the known history of Paleolithic parietal art?

Indeed, and to shorten a longer story, the discovery and dating of Chauvet cave in 1994, abetted by improvements in radiocarbon methods that provide accurate results from tiny samples, have now disproven the controlling gradualist and progressivist assumption in an entire tradition of research. The paintings at Chauvet, dated as the oldest of all known sites (30,000 to 34,000 years BP), include all features previously regarded as identifying the highest, and latest, stage of achievement in a sequence of increased artistry (as found in the much younger caves of Lascaux and Altamira). In other words, the full range of styles extended throughout the entire interval of dated caves, with the most sophisticated forms fully present at the oldest site now known.

Bahn and Vertut (1988) invoked punctuated equilibrium in their prescient anticipation of the disproof that would soon follow. They also made an astute argument—based on punctuated equilibrium's concept of species as discrete individuals with considerable capacity for spatial variation at any one time vs. the tendency of anagenetic thinking to regard the phenotype of any moment as a uniform stage in a temporal continuum—that geographic variation, in itself, should have precluded any expectation of a simple chronological sequence, even if a general trend did pervade the entire series. After all, why should areas as distant as southern Spain, northeastern France and southeastern Italy go through a series of progressive stages in lockstep over 20 thousand years? Regional and individual variation can swamp general trends, even today in our globally connected world of airplanes and televisions. Why did we ever think that evolution should imply a pervasive signal of uniform advance? Bahn and Vertut (1988) write:

> The development of Paleolithic art was probably akin to evolution itself: not a straight line or ladder, but a much more circuitous path—a complex growth like a bush, with parallel shoots and a mass of offshoots; not a slow, gradual change, but a "punctuated equilibrium," with occasional flashes of brilliance . . . Each period of the Upper Paleolithic almost certainly saw the coexistence and fluctuating importance of a number of styles and techniques, . . . as well as a wide range of talent and ability . . . Consequently, not every apparently "primitive" or "archaic" figure is necessarily old . . . and some of the earliest art will probably look quite sophisticated.

A similar reconceptualization and corrective, for a more restricted region at a smaller scale of centuries rather than millennia, has been offered, citing punctuated equilibrium as a source of ideas, in Berry's (1982) treatise on the history of the Anasazi people of western North America. Berry treats the Anasazi as a geographically variable cultural entity, in many ways akin to a biological species under punctuated equilibrium, and not, as in most previous writing, as a group in continuous flux, with nearly all variation expressed temporally. Eldredge and Grene (1992, p. 118) write of Berry's work:

> Rather than interpreting the pattern as a linear history, in which change sometimes occurred rapidly and at other times at a more leisurely pace, Berry argues that the patterns of stasis interrupted by spurts of rather profound cultural change do not represent linear evolution, but rather a sequence of habitation and replacement. The Anasazi are a historical whole, as regionally diverse and as temporally modified as they were. They were replaced by another cultural system, not as a smooth evolutionary outgrowth but because the Anasazi were eventually (and rather abruptly) no longer able to occupy their territory.

Several social scientists have used the model of punctuated equilibrium as a guide to reconstructing patterns in social and technological development as punctuationally disrupted and then reformulated, rather than gradually altering—as in Weiss and Bradley (2001) on climatic forcing as a prod to rapid societal collapse in early civilizations throughout the world, over several millennia of time and types of organization. Adams (2000) has generalized this argument about "accelerated technological change in archaeology and ancient history." He explicitly cites the Lyellian tradition as a former impediment to recognizing and resolving such social punctuations (2000, pp. 95–96): "Built into traditional Darwinian 'descent with modification' was an acceptance of the standpoint of Lyell's geological gradualism. In its time, his assumption of uniformitarian change in the earth's geological history carried the day against competing doctrines of catastrophism. Today, however, there is increasing recognition of great diversity in rates of evolutionary change . . . Accelerated phases of change, often referred to in evolutionary biology as punctuations, invite closer study by students of human as well as natural history."

Finally, as a generality for the key transition to agriculture that marks (through the accumulation of wealth leading to social stratification, and the initiation of fixed-placed dwelling leading to towns and cities) the multiple inception of what, for better or worse, we generally call "civilization"—Boulding (1992, p. 181) cites active stasis and rapid punctuation as the predominant pattern, in opposition to a uniformitarian tradition most famously promulgated in Alfred Marshall's *Principles of Economics,* one of the most influential textbooks ever written, and a volume that, through decades and numerous editions, bore on its title page the Leibnizian, Linnaean, and Darwinian maxim: *Natura non facit saltum.* Boulding writes: "In the economy we certainly find periods of relative stability, in which society is getting neither much richer nor much poorer, but these periods of stability do seem to be punctuated by periods of very rapid economic development. The transition from hunting-gathering societies to agriculture at any particular locality seems to set off a period of rapid economic growth. This transition was usually rather rapid and, it would seem, irreversible."

Examples From Human Institutions and Theories about the Natural World
If relatively prolonged periods of actively maintained stability, followed by episodic transition to new positions of repose, mark the most characteristic style of change across nature's scales, and if we have generally tried to impose a gradualistic and progressivistic model of change upon this different reality, then we must often face anomalies that engender confusion and frustration in our personal efforts to improve our lives or to master some skill. To cite two mundane examples from my own experience, I spent several, ultimately rather fruitless years learning to play the piano. Whenever I tried to master a piece, I would become intensely frustrated at my minimal progress for long periods, and then exhilarated when everything "came together" so

quickly, and I could finally play the piece. I also liked to memorize long passages of poetry and great literature, primarily Shakespeare and the Bible, an activity then practiced and honored in the public primary and secondary schools of America. I would get nowhere forever, or so it seemed—and then, one fine day, I would simply know the entire passage.

Only years later—and perhaps serving as a spur to my later interest in punctuated equilibrium—did I conceptualize the possibility that plateaus of stagnation and bursts of achievement might express a standard pattern for human learning, and that my previous frustration (at the long plateaus), and my exhilaration (at the quick and rather mysterious bursts), might only have reflected a false expectation that I had carried so long inside my head—the idea that every day, in every way, I should be getting just a little bit better and better.

I don't know that explicit instruction in the higher probability of punctuational change, and the consequent appeasement of frustration combined with a better understanding of exhilaration, would improve the quality of our lives. (For all I know, the frustration and exhilaration yield important psychological benefits that outweigh their inadequate mapping of nature.) But I do suspect that a general recognition of the principles of punctuational change—leading us to understand that learning generally proceeds through plateaus of breakthroughs, and that important changes in our lives occur more often by rapid transition than by gradual accretion—might provide some distinct service in our struggles to fulfill the ancient and honorable Socratic injunction: know thyself.

I also think that an explicit application of punctuational models to many aspects of change in human institutions and technologies might improve our grasp and handling of the social and political systems that surround and include us. For example, in a stimulating paper emanating from her own research on "project groups" formed to study and initiate organizational change, Gersick (1991) explored the commonalities of punctuational change at six distinct levels of increasing scope—in the lives of individuals, the structures of groups (her own work), the history of human organizations, the history of ideas, biological evolution (our work on punctuated equilibrium), and general theory in physical science (Prigogine on bifurcation theory). Her paper, published in the *Academy of Management Review*, bears the title, "Revolutionary change theories: a multilevel exploration of the punctuated equilibrium paradigm," and begins by stating: "Research on how organizational systems develop and change is shaped, at every level of analysis, by traditional assumptions about how change works. New theories in several fields are challenging some of the most pervasive of these assumptions, by conceptualizing change as a punctuated equilibrium: an alternation between long periods when stable infrastructures permit only incremental adaptations, and brief periods of revolutionary upheaval." (See also Wollin, 1996, on the utility of punctuated equilibrium for resolving the dynamics of growth of complex human and social systems in general: "A hierarchy based approach to punctuated equilibrium: an alternative to thermodynamic self-organization in explaining complexity.")

In her own work on "task groups," Gersick (1988) reached a surprising conclusion. These associations did not proceed incrementally towards their assigned goals, but rather tended to hem, haw and dither until they reached a particular, and temporally definable, point of quick transition towards a solution. "Project groups," she writes (1991, p. 24), "with life spans ranging from one hour to several months reliably initiated major transitions in their work precisely halfway between their start-ups and expected deadlines. Transitions were triggered by participants' (sometimes unconscious) use of the midpoint as a milestone, signifying 'time to move.'"

These particular results inspired Gersick's more general consideration of punctuational models. In so doing, she explicitly follows both approaches previously advocated (p. 236) as strategies for transcending metaphor and discovering causally meaningful connections among punctuational phenotypes of change across levels and disciplines: the identification of "conjoints," or properties correlated with the basic punctuational pattern (the basic strategy of documenting a complexity in number and interaction of parts too high to attribute to causally accidental resemblance, and therefore necessarily based on homology); and the proposal of a general rationale, transcending the particular of any scale of analysis or class of objects, for the punctuational character of change.

For example, Gersick emphasizes the need for active resistance towards change as a validation of stasis, and she makes the same link that Eldredge and I have stressed for biological evolution between punctuational change in hierarchical systems and structural constraint viewed as partly contrary to an adaptationist paradigm. She writes (1991, p. 12): "Gradualist paradigms imply that systems can 'accept' virtually any change, any time, as long as it is small enough; big changes result from the insensible accumulation of small ones. In contrast, punctuated equilibrium suggests that, for most of systems' histories, there are limits beyond which change is actively prevented, rather than always potential but merely suppressed because no adaptive advantage would accrue."

Gersick's lists of commonalities among her six levels, both for periods of stasis and for episodes of punctuation, satisfy the strategy of conjoints, while her ranked list of scales, and especially her linkages of particular categories to the two most overarching theories of punctuational change—Kuhn's for human thought and Prigogine's for the natural world—meet the criterion of generalization. For example, her chart of comparison among the six levels for "equilibrium periods" cites both commonalities and conjoints, with an interestingly different emphasis (from our concerns with biological systems) upon the potential for strong limitation placed upon incremental pathways within a plateau—an important theme for the Lamarckian character of human cultural change. She lists as "commonalities" (p. 17): "During equilibrium periods, systems maintain and carry out the choices of their deep structure. Systems make adjustments that preserve the deep structure against internal and external perturbations, and move incrementally along paths built into the deep structure. Pursuit of stable deep structure choices may result in behavior that is turbulent on the surface."

Similarly, her chart for punctuational episodes stresses the unpredictability and potential nonprogressionism of outcomes, a surprising theme for human systems based on supposed and explicit goals, but a notion that we did not emphasize in formulating punctuated equilibrium because biological evolution proceeds in a highly contingent manner for so many other reasons (see Gould, 1989a), some recognized and emphasized by Darwin himself. Thus, this important theme, while equally central within the structure of our theory, did not have similar salience for us—and we thank Gersick for her insight and generalization. Gersick writes in her heading (p. 20): "Revolutions are relatively brief periods when a system's deep structure comes apart, leaving it in disarray until the period ends, with the 'choices' around which a new deep structure forms. Revolutionary outcomes, based on interactions of systems' historical resources with current events, are not predictable; they may or may not leave a system better off. Revolutions vary in magnitude."

As another example of fruitful borrowing across disciplines, Mokyr (1990, p. 350) begins his study of technological change by noting that Alfred Marshall's advocacy of gradualism in economics played a similar role in the human sciences to Darwin's impact upon the natural sciences: "Charles Darwin and Alfred Marshall were both extremely influential men . . . Darwin and Marshall both believed that nature does not make leaps. Both were influenced by a long and venerable tradition that harked back to Leibniz rooted in the Aristotelian notion of the continuity of space and time." (On the same subject, see also Loch, 1999, on "A punctuated equilibrium model of technology diffusion.")

The development of improved systems for human communication in the past century must represent one of the most goal-directed and clearly progressionist sequences identifiable in either the human or natural sciences—hence the apparent consonance with gradualist models, and the unpromising character, at first glance, of punctuationalist alternatives. But Mokyr defends a punctuational reformulation by centering his descriptions and explanations on a comparison of stages in this technological trend with the discrete origin of biological species as genuine entities, and then citing punctuated equilibrium for emphasizing this theme as a reform within Darwinian biology (Mokyr, 1990, p. 351).

While not denying the clearly goal-oriented and progressive nature of the trend—and while also (along with Gersick) noting the Lamarckian capacity of human culture to change directionally and incrementally within plateaus, but also stressing the qualitative differences between this limited gradualism and the much larger and quicker transitions of goal-directed punctuations—Mokyr (p. 354) describes the basic outline of this history as four stages separated by punctuational breaks. He also, again as with Gersick, stresses the nonpredictability of outcomes, and then notes, continuing the analogy with speciation, that punctuational models enjoin the study of particular conditions favorable to leaps of change, an issue that does not arise in explanations based upon gradualistic anagenesis:

> Long-distance communications thus illustrate the abruptness of technological change. There was no natural transition from semaphore to the electrical telegraph, nor a gradual movement from the telegraph to the first radio transmission by Marconi in 1894, nor a smooth natural development from the long-wave radio used in the first thirty years to the shortwave systems of the later 1920s. Each of the three systems was subsequently perfected by a long sequence of microinventions, but these would not have occurred without the initial breakthroughs. Their concept was novel, they made things possible that were previously impossible, and they were pregnant of more to come. Therein lies the essence of a macroinvention. . . . Many macroinventions, just like the emergence of species, were the result of chance discover-

ies, luck, and inspiration. Biologists agree that certain environments are more conducive to speciation than others.

Just as for technological change, we tend to view the history of scientific ideas on particular subjects as, in principle, the most incrementally progressionist of all human activities by the empiricist paradigm of ever-closer approximation to natural truth through objective accumulation of data under unchanging principles of "the scientific method"—an idea famously challenged by Kuhn (1962) in the most influential punctuationalist theory in 20th-century scholarship. Thus, punctuationalist reformulations in this domain tend to strike people as especially surprising. The distinguished Dutch petrologist E. Den Tex (1990), in summing up his career of study on the nature of granites, credited punctuated equilibrium with reorganizing his lifelong attempt to make sense of a complicated history, stretching back to the late 18th century (the neptunist-plutonist debate between the schools of Werner and Hutton), and featuring fluctuations between poles of two dichotomies (sedimentary vs. igneous formation, and recruitment from deep magmas vs. transformation from existing crustal rocks), complicated by shifting allegiances and amalgamations, followed by breakages, of separable aspects of all end-member theories.

In his fascinating and highly personal paper, entitled "Punctuated equilibria between rival concepts of granite genesis," Den Tex (1990) notes that he had first tried to apply other models of noncontinuous and progressive change, especially the celebrated Hegelian notion of successive syntheses reached by opposition between a thesis and its antithesis. He then found a better general description, with new and fruitful hints for explanation, in our model of punctuated equilibrium, particularly in the parsing of history as discrete steps, analogous to individuated species with definite sets of properties—a process "in dynamic equilibrium . . . punctuated from time to time by allopatric speciation, *i.e.* by rapid, random, discrete steps taking place in locations isolated from the main stem" (1990, p. 216).

Finally, since punctuated equilibrium arose as a theory about change in the natural world (not in the history of human understanding thereof), Eldredge and I have been gratified by the utility of our theory in suggesting structurally homologous modes of change in other branches of natural science. I have been especially pleased by geological examples distant from our own paleontological concerns, because no other field can match anglophonic geology—resulting largely from the legacy of Lyellian uniformitarianism (see *SET*, Chapter 6)—in explicit, and often exclusive, fealty to strictly gradualistic models.

Lawless (1998, and a good name for iconoclasts), in an article entitled "Punctuated equilibrium and paleohydrology," notes the hold that gradualistic models have imposed on the history of ideas about hydrothermal ore deposits, particularly of gold. He begins by expressing a paradox: if ores accumulate gradually in such systems, and given the average amount of gold carried in most percolating waters, minable deposits should be much more common—indeed almost ubiquitous in hydrothermal systems that persist for at least 25,000 years. But the much rarer distribution of such deposits suggests to Lawless that periods of accumulation must be limited to brief episodes "which cause vigorous boiling through a restricted volume of the reservoir" (p. 165). Lawless views the general history of hydrothermal systems—including the development of ore deposits as just one feature among many—as punctuational, and caused by rapid, intense, rarely-acting forces: "Such disturbances caused by tectonic activity, magmatism, volcanic activity, erosion, climatic changes or other processes may occur at long intervals, but be responsible for producing some of the most significant characteristics of the system, including economic mineral deposits."

As a matter of potential practical importance, Lawless recognizes that, just as punctuated equilibrium must not be construed as an argument against predictable trends, but rather as a different mechanism for the episodic production of such trends, so too might the punctuational origin of hydrothermal ore deposits be reconceptualized as directional but episodic. He writes (p. 168): "Recognition of the quasi-cyclic and episodic nature of these events within the lifetime of a hydrothermal system could be described as leading to more 'catastrophic' models. There are similarities to the concepts of punctuated equilibrium recently proposed in paleontology and biological evolution . . . these concepts emphasize the importance of specific events which are of random occurrence on a short time scale but statistically predictable on a longer scale."

Two Concluding Examples, a General Statement, and a Coda

As final examples in this chapter, two recent authors have used punctuated equilibrium as the central organizing principle for books on subjects of different scale, but of great importance in human life and history—Kilgour (1998) on *The Evolution of the Book*, and Thurow (1996) on *The Future of Capitalism*. Moreover, each author uses punctuated equilibrium not as a vague metaphor, but as a specific model of episodic change offering casual insights through the identification of structural homologies as defined in this chapter.

Kilgour notes that a theme of greater efficiency marks the history of bookmaking (and might be misread as evidence for anagenetic gradualism, just as trends in the evolutionary history of clades have often been similarly misconstrued when a punctuational model of successive plateaus defined by discrete events of branching actually applies). He writes (p. 4): "Form aside, the major change throughout the entire history of the book has been in the continuous increase in speed of production: from the days required to handwrite a single copy, to the minutes to machine-print thousands of copies, to the seconds to compose and display text on an electronic screen."

But, as Kilgour knows, and adopts as the major theme of his book, form cannot be put "aside." When one probes through these progressive improvements in function to underlying bases in form, the history of the book becomes strongly punctuational. In a pictorial summary for his central thesis (Fig. 5-8), Kilgour views the evolution of the book—defined (p. 3) as "a storehouse of human knowledge intended for dissemination in the form of an artifact that is portable, or at least transportable, and that contains arrangements of signs that convey information"—as a sequence of four great punctuations: the clay tablet, the papyrus roll, the codex (modern book), and the electronic "book" (with no canonical form as yet since we are now enjoying, or fretting our way through, the rare privilege of living within a punctuation), with three "subspeciational" punctuations within the long domination of the codex (Gutenberg's invention of printing with movable type in the mid 15th century, and the enormous additional increases in production made possible first by the introduction of steam power at the beginning of the 19th century, and then by the development of offset printing in the mid 20th century).

As I have emphasized throughout this discussion of human cultural and technological change, the Lamarckian nature of inheritance for these processes permits more directional accumulation within periods of overall stasis in basic design than analogous chronologies of biological evolution can probably exhibit. Thus, while the codex (that is, the familiar "bound book") enjoyed its millennium and a half of domination in fundamentally unchanged form, several innovations both in design (the introduction of pagination and indexes) and in human practice and collateral discovery (the invention of eyeglasses at the end of the 13th century, and the spread of the "newfangled" practice of silent reading in the 15th century) greatly expanded the utility of a product that remained stable in form. (While books remained scarce, people read aloud and, apparently, did not even imagine a possibility that seems obvious to us—namely, that one might read without speaking the words. By reading aloud, one copy could be shared with many, but silent reading demands a copy for each participant.)

But the four great designs (or at least the three we know, for the fourth has not yet stabilized) have experienced histories strikingly akin, in more than vaguely metaphorical ways, to the origin and persistence of biological species treated as discrete individuals. In the first of three major similarities,

Clay Tablet	**First Punctuation**	**2500 BC**
Papyrus Roll	**Second Punctuation**	**2000 BC**
Codex	**Third Punctuation**	**AD 150**
Printing	**Fourth Punctuation**	**1450**
Steam Power	**Fifth Punctuation**	**1800**
Offset Printing	**Sixth Punctuation**	**1970**
Electronic Book	**Seventh Punctuation**	**2000**

5-8. Kilgour, in his 1998 work on the evolution of the book, presents a punctuational model of clay tablet, to papyrus roll, to codex, to printing, and onward to the undetermined future of the electronic book. All major features of the biological model apply here in adequate isomorphism for causal insight, including the survival of ancestors after the branching of descendants.

each principal form persisted in effectively unchanged design for long periods of time by standards of human technological innovation. Moreover, each transition introduced a great improvement by solving an inherent structural problem in the previous design. Therefore, the extended persistence of each flawed design illustrates an important reason, more "environmental" than structural, for the existence of stasis in natural systems: the advantages of incumbency. "The extinction of clay tablets," Kilgour writes (pp. 4–5) "was ensured by the difficulty of inscribing curvilinear alphabet-like symbols on clay"; while "the need to find information more rapidly than is possible in a papyrus-roll-form book initiated the development of the Greco-Roman codex in the second century A.D." Of this predominant stasis, Kilgour writes (p. 4): "Extremely long periods of stability characterize the first three shapes of the book; clay tablets and papyrus-roll books existed for twenty-five hundred years, and the codex for nearly two thousand years. An Egyptian of the twentieth century B.C. would immediately have recognized, could he have seen it, a Greek or Roman papyrus-roll book of the time of Christ; similarly, a Greek or Roman living in the second century A.D. who had become familiar with the then new handwritten codex would have no trouble recognizing our machine-printed book of the twentieth century."

Secondly, the successive stages do not specify segments of an anagenetic flow (whatever the punctuational character of each introduction), but rather arise as discrete forms in particular areas—thus following the pattern of branching speciation so vital to the validation of punctuated equilibrium, and also meeting the chief operational criterion for distinguishing punctuated equilibrium from punctuated anagenesis: the survival of ancestral forms after the origin of new species. Kilgour notes (p. 158) that "clay tablets and papyrus-roll books coexisted for two thousand years, much as two biological species may live together in the same environment." He also notes, both wryly and a bit ruefully, "that books on paper and books on electronic screens, will, like clay tablets and papyrus books, coexist for some time, but for decades rather than centuries" (p. 159).

Third, each form—at least before improved communication of the past two centuries made such localization virtually inconceivable—originated in a particular time and place, and in consonance with features of the immediately surrounding environment, a meaningful analog to the locally adaptive origin of biological species. Kilgour writes (p. 4): "the Sumerians invented writing toward the end of the fourth millennium B.C. and from their ubiquitous clay developed the tablet on which to inscribe it. The Egyptians soon afterward learned of writing from the Mesopotamians and used the papyrus plant, which existed only in Egypt, to develop the papyrus roll on which to write."

I know Lester C. Thurow as a colleague from another institution in Cambridge, MA, but I had never discussed punctuated equilibrium with him, and was surprised when he used our theory as one of two defining metaphors in his book, *The Future of Capitalism.* In distinction to most examples in this chapter, Thurow does invoke punctuated equilibrium in a frankly metaphorical and imagistic manner, but he also shows a keen appreciation for the conjoints of punctuated equilibrium applied to his subject of macroeconomics. Thurow writes in the context of the collapse of communism in the Soviet Union, and in the survival of capitalism as a distinctive, effectively universal, and perhaps uniquely workable system of human economic organization within a highly technological matrix. Yet capitalism now faces a crisis of reorganization in a world where major change (both natural and social) occurs by punctuated equilibrium and not by slow incrementation. Thurow therefore presents two metaphors from the natural sciences to ground his argument: "To understand the dynamics of this new economic world, it is useful to borrow two concepts from the physical sciences—plate tectonics from geology and punctuated equilibrium from biology" (p. 6).

In an obviously imagistic metaphor with no causal meaning, Thurow identifies the five "economic tectonic plates" that will incite the next punctuation by crunching and grinding as their rigid borders crash: "the end of communism," "the technological shift to an era dominated by man-made brainpower industries," "a demography never before seen" (increasing average age, greater movement of populations to cities, etc.), "a global economy," and "an era where there is no dominant economic, political or military power" (pp. 8–9).

But his invocation of punctuated equilibrium shows more structural and causal connection to the "parent" phenomenon from the natural sciences. Thurow (p. 7) notes both the long plateaus and the tendency for rapid historical shifts in transitions between macroeconomic systems that organize entire societies:

> Periods of punctuated equilibrium are equally visible in human history. Although they came almost two thousand years later, Napoleon's armies could move no faster than those of Julius Caesar—both depended upon horses

> and carts. But seventy years after Napoleon's death, steam trains could reach speeds of over 112 miles per hour. The industrial revolution was well under way and the economic era of agriculture, thousands of years old, was in less than a century replaced by the industrial age. A survival-of-the-fittest social system, feudalism, that had lasted for hundred of years was quickly replaced by capitalism.

More notably, and marking Thurow's fruitful use, he stresses important conjoints of punctuated equilibrium as the most relevant—and practical—themes for our current and dangerous situation. First, the common phenotype of punctuation leads him to recognize that general structural rules must underlie both the maintenance of stasis and, through their fracturing, the episodes of punctuation as well. The rules will differ in social and natural systems, but the general principle applies across domains. Thurow argues that stasis requires a meshing of technology and ideology, while their radical divergence initiates punctuation, a situation that we face today. (Marx, in an entirely different context, held a similar view about both the character of the rule and the punctuational outcome.)

> Technology and ideology are shaking the foundations of twenty-first-century capitalism. Technology is making skills and knowledge the only sources of sustainable strategic advantage. Abetted by the electronic media, ideology is moving toward a radical form of short-run individual consumption maximization at precisely a time when economic success will depend upon the willingness and ability to make long-run social investments in skills, education, knowledge, and infrastructure. When technology and ideology start moving apart, the only question is when will the "big one" (the earthquake that rocks the system) occur. Paradoxically, at precisely the time when capitalism finds itself with no social competitors—its former competitors, socialism or communism, having died—it will have to undergo a profound metamorphosis (p. 326).

Second, Thurow continually stresses the contingency implied by the scale change between causes of alteration in "normal" times (with greater potential and range in social than in natural systems due to the Lamarckian character of cultural inheritance) and the different modes and mechanisms of punctuational episodes. Thus, the rules we may establish from our experience of ordinary times cannot predict the nature or direction of any forthcoming punctuation.

> In a period of punctuated equilibrium no one knows that new social behavior patterns will allow humans to prosper and survive. But since old patterns don't seem to be working, experiments with different new ones have to be tried (p. 236) . . . How is capitalism to function when the important types of capital cannot be owned? Who is going to make the necessary long-run investments in skills, infrastructure, and research and development? How do the skilled teams that are necessary for success get formed? In periods of punctuated equilibrium there are questions without obvious answers that have to be answered (p. 309).

Thurow therefore stresses the virtues of flexibility, ending his book (pp. 326–327) with an appropriate image for "the period of punctuated equilibrium" that "the tectonic forces altering the economic surface of the earth have created":

> Columbus knew that the world was round, but he . . . thought that the diameter of the world was only three quarters as big as it really is. He also overestimated the eastward land distance to Asia and therefore by subtraction grossly underestimated the westward water distance to Asia . . . Given the amount of water put on board, without the Americas Columbus and all his men would have died of thirst and been unknown in our history books. Columbus goes down in history as the world's greatest explorer . . . because he found the completely unexpected, the Americas, and they happened to be full of gold. One moral of the story is that it is important to be smart, but that it is even more important to be lucky. But ultimately Columbus did not succeed because he was lucky. He succeeded because he made the effort to set sail in a direction never before taken despite a lot of resistance from those around him. Without that enormous effort he could not have been in the position to have a colossal piece of good luck.

To end on a more personal note, if I were to cite any one factor as probably most important among the numerous influences that predisposed my own mind toward joining Niles Eldredge in the formulation of punctuated equilibrium, I would mention my reading, as a first-year graduate student in 1963, of one of the 20th century's most influential works at the interface of philosophy, sociology and the history of ideas: Thomas S. Kuhn's *The Structure of Scientific Revolutions* (1962). (My friend Mike Ross, then studying with the eminent sociologist of science R. K. Merton in the building next to Columbia's geology department, ran up to me one day in excitement, saying "you just have to read this book right away." I usually ignore such breathless admonitions, but I respected Mike's judgment, and I'm surely glad that I followed his advice. In fact, I went right to the bookstore and bought a copy of Kuhn's slim volume.)

Of course, Kuhn's notion of the history of change in scientific concepts advances a punctuational theory for the history of ideas—going from stable "paradigms" of "normal science" in the "puzzle solving" mode, through accumulating anomalies that build anxiety but do not yet force the basic structure to change, through rapid transitions to new paradigms so different from the old that even "conserved" technical terms change their meaning to a sufficient extent that the two successive theories become "incommensurable." The book has also served, I suspect, as the single most important scholarly impetus towards punctuational thinking in other disciplines.

Since the appearance of our initial paper on punctuated equilibrium in 1972, several colleagues have pointed out to me that Kuhn himself, in a single passage, used the word "punctuated" to epitomize the style of change described by his theory. These colleagues have wondered if I borrowed the term, either consciously or unconsciously, from this foundational source. But I could not have done so. (I do not say this in an exculpatory way—for if I had so borrowed, I would be honored to say so, given my enormous respect and personal affection for Kuhn, and the pleasure I take in being part of his intellectual lineage. We did, but in an entirely different discussion about the definition of paradigms, cite Kuhn's book in our 1972 paper.) Kuhn used the word "punctuated" in the 1969 "postscript" that he added to the second edition of his book (see quotation below). In 1972, I had only read the first edition.

I mention this final point not as pure self-indulgence, but largely because Kuhn's single use of the word "punctuated," located in the closing paragraphs of the seventh and last section (entitled "the nature of science") of his postscript, expresses a surprising opinion that seems eminently exaptable as an appropriate finale to this chapter. Like all scholars whose works become widely known through constantly degraded repetition that strays further and further from unread original sources, Kuhn could become quite prickly about fallacious interpretations, and even more perturbed by bastardized and simplistic readings that caricatured his original richness.

He therefore ended his postscript by discussing two "recurrent reactions" (1969, p. 207) to his original text. He regarded the first reaction (irrelevant to this chapter) as simply unwarranted, so he just tried to correct his critics.

But the second reaction bothered him more—as "favorable" (p. 207) to his work but, in his view, "not quite right" in granting him too much credit. He states that, whereas he had certainly hoped to catch the attention of scientists, he cannot understand why scholars in nonscientific disciplines should have found his work enlightening. After all, he claims (p. 208), he had only tried to explain the classical punctuational views of the arts and humanities to practitioners of another field—science—where social and philosophical tradition had long clouded this evident point. (The word "punctuated" appears in this passage):

> To one last reaction to this book, my answer must be of a different sort. A number of those who have taken pleasure from it have done so less because it illuminates science than because they read its main theses as applicable to many other fields as well. I see what they mean and would not like to discourage their attempts to extend the position, but their reaction has nevertheless puzzled me. To the extent that the book portrays scientific development as a succession of tradition-bound periods punctuated by noncumulative breaks, its theses are undoubtedly of wide applicability. But they should be, for they are borrowed from other fields. Historians of literature, of music, of the arts, of political development, and of many other human activities have long described their subjects in the same way. Periodization in terms of revolutionary breaks in style, taste, and institutional structure have been among their standard tools. If I have been original with respect to concepts like these, it has mainly been by applying them to the sciences, fields which had been widely thought to develop in a different way.

To this passage, I can only respond that I see what *he* means, but I also think that Kuhn undercuts the range of his own originality and influence by misreading the very social context that inspired his work. Indeed, the ethos of science—the conviction that our history may be read as an ever closer approach to an objective natural reality obtained by making better and better factual observations under the unvarying guidance of a timeless and rational procedure called "the scientific method"—does establish the most receptive context imaginable for mistaken notions about gradualistic and linear progressionism in the history of human thought. Indeed also, the traditions of disciplines that practice or study human artistic creativity—with their concepts of discrete styles and revolutionary breaks triggered by "genius" innovators—should have established punctuational models as preferable, if not canonical.

But, as Kuhn acknowledges, his book enjoyed a substantial vogue among artists and humanists, who also felt surprised and enlightened by his punctuational theory for the history of ideas. I think that Kuhn underestimated the "back influence" of science upon preexisting fields in humanistic study. Eiseley (1958) labeled the formative era of evolutionary theory as "Darwin's century," and we must never underestimate the influence of the Darwinian revolution, and of other 19th century notions, particularly the uniformitarian geology of Lyell, upon reconceptualizations of modes of discovery and forms of content in other disciplines, no matter how distant. Progressivistic gradualism—so central to most late 19th century versions of biological and geological history, and so strongly abetted by the appearance of "progress" (at least to Westerners) in the industrial and colonial expansion of Western nations, at least before the senseless destruction of the First World War shattered all such illusions forever—became a paradigm for all disciplines, not just for the sciences. Kuhn may have called upon some classical notions from the arts and humanities to construct a great reform for science; but his corrective also, and legitimately, worked back upon a source that had strayed from a crucial root idea to become beguiled by a contrary notion about change that seemed more "modern" and "prestigious."[4]

In addition, and finally, I think that Kuhn underestimated the potential role of scientific ideas in resolving old puzzles that have long stymied humanistic understanding of artistic creativity, and that remain seriously burdened by the hold of theories as ancient as the Platonic notion of essences and universals. In a lovely passage, directly following the "punctuational" quotation just cited, Kuhn acknowledges that the Darwinian concept of species as varying populations without essences (and even without abstractions, like mean values, to act as preferred or defining states) might break through a powerful and constraining prejudice, ultimately rooted in Platonic essentialism, that leads us to search for chimaerical idealizations as ultimate standards of comparison in the definition and evaluation of artistic "style." Kuhn writes, now viewing styles as paradigms and paradigms as higher level individuals like biological species (pp. 208–209): "Conceivably the notion of a paradigm as a concrete achievement, an exemplar, is a second contribution [of my book]. I suspect, for example, that some of the notorious difficulties surrounding the notion of style in the arts may vanish if paintings can be seen to be modeled on one another rather than produced in conformity to some abstracted canons of style."

Punctuated equilibrium represents just one localized contribution, from one level of one discipline, to a much broader punctuational paradigm about the nature of change—a worldview that may, among scholars of the new millennium, be judged as a distinctive and important movement within the intellectual history of the later 20th century. I am pleased that our particular formulation did gain a hearing and did, for that reason, encourage other scholars over a wide range of scientific and nonscientific disciplines (as illustrated in this book) to consider the larger implications of the more general punctuational model for change. I am especially gratified that many of these scholars did not just borrow punctuated equilibrium as a vague metaphor, however useful, but also understood, and found fruitful, some of the more specific "conjoints" distinctive to the level and phenomenon of punctuated equilibrium, but also applicable elsewhere. For the punctuational paradigm encompasses much more than a loose and purely descriptive claim about phenotypes of pulsed change, but also embodies a set of convictions about how the structures and processes of nature must be organized across all scales and causes to yield this commonality of observed results. Only in this sense—punctuated equilibrium as a distinctive contribution to a much larger and ongoing effort—can I understand Ruse's gracious reappraisal of his initial negativity toward punctuated equilibrium: "Grant then that there is indeed something going on that looks like a paradigm (or paradigm difference) in action. People (like my former self) who dismissed the idea were wrong—and missing something rather interesting to boot" (Ruse, 1992, p. 162).

From the more restricted perspective of the aims of this book and of the larger work, *The Structure of Evolutionary Theory*, I can at least assert that punctuated equilibrium unites three definitive themes—the three legs of my tripod of support for an expansion of Darwinian theory, thereby leading me to conclude that an empirically legitimate and logically sound structure does encompass and unite these three arguments into a coherent and general reformulation and extension of the Darwinian paradigm: the hierarchical theory of selection on leg one, the structuralist critique of Darwinian functionalism and adaptationism on leg two, and the paleontologist's conviction (leg three) that general macroevolutionary processes and mechanisms cannot be fully elucidated by uniformitarian extrapolation from the smallest scale of our experiments and personal observations. Punctuated equilibrium has proven its mettle in:

1. Elucidating and epitomizing what may be the primary process of a distinctive level in the evolutionary hierarchy: the role of species as Darwin-

ian individuals, and the speciational reformulation of macroevolution—for leg one.

2. Defining (and, in part, thereby creating) the issue of stasis as a subject for study, and in helping to explicate the structural rules that hold entities in active stasis at various levels, but then permit rapid transition to qualitatively different states—for leg two.

3. Stressing that level-bound punctuational breaks preclude the prediction or full understanding of extensive temporal change from principles of anagenetic transformation at the lowest level (a mode of evolution, moreover, that punctuated equilibrium regards as rare in any case), thus emphasizing contingency and denying extrapolationist premises and methodologies—for leg three.

In developing this set of implications, I do hold, in my obviously biased way, that punctuated equilibrium has performed some worthy intellectual service. The relative frequency of its truth value, of course, must be regarded as another matter entirely, and an issue that only time can fully resolve. But I would maintain that, in the quarter century following its original formulation, punctuated equilibrium has at least prevailed, against an initial skepticism of active and general force and frequency, in three central empirical claims (quite independent of any theoretical weight that evolutionary biology may ultimately wish to assign): (1) documentation of the basic mechanism in cases now too numerous and too minutely affirmed to deny status as an important phenomenon in macroevolutionary pattern; (2) validation of stasis as a genuine, pervasive, and active phenomenon in the geological history of most species; and (3) establishment of predominant relative frequency in enough comprehensive and well-bounded domains to assure the control of punctuated equilibrium over substantial aspects of the phyletic geometry of macroevolution. A fourth, and ultimately more important, issue for evolutionary theory remains unresolved: the implications of these empirical findings for the role of genuine selection among species-individuals (rather than merely descriptive species-sorting as an upwardly cascading expression of conventional Darwinian selection acting at the organismic level) as the causal foundation of macroevolutionary pattern.

Let me therefore end this book by restating the last paragraph of the review article for *Nature* that Eldredge and I wrote (Gould and Eldredge, 1993, p. 227) to celebrate the true majority, or coming of age—that is, the 21st birthday—of punctuated equilibrium. We wrote this paragraph to assess the role of punctuated equilibrium within a larger and far more general intellectual (and cultural) movement that, obviously, punctuated equilibrium did not create or even instigate, but that our theory didn't simply or slavishly follow either. We did, I think, contribute some terms and concepts to the larger enterprise, and we did encourage scholars in distant fields to apply a mode of thinking, and a model of change, that had formerly been as unconventional (or even denigrated) in their fields as in ours. But we do find ourselves in the paradoxical, and at least mildly uncomfortable, position—as we tried to express in these closing words of our earlier article—of having developed a theory with empirical power, and at least some theoretical interest, in its own evolutionary realm, but that must largely depend, for any ultimate historical assessment, upon the fate and efficacy of more general intellectual currents (including both dangerous winds of fashion and solid strata of documentation) well beyond our control of competence.

> In summarizing the impact of recent theories upon human concepts of nature's order, we cannot yet know whether we have witnessed a mighty gain in insight about the natural world (against anthropocentric hopes and biases that always hold us down), or just another transient blip in the history of correspondence between misperceptions of nature and prevailing social realities of war and uncertainty. Nonetheless, contemporary science has massively substituted notions of indeterminacy, historical contingency, chaos and punctuation for previous convictions about gradual, progressive, predictable determinism. These transitions have occurred in field after field. Punctuated equilibrium, in this light, is only paleontology's contribution to a *Zeitgeist*, and *Zeitgeists*, as (literally) transient ghosts of time, should never be trusted. Thus, in developing punctuated equilibrium, we have either been toadies and panderers to fashion, and therefore destined for history's ash-heap, or we had a spark of insight about nature's constitution. Only the punctuational and unpredictable future can tell.

APPENDIX

NOTES

BIBLIOGRAPHY

ILLUSTRATION CREDITS

INDEX

APPENDIX: A LARGELY SOCIOLOGICAL (AND FULLY PARTISAN) HISTORY OF THE IMPACT AND CRITIQUE OF PUNCTUATED EQUILIBRIUM

The Entrance of Punctuated Equilibrium into Common Language and General Culture

As a personal indulgence, after nearly 20 years' work on this book, I wish to present an unabashedly subjective, but in no sense either consciously inaccurate or even incomplete, account of the extra-scientific impact and criticism of punctuated equilibrium during its first quarter century. As extra-scientific, I include both the spread and influence of punctuated equilibrium into nonbiological fields and into general culture, and also the subset of opinions voiced by biological colleagues, that, in my judgment, are not based on logical or empirical argument, but rather on personal feelings spanning the gamut from appreciation to bitter jealousy and anger. I realize that such an effort, which I do regard as self-indulgent, may be viewed as unseemly by some colleagues. I would only reply, first—speaking personally—that I have, perhaps, earned the right after so much deprecation (matched or exceeded, to be sure, by a great deal of support); and second—speaking generally—that such an effort may have value for people interested in metacommentary upon science (historians, sociologists, scientific colleagues with an introspective bent), if only because few scientific theories garner so wide a spate of reactions, both popular and professional (and for reasons both worthy and lamentable). Moreover, the comments, while necessarily and admittedly partisan, of an originator of the theory (who has also kept a chronological file, as complete as he could manage, on the developing discussion) might have some worth as primary source material (in contrast with more objective, but secondary, analysis and interpretation). Therefore, I do not, in this section, include any overt discussion of the rich and numerous scientific critiques, issues, extensions, and arguments inspired by punctuated equilibrium. These subjects have already been treated in the main body of this book.

Needless to say, density and intensity of discussion bear no necessary correlation with the worth or validity of a subject; after all, many theological phenomena that have provoked wars, filled libraries, and consumed the lives of countless scholars, may not exist at all. Still, if only for naïve reasons, I take a generally hopeful view about human intelligence and discernment—at least to the extent of believing that when large numbers of thoughtful people choose to devote substantial segments of careers to the consideration of a new idea, this expenditure probably records the idea's genuine value and interest, and does not represent a pure snare or delusion.

Thus, above all else, I take pleasure in the perceived and expressed utility of punctuated equilibrium in altering a field that had largely languished in doldrums of little to do (as gradualism had defined the domain of recognized empirics for fine-scale evolution, and very few cases of paleontological gradualism could ever be documented)—and in providing an operational base for fruitful study by showing that the primary empirical signals of stasis and punctuation represented meaningful data on the tempo and mode of evolution, and not just a mocking signal from nature about the discouraging imperfection of the fossil record. The greatest success of punctuated equilibrium lies not in any torrent of words provoked by the theory, but in the volume of empirical study pursued under its aegis by paleontologists throughout the world.

While this professional debate unfolded in full force, the name and concept of punctuated equilibrium also moved from the scientific literature into general culture, at least on the intellectual edges, but often into more popular consciousness as well. Consider five categories recording this spread:

1. ESTABLISHMENT OF THE NAME IN DICTIONARIES AND LITERATURE. Punctuated equilibrium has won entry to latest editions of standard general dictionaries of the English language, including the Addenda to *Webster's Third New International Dictionary* (1986), where it shares a page with such other neologisms as *psychedelic, psychobabble, pump iron, putz, quark, rabbit ears* (which, as a name for those old indoor TV antennas, will no doubt pass quickly into oblivion along with the item itself), and *race walk-*

ing. I congratulate *Webster's* on a better and more accurate definition than many professional colleagues have misdevised for their denigrations. *Webster's* suggests: "a lineage of evolutionary descent characterized by long periods of stability in characteristics of the organism and short periods of rapid change during which new forms appear esp. from small subpopulations of the ancestral form in restricted parts of its geographic range." Punctuated equilibrium also occurs in most alphabetical compendia of scientific terms and concepts, including the *World Information Systems Almanac of Science and Technology* (Golob and Brus, 1990), *The Penguin Dictionary of Biology* (Thain and Hickman, 1990), and the *Oxford Dictionary of Natural History* (Allaby, 1985).

As a further mark of general recognition, several novelists have made casual references to punctuated equilibrium (in works for mass audiences, not arcana for the literati). Stephen King mentions punctuated equilibrium in chapter 30, "Thayer gets weird," of *The Talisman.* The celebrated English novelist John Fowles included the following passage in his novel, *A Maggot.* (Fowles, a distinguished amateur paleontologist himself, also created fiction's outstanding paleontologist, the hero of his novel, *The French Lieutenant's Woman*): "This particular last day of April falls in a year very nearly equidistant from 1689, the culmination of the English Revolution, and 1789, the start of the French; in a sort of dozing solstitial standstill, a stasis of the kind predicted by those today who see all evolution as a punctuated equilibrium." In his 1982 crime novel, *The Man at the Wheel,* Michael Kenyon doesn't cite the name, and does veer towards the common saltational confusion, but presumably wrote this passage in the light of public discussion about punctuated equilibrium. "Now there're biologists saying Darwin got it wrong, or at any rate not wholly right, because evolution isn't slow, continuous change, it's sudden bursts of change after millions of years of nothing, so if the polar bear happened suddenly, why not the world?"

2. LISTING THE THEORY AS AN EVENT IN CHRONOLOGICAL ACCOUNTS OF THE GROWTH OF 20TH-CENTURY KNOWLEDGE. Isaac Asimov cited punctuated equilibrium among the seven events of world science chosen to characterize 1972 in his book, *Asimov's Chronology of Science and Technology* (1989). Rensberger (1986) included punctuated equilibrium in his alphabetical compendium *How the World Works: A Guide to Science's Greatest Discoveries. In Our Times* (Glennon, 1995), a lavishly illustrated "coffee table" book on the cultural history of the 20th century, used the Darwinian centennial of 1982 for discussing the strength of evolution (contra

creationism) and the status of the field. Among three "takes" to mark the year (evolution shared space with acid rain and Madonna's first single recording, while a sidebar list of "new in 1982" includes liposuction and Halcion sleeping pills), the column on evolution bears the title "Darwin refined," and concentrates exclusively on punctuated equilibrium. In a refreshing departure from common journalistic accounts, the epitome is both incisive and generally accurate: ". . . The theory of punctuated equilibria reconciled Darwinism with paleontological reality . . . The debate could be traced back to Darwin, who'd candidly admitted that gradual evolution did not square with the fossil record. Gould emphasized that Darwinism was 'incomplete, not incorrect.' The theory of punctuated equilibria, however, proved a crucial refinement of Darwinian thought, as well as a useful model for other disciplines from anthropology to political science."

In a long article for *The New Yorker* on the history of the American Museum of Natural History, Traub (1995) emphasized the frustration of staff scientists when public supporters only recognize exhibits and have no inkling that the Museum also operates as a distinguished institute of scientific research. In listing accomplishments from Boaz to Mead in anthropology, and from Osborn to Simpson and Mayr for paleontology and evolution, Traub mentioned only punctuated equilibrium to mark the continuation of this tradition into recent years: "And in the early seventies, Stephen Jay Gould and Niles Eldredge accounted for anomalies in the fossil record by arguing that evolution proceeded not steadily but sporadically—a theory known as 'punctuated equilibria.'"

I acknowledge, of course, the blatant unfairness of this selectivity, especially the legitimate grievances prompted thereby among those who built cladistic theory at the Museum during these same years. I cite this example (with some mixture of embarrassment and, to be honest, a tinge of pride as well) to point out how punctuated equilibrium became, in popular culture, a synecdoche for professional discussion about evolution. I also recognize that this journalistic ploy rightly angers and inspires jealousy among colleagues; I merely point out that Eldredge and I cannot fairly be blamed for this cultural phenomenon. Neither of us ever organized a symposium, or even called a reporter, to discuss punctuated equilibrium in public (and neither of us was interviewed by the author of this *New Yorker* article).

3. INTERNATIONAL SPREAD. Punctuated equilibrium has been prominently discussed in newspaper and magazine articles of nearly all major Western nations—in France as *équilibres intermittents* (Blanc, 1982) or *équi-*

libres ponctués (the leading newspaper *Le Monde* of May 26, 1982; Devillers and Chaline, 1989; Courtillot, 1995); in Spain and Latin America as *equilibrio interrumpido* (Sequieros, 1981) or *equilibrio punctuado* (Valdecasas and Herreros, 1982; Franco, 1985); in Italy as *equilibri punteggiati* (Salvatori, 1984); and in Germany as *Unterbrochenen Gleichgewichts* (Glaubrecht, 1995). (All these citations come from popular articles about the theory, not from biopic pieces about the authors, or from technical literature.)

Punctuated equilibrium has also been featured in Western but non Indo-European, accounts in Hungary, Finland and Turkey, and in several articles of the non-Western press, notably in India, Japan, Korea and China. The theory even penetrated the strongest of political iron curtains to emerge, on March 21, 1983, as a feature article in Maoist Mainland China's major newspaper *Ren Min Ri Bao (The People's Daily).* They wrote, in a commentary not notable for accuracy: "Theories against Darwin have taken the opportunity to make their appearances. The most typical of all this is the theory of 'punctuated equilibrium' . . . This theory holds that organic evolution proceeds by leaps and bounds and not through continuous change."

4. TEXTBOOKS. This criterion may be viewed as the most unenviable of all, but when a new idea enters textbooks as a "standard," almost obligatory item (remember that no other written genre ranks as more conservative or more cloned through endless copying and regimentation by publishers' requirements), then we may affirm that the notion has flowed into a cultural mainstream. As I shall document on pages 319–325, punctuated equilibrium has become a standard entry in textbooks at both the college and high school levels in America.

5. AN ITEM IN GENERAL CULTURE. When the National Center for Science Education, America's leading anti-creationist organization, put out two bumper stickers as sardonic comments upon the favored evangelical "Honk if you love Jesus," they chose "Honk if you love Darwin" and "Honk if you understand punctuated equilibrium." (Niles Eldredge tells me that, in his car one day, he became frightened by a persistent honker; when he ventured a sheepish glance, fearing an encounter with a gun, or at least an upraised third finger, he noted only a smile on the other driver's face, and a finger pointing downward to the bumper sticker.) My colleagues may be satirizing punctuated equilibrium as terminological mumbo jumbo, but at least they thought they could raise some money (and some laughs) with this item!

Although not always understood or properly employed (but often, to my surprise and gratification, excellently epitomized and tactfully used), punctuated equilibrium has become a recognized term and concept both in scholarship of widely disparate fields, and in popular culture. I first noted this spread in 1978, a few years before punctuated equilibrium splashed into public recognition, when nationally syndicated columnist Ellen Goodman featured punctuated equilibrium in an op-ed piece entitled "Crisis is a way of life bringing sudden change." I feel that Goodman, then unknown to me but now a respected colleague and friend, captured the essence of punctuated equilibrium's suggestions about the general nature of change, and did so with clarity and insight—a good beginning, not always followed in later commentary. Goodman wrote (in part):

> I am not normally the sort of person who curls up in front of the fire with a good science book. The last time I found Charles Darwin interesting was in "Inherit the Wind." But I was still intrigued by Stephen Ray [sic] Gould's thoughts about evolution . . . [for he] has written about natural change in a way that makes sense out of our current lives and not just out of fossils. Gould thinks Darwin's view of evolution . . . was actually "a philosophy of change, not an indication from nature." he says that "gradualism" was part of the 19th century prejudice in favor of orderliness . . . In that sense, I suppose we are all still Darwinians. How many of us harbor the hope that the change in our lives will be gradual, rather like being promoted from the seventh grade to the eighth. We would like our lives to be an accumulation of skills and wisdom . . .
>
> [But] people may go through the greatest changes in their lives in the shortest chunks of time. I have known someone who, after years of stagnation, raced through a decade of personal growth in the first year of a new career. I have known others who experienced a generation's worth of change in six post-divorce months . . .
>
> We often underestimate the suddenness, even the randomness, of the change itself. I suppose that our observations are no more colored than Darwin's. We see gradual change, in part, because we go looking for it. We find it because we need it. Our research into the past reflects our fear of the future . . . Natural history is, as [Gould] puts it, "a series of plateaus punctuated by rare and seminal events that shift systems from one level to another." In that way, I suspect, people have a lot in common with rocks.

Punctuated equilibrium has often been the explicit focus of scholarship in distant fields (see the concluding section of Chapter 5 for a technical discussion with extended examples; I only mention the range of invocations here)—including a lead article by Gans (1987) on "punctuated equilibria and political science" with five commentaries by other scholars and a response by Gans in the journal *Politics and the Life Sciences;* an analysis of my rhetorical style by Lyne and Howe (1986) in the *Quarterly Journal of Speech;* and an exchange between Thompson (1983) and Stidd (1985) in the journal *Philosophy of Science.* However, the general spread of punctuated equilibrium into vernacular culture will not be best illustrated by such explicit treatments (which may be read as didactic efforts to instruct), but rather by more casual comments implying a shared context of presumed understanding before the fact. I therefore present a partial and eclectic list, united only by the chancy criterion that someone called the items to my attention:

1. In economics, the distinguished columnist David Warsh used punctuated equilibrium to illuminate episodic change and long plateaus in the history of markets and prices ("What goes up sometimes levels off," *Boston Globe,* 1990, and a good epitome for stasis vs. progressivism), and also to support the general concept of punctuational change at all levels, in a defense of capitalism with the ironic title "Redeeming Karl Marx" (*Boston Globe,* May 3, 1992). In a recent bestselling book, *The Future of Capitalism,* MIT economist Lester Thurow centered his argument upon two concepts borrowed from the evolutionary and geological sciences—punctuated equilibrium and plate tectonics (see further comment on pp. 281–283).

2. In political theory, the scholarly book of Carmines and Stimson (1989) argues for an episodic model of change based on case studies of the New Deal and race relations in America. "Dynamic evolutions," they write (1989, p. 157), "thus represent the political equivalent of biology's punctuated equilibrium."

3. In sociology, Savage and Lombard's (1983) model "of the process of change in the structure of work groups" cites a key prod from punctuated equilibrium:

> Social scientists, at least in small-group studies, generally follow the uniformitarians' view. In recent years studies in several fields have led to revisions in arguments about these classic views. Paleobiology . . . continues to provide some of the most specific and convincing of the newer studies. Even though the field is far removed from the study of changes in work groups in South America, it is informative to examine some of them. Writing in 1972

about the fossil record of mollusks, Eldredge and Gould concluded that in the development of a new species "the alternative picture [to gradual and continuous change is] of stasis punctuated by episodic events."

4. In history, Levine (1991) used our term and concept to center his argument about the history of working-class families in an article entitled: "Punctuated Equilibrium: The modernization of the proletarian family in the age of ascendant capitalism."

5. In literary criticism, Moretti (1996) cited punctuated equilibrium to epitomize the history of the epic as a literary genre, the principal subject of his book: "It is an undulating curve; a discontinuous history that soars, then gets stuck. Overall, it is the conception illustrated by Gould and Eldredge with the theory of 'punctuated equilibria'" (Moretti, 1996, p. 75).

6. In art history, Bahn and Vertut (1988) used punctuated equilibrium to refute the standard gradualist and progressivist views of the greatest scholars of Paleolithic cave painting, the Abbé H. Breuil and André Leroi-Gourhan (see further comments on pp. 266–271).

7. In the dubious, but popular, literature of "self help" Connie Gersick's fascinating and thought-provoking work (see pp. 271–275) links individual and organizational growth to patterns of punctuated equilibrium. But her subtlety was badly sandbagged in the news bulletin of the University of California (where she teaches at UCLA) for February 21, 1989: "Gersick likened this transition to a midlife crisis, which, she said, is part of a phenomenon known as 'punctuated equilibrium' . . . For organizations which rely on the results of creative efforts, Gersick notes that understanding the transitions within the creative process can help groups to work more effectively. 'Managers may be able to build more punctuation points into the process.'"

8. In humor (and to restore equilibrium after the last quotation), Weller properly situates punctuated equilibrium between gradualism and true saltationism in his book *Science Made Stupid* (for another example, see Fig. A-1).

These citations obviously vary greatly in cogency and utility, but they do indicate that punctuated equilibrium has struck a chord of consonance with themes in contemporary culture that many analysts view as central and troubling. Some usages amount to mere misguided metaphorical fluff, but others may direct and focus major critiques. In any case, since people are not stupid (at least not consistently so over such a broad range of disciplines), I must conclude that punctuated equilibrium has something general, perhaps even important, to say.

Poorly punctuated equilibrium.

A-1. A humorous perspective on punctuated equilibrium.

An Episodic History of Punctuated Equilibrium

EARLY STAGES AND FUTURE CONTEXTS

I have never enjoyed a reputation for modesty, so I believe that the following introductory comments represent a genuine memory, not a bias following a bent of personality. I was proud of our 1972 paper, and of my initiating oral presentation at the 1971 meeting of the Geological Society of America in Washington, D.C. I hoped that punctuated equilibrium would influence the practice of paleontology by showing that the fossil record, read literally, might depict the process of evolution as understood by neontologists, and not only reflect an absence of evidence pervasive and discouraging enough to make the empirical study of macroevolution virtually impossible at fine scale. Among the ordinary run of papers, this goal cannot be called modest, so I maintained some hope for punctuated equilibrium from the start. But I had no premonition about the hubbub that punctuated equilibrium would generate—for two reasons internal to the theory and to professional life, and also for our general inability to know the contingencies of external history.

For the internal reasons, I simply did not grasp, at first, the broader implications of punctuated equilibrium for evolutionary theory, as embodied in our proposals about stasis and the necessary explanation of macroevolutionary pattern by species sorting. I do clearly remember—and this recollection continues to strike me as viscerally eerie—that I felt something significant lurking in the short section on trends (1972, pp. 111–112) that Eldredge had written. I somehow knew that this section included the most important claim in the paper, but I just couldn't articulate why. Second, I never imagined that the paper would generate any readership beyond the small profession of paleontology.

Two incidents reinforce my memory of modest expectations. My father, a brilliant and self-taught man who never had much opportunity for formal education and therefore grasped the logic of arguments far better than the sociological realities of the academy, got excited when he read the 1972 paper in manuscript, and said to me: "this is terrific; this will really make a splash; this will change things." I replied with mild cynicism, and with the distinctive haughtiness of an "overwise" youngster who views his parents as naive, that such a hope could not be fulfilled because so few scientists ever bother to read papers carefully, or to mull over the implications of an argument not rooted exclusively in graphs and tables. As a second example, my former thesis advisor John Imbrie stopped to congratulate me on my "well argued non-Neodarwinian argument about paleontology and evolution" after my original oral presentation in 1971. I appreciated the praise, but remained mystified by why he thought that an argument for operational paleobiology, based on proper scaling of allopatric speciation, could be viewed as theoretically iconoclastic as well.

The early history of punctuated equilibrium unfolded in a fairly conventional manner for ideas that "catch on" within a field. The debate remained pretty much restricted to paleontology (and largely pursued in the new journal founded by the Paleontological Society to publish research in the growing field of evolutionary studies—*Paleobiology*). Theoretical implications received an airing, but most discussion, to our pride and delight, arose from empirical and quantitative studies done explicitly to test the rival claims of gradualism vs. punctuation and stasis in data-rich fossil sequences. Most important were the critical studies of Gingerich (1974, 1976) on putative gradualism in Tertiary mammalian sequences from the western United States. In any case, our hopes for a fruitful unleashing of empirical studies based on new respect for the power and adequacy of the fossil record were surely fulfilled.

Enough data, argument, and misconception as well had accumulated by the summer of 1976 that Eldredge and I decided to write a retrospective and follow-up—a longer article dedicated mostly to the detailed analysis of published data, and appearing in *Paleobiology* under the title: "Punctuated equilibria: the tempo and mode of evolution reconsidered" (Gould and Eldredge, 1977). Meanwhile, we couldn't fail to note that the arguments of punctuated equilibrium, substantially aided by the support and extension of our colleague S. M. Stanley in a widely discussed *PNAS* article of 1975 that introduced the term "species selection" in a modern context (and developed the implications that I had been unable to articulate from our original section on evolutionary trends), were now beginning to attract attention in the larger field of neontological evolutionary studies. Stanley then followed with an important book on macroevolution (1979).

From an isolated South Africa, Elisabeth Vrba published an astonishing paper (1980) that gave an even more cogent and comprehensive voice to the macroevolutionary implications of punctuated equilibrium. (Following British custom from a former colony, she published as E. S. Vrba; Eldredge and I had never heard of her work and didn't even know her gender. The paper burst upon us as a most wonderful surprise.) In 1980, to fulfill an invitation from the editors to celebrate the 5th anniversary of our new journal *Paleobiology,* I then published a general article on the potential reform of evolutionary theory, a pretty modest proposal I thought, but, oh my, did neo-Darwinian hackles rise (see pp. 329–331).

At this point, the story becomes more like ordinary history in the crucial sense that predictable components, driven by the internal logic of a system, interact with peculiar contingencies to yield a result that no one could have anticipated. Punctuated equilibrium did begin to receive general commentary in professional journals (with Ridley's 1980 *News and Views* piece for *Nature* as a first example), but I am sure that our theory would never have become such a public spectacle if this interest had not coincided with two other events (or rather one event and a surrounding political context).

In October 1980, Chicago's Field Museum of Natural History held a large international conference on Macroevolution. This meeting, inspired in good part (but by no means entirely, or even mainly) by the developing debate over punctuated equilibrium, would have been a major event in our profession in any case. But the Chicago meeting escalated to become something of a cultural *cause célèbre* because, and quite coincidentally, the symposium occurred at the height of renewed political influence for the creationist movement in America.

This fundamentalist movement, dedicated (as a major political goal) to suppressing the teaching of evolution in America's public schools, had flourished in the early 1920's under the leadership of William Jennings Bryan, had culminated in the famous Scopes trial in Tennessee in 1925, but had then petered out and become relatively inactive, especially following the 1968 Supreme Court decision, *Epperson* vs. *Arkansas,* that finally overturned the anti-evolution laws of the Scopes era on First Amendment grounds.

But creationism surged again in the 1970s, largely in response to an increasingly conservative political climate, and to the growing political savvy and organizational skills of the evangelical right. Creationists enjoyed a second round of success in the late 1970s, culminating in the passage of "equal time" laws for creationism and evolution in the states of Arkansas and Louisiana. We would eventually win this battle, first by overturning the Arkansas law in early 1982 (see pp. 309–314 for the role of punctuated equilibrium in this trial), and then by securing a resounding Supreme Court victory in 1987 in *Edwards* vs. *Aguillard.* But, in 1980 as the Chicago meeting unfolded, creationists were enjoying the height of their renewed political influence, and evolutionists were both justly furious and rightly worried.

Even with this temporal conjunction, the Chicago meeting wouldn't have attracted public attention if the press had not been alerted by accidental circumstances (neither the participants nor the organizers invited general journalists to the meeting). At most, reports would have appeared in the *News and Views* sections of *Nature* and *Science,* and professional history might have been tweaked or even altered a bit.

But the general press caught on and grossly misread the forthcoming meeting as a sign of deep trouble in the evolutionary sciences (rather than the fruitful product of a time of unusual interest and theoretical reassessment for a factual basis that no one doubted), and therefore as an indication that creationism might actually represent a genuine alternative, or at least a position that stood to benefit from any perceived confusion among evolutionists.

No single source can be blamed for thus alerting and misinforming the press, but an unfortunate article by James Gorman appeared in the popular magazine *Discover* just a month before the meeting ("The tortoise or the hare," October, 1980), leading with the following confused and unfortunate paragraph:

> Charles Darwin's brilliant theory of evolution, published in 1859, had a stunning impact on scientific and religious thought and forever changed

> man's perception of himself. Now that hallowed theory is not only under attack by fundamentalist Christians, but is also being questioned by reputable scientists. Among paleontologists, scientists who study the fossil record, there is growing dissent from the prevailing view of Darwinism. Partly as a result of the disagreement among scientists, the fundamentalists are successfully reintroducing creationism into textbooks and schoolrooms across the U.S. In October, a hundred or so scientists from half a dozen different disciplines will gather at Chicago's Field Museum . . .

This misconstruction yielded two unfortunate consequences—first, in inspiring a substantial contingent of the general press to attend the Chicago meeting under the false assumption that these technical proceedings would yield newsworthy stories about the success and status of creationism; and, second, by creating a blatantly false taxonomy that dichotomized natural historians into two categories: true-blue Darwinians vs. anyone with any desire to revise anything about pure Darwinism (including the strangest bedfellows of evolutionary revisionists and creationist ignoramuses). We must never doubt the potency of such false taxonomies, especially when promulgated by a general press that grasps the true issues poorly, and also plays to an audience too prone to read any dispute as a dichotomous pairing of good and evil. (Consider, for example, the harm done when scientific fraud, the worst of conscious betrayals for all we hold dear as a profession, gets linked with scientific error, a correctable and unavoidable consequence of any boldness in inquiry, because both lead to false conclusions. The pairing of punctuated equilibrium and creationism because both deny pure Darwinian gradualism, falls into the same category.)

The Chicago meeting also produced many good and responsible commentaries in the general press (Rensberger in the *New York Times,* November 4, 1980, for example) and in professional journals (Lewin in *Science* for November 21, 1980, for example). But some very bad accounts also appeared, especially unfortunate in their linkage of success for punctuated equilibrium with the spread of creationism. For example, a lead article in *Newsweek* (November 2, 1980), perhaps the most widely read of all reports, did properly brand the link as a confusion, and also stated that punctuated equilibrium represents a revision, not a refutation, of evolution, but such passing "subtleties" can easily be missed when subjects become so tightly juxtaposed, as in the *Newsweek* story: "At a conference in mid-October at Chicago's Field Museum of Natural History, the majority of 160 of the world's top paleontologists, anatomists, evolutionary geneticists and developmental biologists supported some form of this theory of 'punctuated equilibria.' While the scientists have been refining the theory of evolution in the past decade, some nonscientists have been spreading anew the gospel of creationism."

This kind of reporting kindled the understandable wrath of orthodox Darwinians and champions of the Modern Synthesis. They became justifiably infuriated by two outrageous claims, both falsely linked to punctuated equilibrium by some press reports. First, some absurdly hyped popular accounts simply proclaimed the death of Darwinism (with punctuated equilibrium as the primary assassin), rather than reporting the more accurate but less arresting news about extensions and partial revisions. For example, the same *Newsweek* article stated that "some scientists are still fighting a rear-guard action on behalf of Darwinism," and "it is no wonder that scientists part reluctantly with Darwin." Moreover, even the best and most balanced articles often carry exaggerated and distorted headlines (most scientists, I suspect, don't know that reporters are not generally permitted to write their own headlines). Boyce Rensberger's *New York Times* story on the Chicago conference could not have been more fair or accurate, but the hyped headline proclaimed: "Recent studies spark revolution in interpretation of evolution." Since the article focused on punctuated equilibrium, some colleagues then blamed Eldredge and me for an exaggeration promulgated neither by ourselves nor by the reporter.

Second, since punctuated equilibrium had served as the most general and accessible topic among the many questions debated at the Chicago Macroevolution meeting, our theory became the public symbol and stalking horse for all debate within evolutionary theory. Moreover, since popular impression now falsely linked the supposed "trouble" within evolutionary theory to the rise of creationism, some intemperate colleagues began to blame Eldredge and me for the growing strength of creationism! Thus, we stood falsely accused by some colleagues both for dishonestly exaggerating our theory to proclaim the death of Darwin (presumably in our own cynical quest for fame), and for unwittingly fostering the scourge of creationism as well. I believe that the strong feelings generated by punctuated equilibrium ever since cannot be divorced from this unfortunate historical context. (I also believe, of course, that the intense interest—as opposed to those intense feelings—arises largely from the challenging intellectual content of the theory itself.)

I don't want to advance the exculpating argument that all unfortunate parts of this debate can be traced to purely external and unpreventable press hyping, and that we and our colleagues, in arguing both for and against punctuated equilibrium, have always walked Simon pure on the intellectual high road. I will discuss on pages 338–341 the extent to which our own actions may have contributed to the unseemly side of the discussion. But I do maintain that this truly uncontrollable external context set the primary reason for extended and unwarranted emotionality over the subjects of punctuated equilibrium and macroevolutionary challenges to conventional Darwinism.

Meanwhile, in a simultaneous unfolding of the tragedy and the farce (in contrast with the famous epigram that historical tragedies generally experience *later* replay as farces), a truly risible episode of intense public discussion about punctuated equilibrium erupted in England. American creationism may not rank as a full tragedy, although any suppression of a cardinal subject in public schools surely qualifies as an academic equivalent of murder. By contrast, the great British Museum debate can only be viewed as comical. In an epitome that risks caricature—although the full story veered as close to pure absurdity, and therefore to unalloyed comedy, as anything I have ever witnessed in the sociology of science—the British Museum (Natural History) opened a new exhibit on dinosaurs, based almost exclusively on the rigid cladism espoused by the Museum curators. Beverly Halstead—a man who might be judged as utterly infuriating and even cruelly meddling had he not been so charming and so personally warm and generous—hated these exhibits with all his heart, for Beverly was an unabashed Simpsonian and a devotee of adaptationist biology. So Beverly, following a uniquely British tradition for generating tempests in teapots by inflationary prose fashioned of pure bombast—just where do you think that Blake's famous lines about seeing the world "in a grain of sand" and heaven "in a wild flower" came from?—decided to float the following blessed absurdity, a guarantee of public attention rather than instant burial, in the letters column of the *Times.* He accused—and I swear that I do not exaggerate—the British Museum of foisting Marxism upon an unwitting public in this new exhibit, because cladism can be equated with punctuated equilibrium, and everyone knows that punctuated equilibrium, by advocating the orthodoxy of revolutionary change, represents a Marxist plot.

Well, the press bit, and a glorious volley of ever more orotund letters appeared, both in the general press and in the professional pages of *Science, Nature,* and the *New Scientist.* Since I don't wish to prolong discussion of this peculiar byway (I doubt that any of the Museum curators had any abiding interest in politics beyond the academy, or personally stood one inch to the left of Harold Wilson), and since this chapter represents my own partisan account, let me simply reproduce my own letter to *Nature*—part of the final volley, just before the editors wisely and forcefully cut off all further fulmination forthwith:

> Sir—I have been following the "great museum debate" in your pages with a profound sense of detached amusement. But as matters are quickly reaching a level of absurdity that may inspire me to write the 15th Gilbert and Sullivan opera, and as I am, in a sense, the focal point for Halstead's glorious uproarious misunderstanding, I suppose I should have my say.
>
> Halstead began all this by charging that the venerable Natural History Museum is now purveying Marxist ideology by presenting cladism in its exhibition halls. The charge is based on two contentions: (1) a supposed link between the theory of punctuated equilibrium, proposed by Niles Eldredge and myself, and cladistic philosophies of classification; and (2) an argument, simply silly beyond words, that punctuated equilibrium, because it advocates rapid changes in evolution, is a Marxist plot. For the first, there is no necessary link unless I am an inconsistent fool; for I, the co-author of punctuated equilibrium, am not a cladist (and Eldredge, by the way, is not a Marxist, whatever that label means, as if it mattered). Under cladism, branching events may proceed as slowly as the imperceptible phyletic transitions advocated by the old school. Punctuated equilibrium does accept branching as the primary mode of evolution, but it is, fundamentally, a theory about the characteristic rate of such branching—an issue which cladism does not address.
>
> For Halstead's second charge, I did not develop the theory of punctuated equilibrium as part of a sinister plot to foment world revolution, but rather as an attempt to resolve the oldest empirical dilemma impeding an integration of paleontology into modern evolutionary thought: the phenomena of stasis within successful fossil species, and abrupt replacement by descendants. I did briefly discuss the congeniality of punctuational change and Marxist thought (*Paleobiology*, 1977) but only to illustrate that all science, as historians know so well and scientists hate to admit, is socially embedded. I couldn't very well charge that gradualists reflected the politics of their time and then claim that I had discovered unsullied truth . . .
>
> I saw the cladistic exhibits last December. I did not care for them. I found

> them one-sided and simplistic, but surely not evil or nefarious. I also felt, as a Victorian aficionado who pays homage to St. Pancras [a wonderously ornate late nineteenth century railroad station] on every visit to London, that most of the newer exhibits are working against, rather than with, the magnificent interior that houses them. But I would not envelop these complaints in ideological hyperbole; Halstead has said enough.

We can best explore the consequences of this historically contingent context by examining the use of punctuated equilibrium in two domains that, in their contrast, span the range of public influence from the ridiculous to the potentially sublime: the political propaganda of creationist literature, and the developing treatment of punctuated equilibrium in journalism, and high school and college textbooks of biology.

CREATIONIST MISAPPROPRIATION OF PUNCTUATED EQUILIBRIUM

Since modern creationists, particularly the "young earth" dogmatists who must cram an entire geological record into the few thousand years of a literal Biblical chronology, can advance no conceivable argument in the domain of proper logic or accurate empirics, they have always relied, as a primary strategy, upon the misquotation of scientific sources. They have shamelessly distorted all major evolutionists in their behalf, including the most committed gradualists of the Modern Synthesis (their appropriations of Dobzhansky and Simpson make particularly amusing reading). Since punctuated equilibrium provides an even easier target for this form of intellectual dishonesty (or crass stupidity if a charge of dishonesty grants them too much acumen), no one should be surprised that our views have become grist for their mills and skills of distortion. I have been told that Duane Gish, their leading propagandist, refers to his compendium of partial and distorted quotations from my work as his "Goulden file."

Standard creationist literature on punctuated equilibrium rarely goes beyond the continuous recycling of two false characterizations: the conflation of punctuated equilibrium with the true saltationism of Goldschmidt's hopeful monsters, and the misscaling of punctuated equilibrium's genuine breaks between species to the claim that no intermediates exist for the largest morphological transitions between classes and phyla. I regard the latter distortion as particularly egregious because we formulated punctuated equilibrium as a positive theory about the nature of intermediacy in such large-scale structural trends—the "stairstep" rather than the "ball-up-the-inclined-plane" model, if you will. Moreover, I have written numerous essays in my popular series, spanning ten printed volumes, on the documentation of this style of intermediacy in a variety of lineages, including the transition to terrestriality in vertebrates, the origin of birds, and the evolution of mammals, whales and humans—the very cases that the usual creationist literature has proclaimed impossible.

To choose a standard example by the movement's "heavies" (Bliss, Parker and Gish, 1980, p. 60), the following text embodies the first standard error, while their accompanying illustration (Fig. A-2) records the second error by equating punctuated equilibrium with the saltational origin of each vertebrate class (if anyone has any lingering doubt about the pseudoscientific character of this movement, try to make any sense at all of this figure, a supposed expression of their proper practice of the graphical and quantitative approach to science): "Gould and Eldredge state that fossils, like living forms, vary only mildly around the average or 'equilibrium' for each kind. But, they say, the appearance of a 'hopeful monster' can interrupt or 'punctuate' this equilibrium. According to the new concept of 'punctuated equilibrium,' fossils are not supposed to show in-between forms. The new forms appeared suddenly, in large steps."

We may at least label creationist Everett Williams as timely in adding the insult of misnaming to the injury of the same distortion in a 1980 newspa-

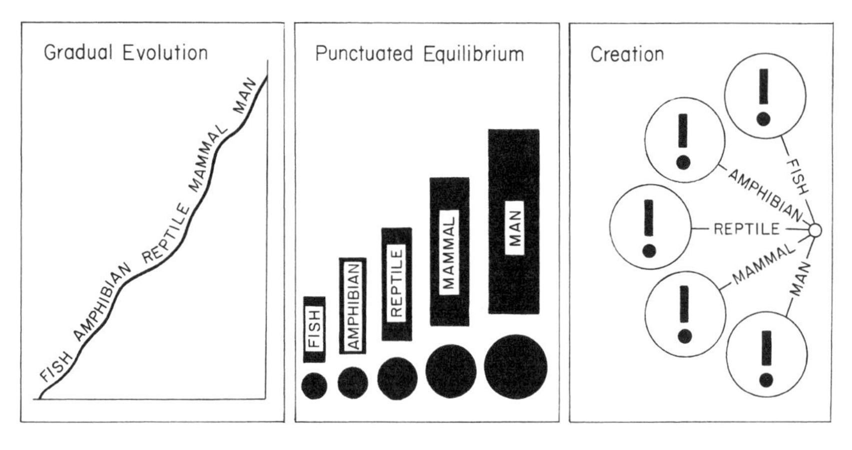

A-2. Creationist distortion of punctuated equilibrium modified from original in "Fossils: key to the present" by Bliss, Parker, and Gish, 1980. They misdepict punctuated equilibrium as a saltationist theory, with all vertebrate classes arising in single steps, all at the same time.

per column: "The latest version of the process is called 'punctual [sic] evolution.' In this version, evolution is seen as moving in giant surges and then becoming stagnant for eons."

A broadsheet from Hillsborough, North Carolina, entitled "Harvard scientists agree: Evolution is a hoax!!!" goes whole hog in assimilating us to its own version of the rock of (small) ages: "The facts of 'punctuated equilibrium' which Gould, Eldredge, Stanley and other top biologists are forcing the Darwinists to swallow fit the picture that Bryan insisted on, and which God has revealed to us in the Bible. Every species of organism was separately created during the six 'days' of creation . . . This is the doctrine taught by Scripture and by Cuvier (the father of paleontology) alike, and modern biology is forcing the Darwinists to accept it."

In *The Genesis Connection,* J. L. Wiester commits the same error of scaling (in maximal degree this time), and then also cites us as hidden supporters of his one true way: "The theory of punctuated equilibrium holds that life did not evolve in the slow uniform method that Darwin envisioned but rather in rapid evolutionary bursts of major change called adaptive radiations. The Cambrian explosion of marine life was such an adaptive radiation . . . The new theory of punctuated equilibrium brings the thinking of science remarkably closer to the biblical view. It is notable that the more evidence scientists discover (or fail to discover), the closer scientific theory moves toward the unchanging biblical pattern."

If observers ever hoped for more accuracy or fairness from "official" publications of the two largest creationist organizations in America (in contrast with the "independents" previously cited), they shall be disappointed. The Jehovah's Witnesses journal *Awake!* reported on the Chicago macroevolution meeting in its issue for September 22, 1981, rooting the story in error number one, the Goldschmidt equation: "This revised view of evolution is called 'punctuated equilibrium,' meaning one species remains for millions of years in the fossil record, suddenly disappears and a new species just as suddenly appears in the record. This, however, is not really a new proposal. Richard Goldschmidt advanced it in the 1930's, called it the 'hopeful monsters' hypothesis, and was much maligned for it. 'Punctuated equilibrium' is a much more impressive designation."

Writing in the September 1982 issue of *Signs,* the leading journal of the Seventh Day Adventists, H. W. Clark discussed the Arkansas creationism trial in terms of error number two, false upward scaling to denial of transitional forms between classes and phyla. Clark equated our punctuations with faunal breaks between geological periods. A sidebar then misnames the author as well as distorting the theory: "Thank you, Dr. Jay Gould: Dr. Jay Gould is the distinguished Harvard Paleontologist who has raised a storm in evolutionary circles with his new theory of 'punctuated equilibrium.' Without intending to do so, he has told the scientific world that Darwin was wrong and the creationists are right. Not that he planned to, of course! Darwin recognized that the theory of evolution needs an unbroken line of gradually changing fossils. Now along comes Dr. Jay Gould and agrees with the creationists: the missing links aren't there and never were. Thank you, Dr. Gould!" You're quite welcome.

On the same theme of shoddiness in supposed creationist "scholarship," I was quite struck by a photograph, supposedly of me, that appeared in M. Bailey's creationist book for children (Greenhaven Press, 1990), *Evolution: Opposing Viewpoints.* The gentleman depicted sports a flowing beard and bald pate—while my head hair has a precisely opposite distribution. He is also considerably older (and, I fancy, a good deal uglier) than I. I finally realized that he is the 19th century robber-baron Jay Gould (no relation, by the way).

While America deals almost exclusively with creationists of the Protestant fundamentalist line (at least among the movement's chief political activists), other religions have their own similar crosses to bear. I was sent a 1983 Hindu version by one Satyaraja dasa entitled "Puncturing the jerk theory." An article by Barbara Sofer in a recent issue of *Hadassah Magazine* reports on the rare phenomenon of Israeli creationism, and cites one adherent: "Schroeder points out that the newer theories of punctuated evolution come closer to the biblical description."

Such inane and basically harmless perorations may boil the blood, but creationist attempts to use punctuated equilibrium in their campaigns for suppressing the teaching of evolution raise genuine worries. Nonetheless, here we can fight back directly—and we have always won. Elijah, after all, taught us how to fight fire (or rather the inability of reprobates to kindle any real flame) with fire—and the splendid man cited by creationists as their own primary hero did promise that truth would make us free.

In the Texas textbook hearings of 1984, for example, Mel and Norma Gabler, the infamous professional propagandists for forcing a right-wing evangelical agenda into textbooks, lobbied for four imposed changes in any evolutionary passages within biology texts. One of the items rested upon the second standard creationist mischaracterization of punctuated equilibrium:

"There are systematic gaps in the fossil record, showing absences of intermediate links. Punctuated equilibrium was devised to explain these gaps. Therefore presentation of evolutionary lineages, such as from amphibians to reptiles to birds and mammals, cannot be supported with evidence. The textbooks should be revised to reflect this understanding."

But our side holds a strong weapon in such public cases, for we can also testify, and therefore expose. We have never failed in these circumstances. The Gablers' proposal lost and the State of Texas endorsed good biology books.

Given the power of the First Amendment, and the fairness and intellectual stature of federal judge William Overton, our success in overturning the Arkansas equal time law, in a trial held in Little Rock in late 1981, seemed assured. I can only regard my own role, as one of six expert witnesses for science and religion, as both minor and probably irrelevant to the inevitable decision. But I was able to speak for paleontology and to add our unique temporal perspective to the documentation.

Clarence Darrow's scientific witnesses were not permitted to testify in the Scopes Trial of 1925. The Louisiana creationism law, the only other statute passed in modern times by a state legislature, and a virtual copy of the Arkansas law, was dismissed by summary judgment following our success in Arkansas, and was therefore never tried in court. The State of Louisiana appealed this case to the Supreme Court, where oral argument consumes but one hour, and only the principal lawyers may testify, with no witnesses called. Therefore, for the first and only time in American history, the Arkansas trial permitted full scale testimony about creationism in a court of law. I feel honored that I had the opportunity to help present the case for evolution as natural knowledge, and for creationism as pseudoscience, in the only legal venue ever provided to experts in the relevant professions throughout this long and important episode in 20th century American history.

My testimony and cross examination at the Arkansas trial lasted for the better part of a day, and focused upon two subjects: the absurdity of attributing the entire fossil record to the single incident of Noah's flood (a favored creationist ploy for cramming the entire geological history of the earth into a mere 6000 years or so), and creationist pseudoscientific practice as illustrated by their clearly willful distortions of the theory of punctuated equilibrium. (We did not, in this trial, try to "prove" evolution—a subject scarcely in need of such treatment, and not for a court of law to adjudicate in any case—but only to expose creationism as a narrow form of dogmatic religion, masquerading as science in an attempt to subvert First Amendment guarantees against the establishment of religion in public institutions.)

Creationists continue to distort punctuated equilibrium, but we continue to win by exposing them in fair forums. For example, in 1997, Rep. Russell Capps of the North Carolina General Assembly used a "standard" misquotation from one of my essays about punctuated equilibrium in arguing before the legislature for a law that would ban the teaching of evolution as a fact (although teachers could still present the subject as a hypothesis). I suspect that Capps simply lifted the quote from Duane Gish's *Evolution: The Fossils Say No!* (I do love that title!) and never read my essay, because his version used exactly the same deletions as Gish's. Rep. Bob Hensley, an opponent of the bill, asked for my aid, and I wrote a letter, which he read to the assembly, detailing this dishonest treatment of my writing. I stated, in part (letter of April 4, 1997):

> [My] article is not an attack on evolution at all, but an attempt to explain how evolution, properly interpreted, yields the results that we actually see in the fossil record. The first part of the quotation is accurate, but about rates of change, not whether or not evolution occurs. The second part of the quotation after the three dots—"it was never seen in the rocks"—seems then to deny that evolution occurs. But if you read my full text and look for the material left out, it is obvious that my word "it" refers to gradualism as a style of evolutionary change, and not to evolution itself. If one reads the rest of the essay, the intent is abundantly clear. For example, I state on page 182: "The modern theory of evolution does not require gradual change. In fact, the operation of Darwinian processes should yield exactly what we see in the fossil record. It is gradualism that we must reject, not Darwinism." . . . Thus you can see that my essay actually says exactly the opposite of the false quotation cited by your colleague. This is so typical of the intellectual level of most creationist literature. Do we really want our students to be taught by this form of dishonest argument?

The counterattack succeeded. Rep. Hensley wrote to me on April 21, 1997: "Because of your efforts, the Bill has now been withdrawn from consideration in the House Education Committee." Shabby and dishonest argument can win a fragile and transient advantage, but so long as we fight back, we will win. God (who, as a self-respecting deity, must honor and embrace empirical truth) really is on our side.

PUNCTUATED EQUILIBRIUM IN JOURNALISM AND TEXTBOOKS

All scientists have read egregiously bad, hyped and distorted press commentaries about the more subtle and nuanced work of their field. I too get annoyed at such stories, but I have also learned to appreciate that most journalists take their job seriously, follow the ethics of the field, and tend to turn out good stories, on balance. When hype occurs, the fault lies just as often with scientists who simplify and overpromote their work, as with reporters who accept what they hear too uncritically. (Journalists should check, of course, and must therefore bear part of the blame, but scientists should begin any general critique of the press with an acknowledgment of our own *in camera* foibles.)

The extensive press coverage of punctuated equilibrium has generally maintained adequate to high quality. Ironically, though, the most common errors—which like the old soldiers, cats and bad pennies of our mottoes, never seem to fade but turn up, however sparsely, again and again with no diminution in frequency—match the mistakes cited by creationists for utterly different purposes. If willful misuse and unintentional, albeit careless, error repeat the same false arguments, then what serves as a common source amidst such different motives (and different frequencies of occurrence, of course—pervasive for creationists, rare for reputable journalists)? Deep constraints on human mentality (common difficulties with concepts of scaling and probability, for example)? Persistent historical and cultural prejudices (about progress and gradualism, for example)? The malfeasance and hyped misleading of original authors (as our severest critics like to claim)? In any case, I am fascinated by the entire issue of commonality in errors across such a maximal range of motives, and I believe that something deep about the nature of mentality and the sociology of knowledge lies exposed therein.

Schemes of oversimplification must rank as the *bête noire* of journalism, at least in the eyes of scientists and other scholars. Since dichotomization stands as our primary mode of taxonomic oversimplification, probably imposed by the deep structure of the human mind, we should not be surprised that journalists have tended to treat the punctuated equilibrium debate as a dichotomous struggle between gradualists and punctuationalists, superimposed upon another false dichotomy (with supposedly perfect mapping between the two) of Darwinians (read gradualists) against anti-Darwinians (read punctuationalists). This struggle then occurs within a political dichotomy—a genuine division this time—of evolution vs. creationism. (The misappropriation of punctuated equilibrium by creationists, as documented in the last section, violates this last dichotomy and can thus be easily grasped as unfair by nearly everyone.)

The error of dichotomy appears most starkly in the minimal length and maximal hype of advertising copy for books. Pergamon's come-on for Nield and Tucker's *Paleontology,* for example, promises that "the approach in the evolutionary discussions is fully in line with the most recent understandings of the punctuated equilibrium/phyletic gradualism debate." The blurb for Oliver Mayo's *Natural Selection and Its Constraints* proclaims: "Among other topical matters, he touches upon the controversial question of 'punctuated equilibrium' or 'phyletic gradualism' as a mechanism for major evolutionary change."

A prominent cultural legend (with "The Emperor's New Clothes" as a prototype) celebrates the young and honest naif as exposer of an evident truth that hidebound adults will not or cannot admit. True to this scenario, the Summer 1993 publicity blurb sheet of Mount Holyoke College reports the happy story of Heather Winklemann, a senior who had just won a prestigious Marshall Fellowship for graduate study in England. The article focuses on her intimidating but successful oral interview, held in San Francisco. As a prospective paleontologist, the committee asked her: "Does evolution work by punctuated equilibrium? Answer yes or no?" Ms. Winklemann replied cogently by exposing the dichotomy as false—and she got her fellowship. The article ends: "'That question took me by surprise,' Winkelmann recalls, 'because if you know anything about the topic you know it can't be answered with a yes or a no. Were they trying to catch me on the question? I told the committee that I couldn't give a yes-or-no answer and why.' Heather Winklemann's answer evidently was what the committee was hoping to hear."

Beyond dichotomy, a failure to recognize the theory's proper scale stands as the most common journalistic error about punctuated equilibrium, in accounts both positive and negative. Many reporters continue to regard yearly or generational changes in populations as a crucial test for punctuated equilibrium. Thus, Keith Hindley reported the fascinating work of Peter Grant and colleagues on changes in population means for species of Darwin's Galápagos finches following widespread mortality due to extreme climatic stresses. Hindley placed the entire story in the irrelevant light of punctuated equilibrium (which cannot even "see" such transient fluctuations in population means from year to year): "Striking new evidence has refuelled the heated scientific debate about the process of evolution . . . The followers of Stephen Gould of Harvard claim that such rapid changes or 'jumps,' caused by environmental pressures, are the key to the emergence of new species . . . This episode has provided Gould's supporters with some of the ammunition their theory has so far lacked: good examples of sudden evolution among species alive today."

Negative accounts of punctuated equilibrium often make the same error. In reviewing a book by Ernst Mayr in the *New York Times*, Princeton biologist J. L. Gould (no relation) discusses the link of punctuated equilibrium to Mayr's views on allopatric speciation. But he then attacks punctuated equilibrium because "its authors seem to believe that species-level changes can occur in one generation, presumably by the production of what the embryologist Richard Goldschmidt called 'hopeful monsters.'"

Among human foibles, our tendency to excoriate a bad job in public, but merely to smile in private at good work, imposes a marked asymmetry upon the overt reporting of relative frequencies in human conduct and intellect. In truth, although I have singled out some "howlers" for quotation in this section, most press reports of punctuated equilibrium have been accurate, while a few have been outstanding. Consequently, I close this section on punctuated equilibrium and the press with two extensive quotations from two leading science writers, one British and one American—with thanks for confirming my faith in the coherence and accessibility of the ideas and implications of punctuated equilibrium. In *The Listener* (magazine of the BBC) for July 19, 1986, Colin Tudge beautifully explains the key concepts and general reforms proposed by punctuated equilibrium, while also giving the critics their due:

> A third modification of the neo-Darwinian orthodoxy is embraced in the hypothesis of punctuated equilibrium, proposed in the early 1970s by the American biologists Niles Eldredge and Stephen Jay Gould. The idea of punctuated equilibrium is not intended to dispute Darwin's central notion that the evolutionary destinies of plants and animals are shaped largely or mostly by natural selection. But it does take issue with two of his subsidiary notions: the idea that evolutionary change brought about by natural selection is necessarily gradual; and the idea that natural selection can operate only at the level of the individual.
>
> No idea in biology has caused more contention and indeed rancor over the past 15 years. Some opponents of Gould and Eldredge argue that their

> observation is just plain wrong—that evolution *is* gradual. Some argue that even if it were true it would be trivial. And some suggest that even if it were true and not trivial, then it is in any case untestable, and therefore not worth considering.
>
> In truth, the paleontological record sometimes seems to show that one form of animal may gradually turn into another, in Darwinian fashion, but often it seems to show precisely the pattern that Gould and Eldredge propose. . . .
>
> It's at this point that some biologists say "So what?" Who ever doubted that evolution can at times proceed more quickly than at others? Even if true (in some cases), the observation is trivial. This, however, is a severe misrepresentation of Gould and Eldredge's idea, for they are not simply making the banal observation that evolution is sometimes fast and sometimes slow. They are suggesting that the "jumps" that can be observed in the fossil record represent the emergence of new species—that is, of groups of organisms that reproduce sexually with each other but not with other groups. . . .
>
> Indeed, Gould and Eldredge go further than this. They suggest that when a species divides to form several new species, this is analogous to the birth of new individuals; and just as natural selection tends to weed out weak individuals in favor of the strong, so it serves to weed out new experimental species. Thus, they suggest, natural selection can operate at the level of the species ("species selection") and not simply at the level of the individual, as Darwin proposed. This is not a trivial observation. . . .
>
> The attacks on punctuated equilibrium seem powerful. But Gould gives as good as he gets, and my own betting is that the theory of punctuated equilibrium, with a bit more buffering from biologists at large, will take its place as an important modification of Darwin's basic ideas.

In an article on Peter Sheldon's claims for extensive gradualism in trilobites, and therefore generally critical of punctuated equilibrium, James Gleick states that our theory has provoked "the most passionate debate in evolutionary theory over the last decade," and then provides a fine summary of our key ideas, and of the intellectual depth of the resulting debate (*New York Times,* December 22, 1987):

> Steady flow or fits and starts—the division between these conceptions of evolution has dominated the debate over evolutionary theory. The punctuated equilibrium model has stimulated much research and drawn many adherents. Some of its central notions have taken firm hold.

> Even the most traditional Darwinians, for example, acknowledge that punctuated equilibrium has become an important part of the picture of evolution. Some species do little of evolutionary interest for millions of years at a time. . . .
>
> But the debate continues to rage, because it concerns far more than speed itself. At stake are the fundamental questions of evolution: when and why does a creature change from one form to another? Is most evolution the slow, unceasing accumulation of the small changes a geneticist sees in laboratory fruit flies, or does it occur in episodes, when a small population, perhaps isolated geographically, suddenly changes enough to give rise to a new species?
>
> Suddenly, in paleontological terms, can mean hundreds of thousands of years. . . . Proponents of punctuated equilibrium take pains to stress that such events rely mainly on the Darwinian principles of natural selection among individuals varying randomly from one another. Even so, to some biologists, punctuated equilibrium seems like a resort to some process apart from the usual rules—"mutations that appear to be magic," Dr. Maynard Smith said.
>
> "They have argued that their results mean that evolution as seen on the large scale is not just the summing up of small events," he said, "but a series of quite special things that people like me"—population geneticists—"don't see. We don't want to be written out of the script."

The movement of scientific ideas into textbooks may provide our best insight into social forces that direct the passage from maximal professional independence into the most conservative of print genres. To be successful, textbooks must sell large numbers of copies to audiences highly constrained by set curricula, teachers who hesitate to revise courses and lessons substantially, and conservative communities that shun scholastic novelty. These external reasons reinforce the internal propensities of publishers who are happy to jazz up or dumb down, but not to innovate, and authors who experience great pressure to follow the conventions of textbook cloning, and not to depart from the standard takes, examples, illustrations, and sequences. Did you ever see a high school biology textbook that doesn't start the evolution chapter with Lamarck's errors, Darwin's truths, and giraffes' necks in that order?

In this context, I delight in the rapid passage of punctuated equilibrium from professional debate to nearly obligatory treatment in the evolution chapter of biology textbooks. I could put a cynical spin on this phenomenon, but prefer an interpretation, in my admittedly partisan manner, based on the successful ontogeny of punctuated equilibrium from a controversial idea to a firm item of natural knowledge, however undecided the issues of relative frequency and importance remain.

But I am also not surprised that textbooks encourage promulgation of standard errors—a tendency arising from pressures to simplify ideas, downplay controversy, favor bland consensus, and generate a fairly uniform treatment from text to text. We often encounter, for example, the same oversimplification by dichotomy that compromises so many press reports. Villee and collaborators (1989) state, for example: "Some scientists believe that evolution is a gradual process, while others think evolution occurs in a series of rapid changes." The headings of entire sections often bear this burden, as in Tamarin's (1986) title for his pages on evolutionary rates: "Phyletic Gradualism Versus Punctuated Equilibrium."

However, the bland consensus favored by textbooks (and euphemistically called "balance") often imposes a peculiar resolution foreign to most journalistic accounts, where controversy tends to be exaggerated rather than defanged to a weak and toothless smile of agreement at a meaningless center. Textbooks therefore tend to present the dichotomy and then to state that "I am right and you are right and everything is quite correct," to quote Pish-Tush in *The Mikado*—as average reality rests upon the blandest version of a meaningless golden mean. The 1996 edition of J. L. Gould and W. T. Keeton proclaims (p. 511) that "the usual tempo of speciation probably lies somewhere between the gradual-change and the punctuated equilibrium models." (But such a various phenomenon as speciation has no "usual tempo," or any single meaningful measure of central tendency at all. Blandness, in this case, reduces to incoherence.)

In another example, Levin (1991, p. 112) concludes with pure textbook boilerplate that could be glued over almost any scientific controversy: "The final chapter on the question of punctuated evolution versus phyletic gradualism has not been written. At present, the proponents of punctuated evolution appear to be more numerous than those of phyletic gradualism. Like most controversies in science, however, the answer need not lie totally in one camp, and it is evident that instances of phyletic gradualism can also be recognized in the fossil record of certain groups of plants and animals."

If we consider dichotomy as a general mental error of oversimplified organizational logic, then the most common scientific fallacy in textbook accounts of punctuated equilibrium resides, once again, in false scaling by application of the theory to levels either below or above the appropriate subject of speciation in geological time. As before, the conflation of punctuated equilibrium (speciation in geological moments) with true saltation (speciation in a single generation, or moment of human perception) persists as the greatest of all scaling errors. I am discouraged by this error for three basic reasons: (1) It has been exposed and explained so many times, both by the authors of punctuated equilibrium and by many others; so continued propagation can only record carelessness. (2) Saltation at any appreciable relative frequency surely represents a false theory, so punctuated equilibrium becomes tied to a patently erroneous idea; whereas misapplication of punctuated equilibrium to higher levels may at least misassociate the name with a true phenomenon (like catastrophic mass extinction). (3) This particular error of scaling embodies our worst mental habit of interpreting other ranges of size, or other domains of time, in our own limited terms.

For example, Mettler, Gregg, and Schaffer's textbook on *Population Genetics and Evolution* (1988, p. 304) states: "The punctuated equilibrium theory, on the other hand, holds that sudden appearance is due to rapid selection, rather than rapid spread, and that stasis results because evolutionary change occurs in large discrete jumps rather than by a series of gene substitutions. There really are no gradual changes or intermediate stages." In their volume on *Sexual Selection* for the prestigious *Scientific American* series (1989, p. 83), Gould and Gould (no relation) write: "The proven ability of selection to operate quickly in at least some cases, has led to the widely publicized theory of punctuated evolution. According to the original version, no intermediate forms are preserved simply because there are no halfway creatures in the first place: new species come into being in single steps."

Turning to misscaling in the other direction, Wessells and Hopson (1988, pp. 1073–1074) equate punctuated equilibrium with the origin of new *Baupläne* and faunal turnovers in mass extinction: "The central tenet of punctuated equilibrium is that a lineage of organisms arises by some dramatic changes—say, the rapid acquisition of body segmentation in annelids—after which there is a lengthy period with far fewer radical changes taking place." They then write of two great evolutionary bursts in the history of sea urchins (following the late Cambrian and Late Triassic mass extinctions). "One might interpret this record to reflect two 'punctuations' in the Ordovician and early Jurassic periods. And the 'equilibrium' times would be from the Ordovician through the Triassic and, perhaps, from the Jurassic to today. This record may be consistent with the punctuated equilibrium hypothesis."

Chaisson's ambitious textbook on nearly everything—*Universe: An Evolu-*

tionary Approach to Astronomy—equates punctuated equilibrium with faunal turnovers in mass extinction. His section entitled punctuated equilibrium (1988, p. 481) begins by stating: "The fossil record of the history of life on Earth clearly documents many periods of mass extinction." He then adds (p. 483): "Punctuated equilibrium merely emphasizes that the rate of evolutionary change is not gradual. Instead, the 'motor of evolution' occasionally speeds up during periods of dramatic environmental change—such as cometary impacts, reversals of Earth's magnetism, and the like. We might say that evolution is imperceptibly gradual most of the time and shockingly sudden some of the time." But Chaisson's "imperceptibly gradual" times—the intervals of so-called "normal" evolution between episodes of mass extinction—build their incremental trends by stair steps based on the true rhythm of punctuated equilibrium in rapid origin and subsequent stasis of individual species.

However, even in this maximally constrained and conservative world of textbooks, some reform has emerged from punctuated equilibrium. Above all, the debate on punctuated equilibrium prodded the authors of nearly all major textbooks to include (often as entirely new sections) substantial and explicit material on macroevolution—in contrast with the appalling absence or shortest shrift awarded to the topic in standard textbooks of the 1950s and 1960s (as documented in *SET*, Chapter 7, pp. 579–584).

At the level of details and content, many textbooks provide gratifyingly accurate (if often critical) definitions and appraisals of punctuated equilibrium. Unsurprisingly, textbooks written by paleontologists have generally provided the clearest treatments. Nield and Tucker (1985, p. 162), for example, stress the role of punctuated equilibrium in rendering the fossil record operational for evolutionary studies: "We usually witness sudden appearance of new species, followed by long static periods and ultimate extinction. Formerly it was supposed that this fact reflected the incompleteness of the fossil record, but the belief now is that it represents something very important about the evolutionary process." Similarly, Dott and Prothero (1994, p. 61) end their section on "The fossil record and evolution" by stating:

> To some paleontologists, species are more than just populations and genes. They are real entities that seem to have some kind of internal stabilizing mechanism preventing much phenotypic change, even when selection forces change. Clearly, the fossil record produces some unexpected results that are not yet consistent with everything we know about living animals and labora-

> tory experiments. This is good news. If the fossil record taught us nothing that we didn't know already by biology, there wouldn't be much point to evolutionary paleontology.

Finally, Dodson and Dodson (1990, p. 520) provide an excellent summary on the implications of punctuated equilibrium for evolutionary theory:

> Most evolutionary biologists are prepared to acknowledge that punctuated equilibrium is an important phenomenon, even if somewhat less so than its more enthusiastic advocates claim. And population geneticists, who have labored mainly to clarify the genetic basis of evolutionary change, may now have to give greater attention to the problem of evolutionary stasis . . . Thus, the question is not whether punctuated equilibria occur, but how general they are and whether they can be absorbed into the modern evolutionary synthesis.

Among the best treatments of punctuated equilibrium in textbooks, I would cite Kraus's book for high school biology (1983), the continuing efforts of Alters and McComas (1994) to design a high school curriculum based on punctuated equilibrium, and the college textbooks by Avers (1989) and Price (1996). Much of the graphical material has also been highly useful—as in Price's ingenious inclusion of both spatial and temporal dimensions to show how allopatric speciation yields both stasis and punctuation in the fossil record—see Fig. A-3.

As a model of excellence, and of clear, accurate, stylish writing as well, the treatment of punctuated equilibrium in the most popular textbook of the 1980s embodies the reasons for this volume's well-deserved status. Helena Curtis was a thoughtful writer, not a professional biologist, but she mastered the material and could write circles around her competition. (I also know, from personal conversation, that she initially felt quite skeptical about the importance of punctuated equilibrium—so her generous treatment records the judgment of a critical observer, not a partisan.) I reproduce below most of Curtis and Barnes's (1985, pp. 556–557) section on "Punctuated Equilibria," the closing topic in their chapter on "evolution." If these authors could be so fair and accurate, then textbooks can achieve excellence as a genre, and punctuated equilibrium lies safely within the domain of the understandable, the informative, and the interesting:

> Although the fossil record documents many important stages in evolutionary history, there are numerous gaps . . . Many more fossils have, of course,

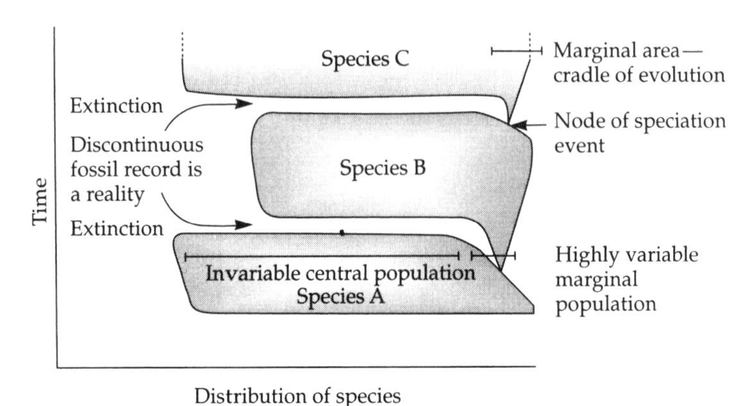

A-3. An excellent textbook figure of punctuated equilibrium from Price, 1996. He includes, in a way that had never occurred to me or Eldredge, both spatial and temporal dimensions to show how allopatric speciation yields both stasis and punctuation in the fossil record. His own caption reads: "A general scenario for the punctuated equilibrium concept of evolutionary change. Visualize a species change pattern that appears when time is measured vertically through a stratigraphic section of rock, and ponder how many such rock sections would be needed to reveal at least the distribution of species in the central population."

been discovered in the 100 years since Darwin's death. Nevertheless, fewer examples of gradual change within forms have been found than might have been expected. Until recently, the discrepancy between the model of slow phyletic change and the poor documentation of such change in much of the fossil record has been ascribed to the imperfection of the fossil record itself.

About a decade ago, two young scientists, Niles Eldredge of the American Museum of Natural History and Stephen Jay Gould of Harvard University, ventured the radical proposal that perhaps the fossil record is not so imperfect after all. Both Eldredge and Gould have backgrounds in geology and invertebrate paleontology, and both were impressed with the fact that there was very little evidence of phyletic change in the fossil species they studied. Typically, a species would appear abruptly in the fossil strata, last 5 million to 10 million years, and disappear, apparently not much different than when it first appeared. Another species, related but distinctly different—"fully formed"—would take its place, persist with little change, and disappear equally abruptly. Suppose, Eldredge and Gould argued, that these long periods of no change ("stasis" is the word they use) punctuated by gaps are not flaws in the record but *are* the record, the evidence of what really happens.

How could it be that a new species would make such a sudden appearance? They found their answer in the model of allopatric speciation. If new species formed principally in small populations on the geographic periphery of the range of the species, if speciation occurred rapidly (by rapidly, paleontologists mean in thousands rather than millions of years), and if the new species then outcompeted the old one, taking over its geographic range, the resulting fossil pattern would be the one observed . . .

As the new model has become more fully developed, particularly by Steven M. Stanley of Johns Hopkins (also a paleontologist), it has become more radical. Its proponents now argue that not only is cladogenesis the principal mode of evolutionary change (as Mayr stated some 30 years ago) but that natural selection occurs among species as well as among individuals. . . . In this new formulation, species take the place of individuals, and speciation and extinction substitute for birth and death. In short, there are two mechanisms of evolution, according to this proposal: in one, natural selection acts on the individual, and in the other, it acts on the species.

Will the punctuated equilibrium model be assimilated into the synthetic theory? Or will some radical new concept of evolutionary mechanisms spread through the scientific strata, outcompeting the old ideas? At this writing, it is too early to tell. All that is clear is that this proposal has stimulated a vigorous debate, a reexamination of evolutionary mechanisms as currently understood, and a reappraisal of the evidence. All of this indicates that evolutionary biology is alive and well and that scientists are doing what they are supposed to be doing—asking questions. Darwin, we think, would have been delighted.

The Personal Aspect of Professional Reaction

Among false dichotomies, the strict division of a professional's reaction into scientific conclusions based on legitimate judgment and personal reasons rooted in emotional feelings represents a particularly naïve and misleading parsing of human motivation. Our analytic schemes do require some heuristic divisions, but the notion that good reason stands in primal antithesis to bad feelings surely caricatures the depth and complexity of human reactions. All scientific critiques arise in concert with a complex and often unconscious range of emotional responses (not to mention a social and cultural context, which scientists, trained to absorb the myth of objectivity, are particularly disinclined to recognize). The fact that we can analyze the pure logic of an argument *a posteriori* tells us little about the ineluctably nonlogical motives and feelings behind any decision to frame such an argument at a given time, and in the hope of a particular outcome.

Nonetheless, because punctuated equilibrium has provoked so much commentary of a personal nature from scientific colleagues, often expressed with unusual intensity both pro and con (for statements published in professional literature), I don't know how else to parse the content in this case. I discussed the rich and numerous intellectual critiques in the main body of this book, but what can be done with the large residuum of unusual personal commentary? I cannot simply ignore it, both because the discussion would then be so selectively incomplete, and also (for personal reasons of course) because I find so much of the most negative commentary so false and unfair—and I do wish to exercise what Roberts Rules calls a "point of personal privilege" as expressed in a basic right of reply. Thus, I have tried to separate personal commentary (in this section) from the critical discourse of ideas, while acknowledging the small psychological sense of such a division. The heuristic advantages of thus splitting each side's clutter from the other's content may justify this procedure.

THE CASE *AD HOMINEM* AGAINST PUNCTUATED EQUILIBRIUM

I should state up front that I regard this discourse as rooted in little more than complex fallout from professional jealousy, often unrecognized and therefore especially potent. I shall, in the next subsection, own what I regard as the share of responsibility that Eldredge and I bear for standard misconceptions about punctuated equilibrium, but I believe that the *ad hominem* literature on this subject primarily records inchoate and unanalyzed feelings and habits of thought among our most negatively inclined colleagues.

The common denominator to all these expressions lies in a charge—the basis of most claims on the low road of accusation *ad hominem*—that punctuated equilibrium is false, empty, or trivial, and that the volume of discussion, both in professional literature and general culture, can only record our trickery, our bombast, our dishonesty, our quest for personal fame, or (in the kindest version) our massive confusion. (But what then must these detractors conclude about the intellectual acumen of so many of their peers who support punctuated equilibrium, or at least find the discussion interesting?) I read the case *ad hominem* as a brief composed of two charges, culminating in what has almost become an "urban legend" equivalent in veracity to those alligators in the sewers of New York City, indefensible in fact or logic, but propagated by confident repetition within the club of true believers. I will respond to each point by analyzing the passages from my writing that have become virtually canonical as supposed confirmation.

1. In the kindest version, we are depicted as merely confused and overly hopeful. We develop a good little modest idea that might help the benighted community of paleontologists, but we then begin to suffer delusions of grandeur, and to believe that we might have something to say about evolution in general. (We really don't of course, for punctuated equilibrium only confirms all the beliefs and predictions of the Modern Synthesis.) We now make the crucial error of deciding that our punctuations must require a new evolutionary mechanism unsuspected by Darwinian gradualism—probably a new style of genetic change directing the process of speciation. But we have only made a fundamental mistake in scaling, for our punctuations are slow enough in microevolutionary time to record the ordinary workings of natural selection.

I cringe when I read characterizations like this because such statements only indicate that the perpetrators haven't read our papers, and must either be expressing their fears or some undocumented gossip that passes for wisdom along academic grapevines. As quotations throughout this chapter amply demonstrate, we have always taken a position contrary to these charges. We didn't err in failing to recognize that a paleontologist's punctuation equals a microevolutionist's continuity. Rather, *we based our theory upon this very idea from the start,* by demonstrating that the conventional allopatric model of speciation scales as a punctuation, not as gradual change through a long sequence of strata, in geological time. Clearly, we could not have located anything theoretically radical in the punctuations of our theory—since we built our model by equating these punctuations with ordinary microevolutionary events of peripatric speciation!

It is true that we staked no unconventional claims for evolutionary theory in our original paper (Eldredge and Gould, 1972)—while urging substantial reform of paleontological practice—but only because we hadn't yet recognized the implications of punctuated equilibrium in this domain. It is also true that we began to urge theoretical reform in subsequent papers (beginning in Gould and Eldredge, 1977, and continuing in Gould, 1982b, 1989b, and Gould and Eldredge, 1993), but we have never based these proposals on the speed or nature of punctuations. Again, as demonstrated by citations throughout this chapter, we locate any revisionary status for punctuated equilibrium in its suggestions about the nature of stasis, and particularly its implications for attributing macroevolutionary phenomena to causes operating on the differential success of species treated as Darwinian individuals. Ordinary speciation remains fully adequate to explain the causes and phenomenology of punctuation.

2. If (as argument one holds) punctuated equilibrium includes no theoretical novelty, and if the theory has enjoyed such intense discussion in both popular and professional literature, then we must have created this anomaly by using rhetorical skills to flog our empty notions in a quest for personal fame. So we hyped, and the media followed like sheep. Dawkins (1986) writes, for example: "Punctuationism is widely thought to be revolutionary and antithetical to neo-Darwinism for the simple reason that its chief advocates have said that it is: said so, moreover, in loud and eloquent voices, making frequent and skillful use of the mass media. The theory, in short, stands out from other glosses on the neo-Darwinian synthesis in one respect only: it has enjoyed brilliant public relations and stage management." (Do I detect a whiff of jealousy in this expostulation?)

I reject this argument about mass media on two grounds: first, for its condescending assumption of such pervasive and universal incompetence within the fourth estate (my section on the press, pages 315–319, includes several examples of highly accurate and critical coverage); and second, for its false and unflattering conjectures about our procedures and integrity. Neither I nor Eldredge has ever engaged in "skillful use" or "stage management" of media.

I have no personal objection to active courting of journalists by scientists, so long as fairness and integrity do not become compromised, especially by caricature, oversimplification or dumbing down. The public is not stupid and can handle scientific material at full conceptual complexity. (Necessary simplification of terminology, and avoidance of jargon, need not imply any sacrifice of intellectual content.) But, as a matter of personal preference, I have never approached the media in this manner. I have never arranged a press conference or meeting, or even placed a phone call to a reporter. I try to be responsive when approached, but I have been entirely reactive in my contact with media on the subject of punctuated equilibrium. Moreover, although I occupied the most "bully pulpit" in America for popular writing about evolution—my monthly column in *Natural History Magazine*, published from January 1974 to January 2001—I never used this forum to push punctuated equilibrium. Of 300 successive essays, I devoted only two to this subject. No ethical or intellectual barrier stood against more extensive treatment, but I preferred to use the great privilege of this forum to learn about new evolutionary byways that I would otherwise not have had time to study, rather than to advocate what I had already treated in the greater depth of professional journals.

So if this second *ad hominem* argument won't even wash for a presumably naive press, how can colleagues regard the attention of a far more sophisticated professional community as nothing but a spinoff from our hype and rhetoric?

When this charge has been laid against me, cited evidence almost always rests upon two supposed claims (and their canonical quotations) expressed in my putatively most radical paper of 1980, entitled: "Is a new and general theory of evolution emerging?" I wrote this paper for the 5th anniversary of *Paleobiology,* as a companion piece to a longer analysis of biological research in our profession: "The promise of paleobiology as a nomothetic, evolutionary discipline" (see Gould, 1980a and b).

The received legend about this paper—I really do wonder how many colleagues have ever based their comments on reading this article with any care, or even at all—holds that I wrote a propagandistic screed featuring two outrageously exaggerated claims: first, the impending death of the Modern Synthesis; and second, the identification of punctuated equilibrium as the exterminating angel (or devil). I do not, in fact and in retrospect (but not in understatement), regard this 1980 paper as among the strongest, in the sense of most cogent or successful, that I have ever written—but neither do I reread it with any shame today. Some of my predictions have fared poorly, and I would now reject them—scarcely surprising for a paper that tried to summarize all major theoretical revisions then under discussion among evolutionists. For example, I then read the literature on speciation as beginning to favor sympatric alternatives to allopatric orthodoxies at substantial relative frequency, and I predicted that views on this subject would change substantially, particularly towards favoring mechanisms that would be regarded as rapid even in microevolutionary time. I now believe that I was wrong in this prediction.

But the relatively short section devoted to punctuated equilibrium (Gould, 1980b, pp. 125–126) presents this subject in a standard and unsurprising manner, and I would not change any major statement in this part of the paper. (My reassessment away from high relative frequency for rapid speciation in microevolutionary time, and back to the peripatric orthodoxy of our original views, represents a rethinking of another section of this 1980 paper, and does not speak to the validity of punctuated equilibrium. As I have emphasized throughout this chapter, punctuated equilibrium was formulated as the expected macroevolutionary expression of conventional allopatric speciation—so a return to this conventional model can scarcely threaten the theory's validity!)

THE SUPPOSED GENERAL DEATH OF THE SYNTHESIS. Given the furor provoked, I would probably tone down—but not change in content—the quotation that has come to haunt me in continual miscitation and misunderstanding by critics: "I have been reluctant to admit it—since beguiling is often forever—but if Mayr's characterization of the synthetic theory is accurate, then that theory, as a general proposition, is effectively dead, despite its persistence as textbook orthodoxy" (Gould, 1980b, p. 120). (I guess I should have written the blander and more conventional "due for a major reassessment" or "now subject to critical scrutiny and revision," rather than "effectively dead." But, as the great Persian poet said, "the moving finger writes, and having writ . . ." and neither my evident piety nor obvious wit can call back the line—nor would tears serve as a good emulsifier for washing out anything I ever wrote!)

Yes, the rhetoric was too strong (if only because I should have anticipated the emotional reaction that would then preclude careful reading of what I actually said). But I will defend the content of the quotation as just and accurate. First of all, I do not claim that the synthetic theory of evolution is wrong, or headed for complete oblivion on the ashheap of history; rather, I contend that the synthesis can no longer assert full sufficiency to explain evolution at all scales (remember that my paper was published in a paleobiological journal dedicated to studies of macroevolution). Two statements in the quotation should make this limitation clear. First of all, I advanced this opinion only with respect to a particular, but (I thought) quite authoritative, definition of the synthesis: "if Mayr's characterization of the synthetic theory is accurate." Moreover, I had quoted Mayr's definition just two paragraphs earlier. The definition begins Mayr's chapter on "species and transspecific evolution" from his 1963 classic—the definition that paleobiologists would accept as most applicable to their concerns. Mayr wrote (as I explicitly quoted): "The proponents of the synthetic theory maintain that all evolution is due to the accumulation of small genetic changes, guided by natural selection, and that transspecific evolution is nothing but an extrapolation and magnification of the events that take place within populations and species."

Second, I talked about the theory being dead "as a general proposition," not dead period. In the full context of my commentary on Mayr's definition, and my qualification about death *as a full generality,* what is wrong with my statement? I did not proclaim the death of Darwinism, or even of the strictest form of the Modern Synthesis. I stated, for an audience interested in macroevolutionary theory, that Mayr's definition (not the extreme statement of a marginal figure, but an explicit characterization by the world's greatest expert in his most famous book)—with its two restrictive claims for (1) "*all* evolution" due to natural selection of small genetic changes, and (2) transspecific evolution as "nothing but" the extrapolation of microevolutionary events—must be firmly rejected if macroevolutionary theory merits any independent status, or features any phenomenology requiring causal explanation in its own domain. If we embrace Mayr's definition, then the synthesis is "effectively dead" "as a general proposition"—that is, as a theory capable of providing a full and exclusive explanation of macroevolutionary phenomena. Wouldn't most evolutionary biologists agree with my statement today?

Nonetheless, I was reviled in many quarters, and in prose far more intemperate and personal than anything I ever wrote, for proclaiming the death of Darwinism, and the forthcoming enshrinement of my own theory as a replacement (see, for example, A. Huxley, 1982; Thompson, 1983; Cain, 1988; Vogel, 1983; Ayala, 1982; Stebbins and Ayala, 1981a and b; Mayr, 1982; and Grant, 1983, under the title: "The synthetic theory strikes back").

Many reasons underlie this error, and I do accept some responsibility for my flavorful prose (but not for any lack of clarity in intended meaning, or for any statement stronger than Mayr's dismissive words about my own profession of macroevolution). One common reason, perhaps the most prominent of all, arises from careless scholarship and cannot be laid at my doorstep. I provided the full quotation that offended so many colleagues, along with Mayr's accompanying words, so necessary to grasp the definition that I used. But my statement is usually quoted in deceptively abridged form, leading to a false reading clearly opposite to what I intended. I usually find my words cited in the following abridgment: "The synthetic theory . . . is effectively dead, despite its persistence as textbook orthodoxy." Much commentary has been based upon this truncated and distorted version, not on my actual words. Fill in those three dots before you fire.

HOMO UNIUS LIBRI. An old and anonymous Latin proverb states: *cave ab homine unius libri*—beware the man of one book. I do appreciate the attention that punctuated equilibrium has received, and, as a fallible mortal, I am not adverse to the recognition that this debate has brought me. But as a curious and general consequence of extensive publicity for a single achievement, the totality of one's work then tends to be read as a long and unitary commentary upon this singular idea or accomplishment. The Latin motto should therefore be read from both ends: we should be wary of a person who has only one good idea, but we should also not automatically assimilate an entire life by synecdoche to the single aspect we know best. Leonardo's war machines bear little relationship to the *Mona Lisa;* Newton's chronology of ancient kingdoms never mentions gravity or the inverse square law; and Mickey Mantle was also the best drag bunter and fastest runner in baseball.

Perhaps I should be flattered by the implied importance thus accorded to punctuated equilibrium, but I do maintain interests, some just as consuming, and some (I hope) just as replete with implications for evolutionary theory. Critics generally complete their misunderstanding of my 1980 paper by first imagining that I proclaimed the total overthrow of Darwinism, and then supposing that I intended punctuated equilibrium as both the agent of destruction and the replacement. But punctuated equilibrium does not occupy a major, or even a prominent, place in my 1980 paper.

This article tried to present a general account of propositions within the Modern Synthesis that, in my judgment, might require extensive revision or enlargement, especially from the domain of macroevolution. I did speak extensively—often quite critically—about the reviled work of Richard Goldschmidt, particularly about aspects of his thought that might merit a rehearing. This material has often been confused with punctuated equilibrium by people who miss the crucial issue of scaling, and therefore regard all statements about rapidity at any level as necessarily unitary, and necessarily flowing from punctuated equilibrium. In fact, as the long treatment in Chapter 5 of this book should make clear, my interest in Goldschmidt resides in issues bearing little relationship with punctuated equilibrium, but invested instead in developmental questions that prompted my first book, *Ontogeny and Phylogeny* (Gould, 1977). The two subjects, after all, are quite separate, and rooted in different scales of rapidity—hopeful monsters in genuine saltation, and punctuated equilibrium in macroevolutionary punctuation (produced by ordinary allopatric speciation). I do strive to avoid the label of *homo unius libri.* I have even written a book about baseball, and another about calendrics and the new millennium.

The section on punctuated equilibrium in my 1980 paper is both short in extent, and little different in content from my treatment of the subject elsewhere. I began with the usual definition: "Our model of 'punctuated equilibria' holds that evolution is concentrated in events of speciation and that successful speciation is an infrequent event punctuating the stasis of large populations that do not alter in fundamental ways during the millions of years that they endure" (p. 125). I then made my usual linkage to ordinary allopatric speciation, not to any novel or controversial mechanism of microevolution. Moreover, I emphasized the scaling error that so often leads people to confuse punctuated equilibrium with saltationism:

> Speciation, the basis of macroevolution, is a process of branching. And this branching, under any current model of speciation—conventional allopatry to chromosomal saltation—is so rapid in geological translation (thousands of years at most compared with millions for the duration of most fossil species) that its results should generally lie on a bedding plane, not through the thick sedimentary sequence of a long hillslope . . . It [gradualism] represents, first of all, an incorrect translation of conventional allopatry. Allopatric speciation seems so slow and gradual in ecological time that most paleontologists never recognized it as a challenge to the style of gradualism—steady change over millions of years—promulgated by custom as a model for the history of life (p. 125).

Finally, I stressed that the radical implications of punctuated equilibrium lay in proposed explanations for such macroevolutionary phenomena as cladal trends, not in any proposal for altered mechanisms of microevolution: "Evolutionary trends therefore represent a third level superposed upon speciation and change within demes . . . Since trends 'use' species as their raw material, they represent a process at a higher level than speciation itself. They reflect a sorting out of speciation events . . . What we call 'anagenesis,' and often attempt to delineate as a separate phyletic process leading to 'progress,' is just accumulated cladogenesis filtered through the directing force of species selection."

IN TRES PARTES DIVISA EST: THE 'URBAN LEGEND' OF PUNCTUATED EQUILIBRIUM'S THREEFOLD HISTORY. The opponents of punctuated equilibrium have constructed a fictional history of the theory, primarily (I suppose) as a largely unconscious expression of their hope for its minor importance, and their jealousy towards its authors. This history even features a definite sequence of stages, constructed to match a classic theme of Western sagas: the growth, exposure and mortification of hubris (try Macbeth as a prototype, but he dies before reaching the final stage of penance; so try

Faust instead, who lusts for the world and ends up finding satisfaction in draining a swamp). This supposed threefold history of punctuated equilibrium also ranks about as close to pure fiction as any recent commentary by scientists has ever generated.

In stage one, the story goes, we were properly modest, obedient to the theoretical hegemony of the Modern Synthesis, and merely trying to bring paleontology into the fold. But the prospect of worldly fame beguiled us, so we broke our ties of fealty and tried, in stage two, to usurp power by painting punctuated equilibrium as a revolutionary doctrine that would dethrone the Synthesis, resurrect the memory of the exiled martyr (Richard Goldschmidt), and reign over a reconstructed realm of theory. But we were too big for our breeches, and the old guard still retained some life. They fought back mightily and effectively, exposing our bombast and emptiness. We began to hedge, retreat, and apologize, and have been doing so ever since in an effort to regain grace and, chastened in stage three, to sit again, in heaven or Valhalla, with the evolutionary elite.

Such farfetched fiction suffers most of all from an internal construction that precludes exposure and falsification among true believers, whatever the evidence. Purveyors of this myth even name the three stages, thus solidifying the false taxonomy. Dawkins (1986), for example, speaks of the "grandiloquent era . . . of middle-period punctuationism [which] gave abundant aid and comfort to creationists and other enemies of scientific truth." In the other major strategy of insulation from refutation, supporters of this "urban legend" about the modest origin, bombastic rise, and spectacular fall of punctuated equilibrium forge a tale that allows them to read any potential disconfirmation as an event within the fiction itself. (Old style gradualism pursued exactly the same strategy in reading contrary data as marks of imperfect evidence within the accepted theory—and thus could not be refuted from within. I am struck by the eerie similarity between the structure of the old theory and the historical gloss invented by opponents of a proposed replacement.)

In particular, and most offensive to me, the urban legend rests on the false belief that radical, "middle-period" punctuated equilibrium became a saltational theory wedded to Goldschmidt's hopeful monsters as a mechanism. I have labored to refute this nonsensical charge from the day I first heard it. But my efforts are doomed within the self-affirming structure of the urban legend. We all know, for so the legend proclaims, that I once took the Goldschmidtian plunge. So if I ever deny the link, I can only be retreating from an embarrassing error. And if I continue to deny the link with force and gusto, well, then I am only backtracking even harder (into stage 3) and apologizing (or obfuscating) all the more. How about the obvious (and accurate) alternative: that we never made the Goldschmidtian link; that this common error embodies a false construction; and that our efforts at correction have always represented an honorable attempt to relieve the confusion of others.

But the urban legend remains too simplistically neat, and too resonant with a favorite theme of Western sagas, to permit refutation by mere evidence. So Dennett (1995, pp. 283–284) writes: "There was no mention in the first paper of any radical theory of speciation or mutation. But later, about 1980, Gould decided that punctuated equilibrium was a revolutionary idea after all . . . [But] it was too revolutionary, and it was hooted down with the same sort of ferocity the establishment reserves for heretics like Elaine Morgan. Gould backpedaled hard, offering repeated denials that he has ever meant anything so outrageous." And Halstead (1985, p. 318) wrote of me (with equal poverty in both logic and grammar): "He seems to be setting up a face-saving formula to enable him to retreat from his earlier aggressive saltationism, having had a bit of a thrashing, his current tack is to suggest that perhaps we should keep the door open in case he can find some evidence to support his pet theories so let us be 'pluralist.'"

I do not, of course, claim that our views about punctuated equilibrium have never changed through the years of debate (only a dull and uninteresting theory could remain so static in the face of such wide discussion). Nor do I maintain a position that would be even sillier—namely, that we made no important errors requiring corrections to the theory. Of course we made mistakes, and of course we have tried to amend them. But I look upon the history of punctuated equilibrium (from my partisan vantage point of course) as a fairly standard development for successful theories in science. We did, indeed, begin modestly and expand outward thereafter. (In this sense, punctuated equilibrium has grown in theoretical scope, primarily as macroevolutionary theory developed and became better integrated with the rest of evolutionary thought—and largely through articulation of the hierarchical model, as discussed in *The Structure of Evolutionary Theory*, Chapter 8).

We started small as a consequence of our ignorance and lack of perspective, not from modesty of basic temperament. As stated before, we simply didn't recognize, at first, the interesting implications of punctuated equilibrium for macroevolutionary theory—primarily gained in treating species as Darwinian individuals for the explanation of trends, and in exploring the extent and causes of stasis. With the help of S. M. Stanley, E. S. Vrba and other colleagues, we developed these implications over the years, and the theory grew accordingly. But we never proposed a radical theory for punctuations (ordinary speciation scaled into geological time), and we never linked punctuations to microevolutionary saltationism.

Of course we made mistakes—serious ones in at least two cases—and the theory has changed and improved by correcting these errors. In particular, and as documented extensively in *SET*, Chapter 8, we were terribly muddled for several years about the proper way to treat, and even to define, selection at the level of species—the most important of all theoretical spinoffs from punctuated equilibrium. We confused sorting with selection (see Vrba and Gould, 1986, for a resolution). We also did not properly formulate the concept of emergence at first; and we remained confused for a long time about emergence of characters vs. emergence of fitness as criteria for species selection (Lloyd and Gould, 1993; Gould and Lloyd, 1999). In retrospect, I am chagrined by the long duration of our confusion, and its expression in many of our papers. But I think that we have now resolved these difficult issues.

Secondly, as discussed on pages 77–79, I think that we originally proposed an incorrect reason for the association of rapid change with speciation. But I believe that we portrayed the phenomenology correctly, and that we have, with the help of Futuyma's (1987) suggestions, now developed a proper explanation. Thus, the theory of punctuated equilibrium has altered substantially to correct these two errors. Interestingly (and ironically), however, these important changes do not figure at all in the deprecating claims of the urban legend about our supposed retreats and chameleon-like redefinitions—for our detractors hardly recognize the existence of punctuated equilibrium's truly radical claim for evolutionary theory: its implications for selection above the species level, and for the explanation of trends.

Punctuated equilibrium, in short, has enjoyed true Darwinian success through the years: it has struggled, survived, changed and expanded. But the theoretical evolution of punctuated equilibrium belongs to the sphere of cultural change with its Lamarckian mode of transmission by direct passage of acquired improvements. Thus, the theory need not remain in Darwinian stasis, but may grow—as it has—in (gulp!) a gradualist and progressive manner.

The saltationist canard has persisted as our incubus. The charge could never be supported by proper documentation, for we never made the link or claim. All attempts collapse upon close examination. Dennett, for example, who insists (1997, p. 64) that "for a while he [Gould] had presented punctuated equilibrium as a revolutionary 'saltationist' alternative to standard neo-Darwinism," documents his supposed best case by assuring readers (1995, p. 285) that "for a while, Gould was proposing that the first step in the establishment of any new species was a doozy—a non-Darwinian saltation." Dennett directly follows this claim with his putative proof, yet another quotation from my 1980 paper, which he renders as follows: "Speciation is not always an extension of gradual, adaptive allelic substitution to greater effect, but may represent, as Goldschmidt argued, a different style of genetic change—rapid reorganization of the genome, perhaps non-adaptive" (Gould, 1980b, p. 119).

I regard Dennett's case as pitiful, but the urban legend can offer no better. First of all, this quotation doesn't even refer to punctuated equilibrium, but comes from a section of my 1980 paper on the microevolutionary mechanics of speciation. Secondly, Dennett obviously misreads my statement in a backwards manner. I am trying to carve out a *small* theoretical space for a style of microevolutionary rapidity at low relative frequency—as clearly stated in my phrase "not always an extension of gradual . . ." But Dennett states that I am proposing this mechanism as a general replacement for gradual microevolutionary change in *all* cases of speciation—"the first step in the establishment of *any* new species" in *his* words. But *my* chosen phrase—"not always"—clearly means "most of the time," and cannot be read as "never." In short, I made a plea for pluralism, and Dennett charges me with usurpation. Then, when I try to explain, I am accused of beating a retreat to save face. When placed in such a double bind, one can only smile and remember Schiller's famous dictum: *Mit Dummheit kämpfen die Götter selbst vergebens.*

Finally, the claim that we equated punctuated equilibrium with saltation makes no sense within the logical structure of our theory—so, unless we are fools, how could we ever have asserted such a proposition? Our theory holds, as a defining statement, that ordinary allopatric speciation, unfolding gradually at microevolutionary scales, translates to punctuation in geological time. Microevolutionary saltation also scales as a punctuation—so the distinction between saltation and standard allopatry becomes irrelevant for punctuated equilibrium, since both yield the same favored result!

Moreover, the chronology of debate proves that we did not issue disclaimers on this subject only to cover our asses as we retreated from exaggerations of our supposed second phase, because we have been asserting this clarification from the very beginning—that is, from the first paper we ever wrote to comment upon published reactions to punctuated equilibrium. Our first response appeared in 1977, long before we issued the supposed clarion call of our false revolution in 1980. We wrote (Gould and Eldredge, 1977, p. 121), under the heading "Invalid claims of gradualism made at the wrong scale": "The model of punctuated equilibria does not maintain that nothing occurs gradually at any level of evolution. It is a theory about speciation and its deployment in the fossil record. It claims that an important pattern, continuous at higher levels—the 'classic' macroevolutionary trend—is a consequence of punctuation in the evolution of species. It does not deny that allopatric speciation occurs gradually in ecological time (though it might not—see Carson, 1975), but only asserts that this scale is a geological microsecond."

We have never changed this conviction, and we have always tried to correct any confusion of scaling between saltation and punctuation, even in papers written during the supposed apogee of our revolutionary ardor, during illusory stage 2 of the urban legend. For example, under the heading of "The relationship of punctuated equilibrium to macromutation," I wrote in 1982 (p. 88): "Punctuated equilibrium is not a theory of macromutation . . . it is not a theory of any genetic process . . . It is a theory about larger-scale patterns—the geometry of speciation in geological time. As with ecologically rapid modes of speciation, punctuated equilibrium welcomes macromutation as a source for the initiation of species: the faster the better. But punctuated equilibrium clearly does not require or imply macromutation, since it was formulated as the expected geological consequence of Mayrian allopatry."

AN INTERLUDE ON SOURCES OF ERROR

With such limited skills in sociology and psychology, and from too close a personal and partisan standpoint, I cannot claim much insight into the general sources of persistent nonscientific errors among professional colleagues. But I wish to offer a few thoughts, at least to separate what Eldredge and I must own from the truly unfair, and often intemperate, charges so often made against us.

Any complex situation arises from multiple causes, with inevitable shortcomings on both sides of any basically dichotomous issue. But when I list our own faults and failures, I find nothing of great depth, and no indication of any sustained stupidity, carelessness, lack of clarity, or malfeasance. Thus, I continue to feel far more aggrieved than intemperate—although I wouldn't give up this lifetime's intellectual adventure for any alternative construction of a scientific career.

For our part, I think that critics can identify three sources of potential confusion that might legitimately be laid at our doorstep, and might have been prevented had our crystal ball been clearer.

1. In our original paper (Eldredge and Gould, 1972), but not subsequently, we failed to explain, in a sufficiently didactic and explicit manner, that when paleontologists use such terms as "rapid," "sudden," or "instantaneous," they refer to expressions of events at geological scales, and not to rates of change in microevolutionary time. But we cannot be blamed for anything more than a failure to anticipate the range of interest that our paper would generate. After all, we wrote this paper for paleontologists, and never expected a wider audience. We used the standard terminology of our profession, well known and understood by all members of the clan. Indeed, few nonpaleontologists ever read this original article, published in an obscure symposium volume with a small press run. From 1977 on, in all papers widely read by neontologists, and serving as a basis for enlarged discussion, we clearly explained the differences in scaling between micro- and macroevolutionary rates.

2. As acknowledged on pages 329–330, I did use some prose flourishes that, in a context of considerable suspicion and growing jealousy, probably fanned the flames of confusion. Although I never stated anything unclearly, and committed no logical errors that could legitimately have inspired a resulting misreading, I should have toned down my style in a few crucial places.

3. We may have sown some confusion by using partially overlapping terminology for a specific theory (punctuated equilibrium), and for the larger generality (punctuational styles of change) in which that theory lies embedded. But this taxonomic usage does stress a legitimate commonality that we wished to emphasize. We also chose and used our terms with explicit consistency and clear definitions—so careful reading should have precluded any misunderstanding.

The testing and development of punctuated equilibrium—a well-defined and circumscribed theory about the origin and deployment of speciation events in geological time—has always been our major concern. But as students of evolution, we have also been interested in the range of applicability for the geometric generalization represented by this theory—the unfolding of change as occasional punctuation within prevailing stasis, rather than as gradualistic continuity—to other scales of space and time, and for other causes and phenomena of life's history. We have called this more general and abstract style of change "punctuational," and have referred to the hypothesis favoring its generality as "punctuationalism."

We have always been careful and clear about the differences between our specific theory of punctuated equilibrium and the general proposition of punctuational change. (In fact, we strove to be explicit, even didactic, about this distinction because we recognized the confusion that might arise otherwise.) But perhaps the words are too close to expect general understanding of the distinction, particularly from hostile critics who have invested their emotional ire in the legend that we have been pursuing an imperialistic, grandstanding quest to enshrine punctuated equilibrium as a new paradigm for all the evolutionary sciences.

Still, as a statement of a basic intellectual principle, why should we allow ourselves to be forced into suboptimal decisions by the least thoughtful and most emotionally driven forms of misunderstanding among critics? Punctuationalism *is* the right and best word for the general style of change expressed by punctuated equilibrium as a specific example at a circumscribed level and phenomenology. As long as we take special care to be clear and explicit about the distinction, why should we sacrifice this most appropriate form of naming? I believe that we have been scrupulous in characterizing and highlighting this point, right from our first introduction in 1977, when we began a section entitled "Towards a general philosophy of change" with these words: "Punctuated equilibria is a model for discontinuous tempos of change at one biological level only: the process of speciation and the deployment of species in geological time. Nonetheless, we believe that a general theory of punctuational change is broadly, though by no means exclusively, valid throughout biology" (Gould and Eldredge, 1977, p. 145). In 1982, in the midst of illusory stage 2, when I was supposedly touting macromutation as the cause of punctuated equilibrium in order to dethrone Darwinism, I explicitly drew the same distinction in order to separate the phenomena, while noting an interesting similarity in abstract geometric style of change across scales (Gould, 1982b, p. 90):

> These legitimate styles of macromutation are related to punctuated equilibrium only insofar as both represent different and unconnected examples of a

> general style of thinking that I have called punctuational (as opposed to gradualist or continuationist thought). I take it that no one would deny the constraining impact of gradualistic biases upon evolutionary theorizing. Punctuational thinking focusses upon the stability of structure, the difficulty of its transformation, and the idea of change as a transition between stable states. Evolutionists are now discussing punctuational theories at several levels: for morphological shifts (legitimate macromutation), speciation (various theories for rapid attainment of reproductive isolation), and general morphological pattern in geological time (punctuated equilibrium). These are not logically interrelated, but manifestations of a style of thought that I regard as promising and, at least, expansive in its challenge to conventional ideas. Any manifestation may be true or false, or of high or low relative frequency, without affecting the prospects of any other. I do commend the general style of thought (now becoming popular in other disciplines as well) as a fruitful source for hypotheses.

However, when I turn to factors that must be laid at our critics' doorstep, I can compile a longer and more serious list, including attitudes and practices that do compromise the ideals of scholarship. (Remember that I deal, in this section, only with personal and nonscientific critiques of punctuated equilibrium. We have also been properly subjected to very sharp, entirely appropriate, and fully welcomed criticism of a technical and scientific nature—and the theory of punctuated equilibrium has only been altered and improved thereby. I have discussed these legitimate criticisms in Chapter 4 of this book.)

1. THE EMOTIONAL SOURCE OF JEALOUSY. Given the vehemence of many deprecations, combined with a weakness or absence of logical or scientific content, I must conclude that the primary motivating factor lies in simple jealousy—that most distressing, yet most quintessentially human, of all destructive emotions ("as cruel as the grave" according to the *Song of Songs;* "the jaundice of the soul," in Dryden's metaphor; and, in the most memorable definition of all, Shakespeare's words of warning to Othello: "It is the green-eyed monster which doth mock the meat it feeds on").

Punctuated equilibrium has generated a large and public volume of commentary. I am confident that genuine interest and content has generated the bulk of this publicity, but I understand the all too human tendency to view achievements of perceived rivals as imposture rooted in base motivation. Moreover, jealousy gains a particularly potent expression in science for the ironic reason that professional norms do not permit us to acknowledge such feelings or motivations, even to ourselves. Our negativities are supposed to arise from perceived fallacies in the logic or empirical content of hypotheses we dislike, not from personal expressions of envy. Thus, if our emotions exude distress and anger, but we cannot admit, or even recognize, jealousy as a source, then we must impute our genuine envy to the supposed intellectual malfeasance of our opponents—and our internal feeling becomes falsely objectified as their failing. This form of transference leads to larger problems in the sociology of science than we have generally been willing to admit.

2. THE PHILOSOPHICALLY INTERESTING ISSUE OF LIMITED CONCEPTUAL SPACE. I have long faced a paradox in trying to understand why many intelligent critics seem unable to understand or acknowledge our reiterated insistence that the radical claim of punctuated equilibrium lies not in any proposal for revised microevolutionary mechanisms (especially not in any novel explanations for punctuations), but rather at the level of macroevolution, in claims for efficacy of higher-level selection based on the status of species, under punctuated equilibrium, as genuine Darwinian individuals.

When smart people don't "get it," one must conclude that the argument lies outside whatever "conceptual space" they maintain for assessing novel ideas in a given area. Many evolutionists, particularly those committed to the strict Darwinism of unifocal causation at Darwin's own organismic level, or below at the genic level, have never considered the hierarchical model, and apparently maintain no conceptual space for the notion of effective selection at higher levels. These scientists then face the following situation: (1) they note correctly that punctuated equilibrium stakes some claim for novelty within evolutionary theory; (2) their concept of "evolutionary theory" does not extend to causation above the organismic level, so they do not grasp the actual content of our claim; (3) they correctly understand that punctuated equilibrium offers no radical statement about microevolutionary mechanics; (4) Q.E.D., the authors of punctuated equilibrium must be grandstanding by asserting a radical claim without content. But the limit lies within the conceptual space of our critics, not in the character of our rhetoric.

3. THE PARTICULAR PREJUDICE THAT FANS THE FLAMES. Certain words embody unusual power, for reasons both practical and emotional—"fire" in a crowded theater, or "communist" at right-wing pep rallies of old. For reasons of impeccable historical pedigree, thoroughly explored in Chapters 2–6 of *The Structure of Evolutionary Theory*, and rooted largely in Darwin's own philosophical preferences, the most incendiary words for dedi-

cated Darwinians (once we get past Lamarckism, creationism, and a few others) must be the various synonyms of "sudden"—"rapid," "instantaneous," "quick," "discontinuous," and the like. Proponents of punctuated equilibrium do use these words—but at an appropriate scale of geological time, to express microevolutionary continuities that translate to punctuations in this larger temporal realm. Nonetheless, some orthodox Darwinians react with knee-jerk negativity towards any claim at all about rapidity. Any invocation of "rapid" must conjure up saltation and Goldschmidt, and must be met by counterattack. How else can we explain such a persistent confusion based on a false construction, then elevated to an urban legend, that the originators of punctuated equilibrium have always tried to identify and dispel?

THE WAGES OF JEALOUSY

The Descent to Nastiness

I treated the general *ad hominem* case against punctuated equilibrium in the last section. But some specific charges against punctuated equilibrium have bordered on the inane, or even the potentially actionable in our litigious world. To mention a few highlights along this low road:

THE CHARGE OF DISHONESTY. The following event unfolds with lamentable predictability in our imperfect world: when a controversy becomes impassioned, someone will eventually try to land the lowest academic blow of all by launching a charge of plagiarism or dishonest quotation. The debate about punctuated equilibrium reached this nadir when Penny (1983) accused us of cooking a quote from the *Origin of Species* by omitting passages without noting the deletion, and thereby changing Darwin's meaning to suit our purposes. Penny quoted from the 6th edition of the *Origin* to back up his claim. We, however, had used the first edition—and had rendered Darwin's words accurately (Gould and Eldredge, 1983). Enough said.

THE CHARGE OF RIP-OFF. A more conventional strategy for those who wish to deny a colleague's originality consists in claiming that a putative novelty really has an old pedigree—a twice told tale, said long before, preferably by a leading scientific light, and not in an obscure source (so that those under question cannot claim forgivable ignorance of minutiae). I suppose, therefore, that when we began to arouse substantial jealousy, someone was bound to argue that Darwin himself had said it all before.

The litany of this claim may hold some sociological interest for the time and energy invested by several commentators (Penny, 1983, 1985; Gingerich,

1984a and b, 1985; Scudo, 1985). These authors did point out some legitimate similarities between certain Darwinian statements and the tenets of punctuated equilibrium—including a significant one-sentence addition to later editions of the *Origin* (which we had indeed missed), acknowledging the occurrence of punctuational tempos, and apparently inspired by Falconer's objections, as highlighted in the introductory section of this chapter.

I regard this case as fundamentally misguided for general historiographic reasons, outlined in Gould and Eldredge (1983, p. 444):

> One simply cannot do history by searching for footnotes and incidental statements, particularly in later editions that compromise original statements. As with the Bible, most anything can be found somewhere in Darwin. General tenor, not occasional commentary, must be the criterion for judging a scientist's basic conceptions. If Darwin historians agree on a single point (for example, see Gruber [1974] and Mayr [1982b]), it is the importance and pervasiveness of Darwin's gradualism—a commitment far stronger than his allegiance to natural selection as an evolutionary mechanism.

Fortunately, one needn't take my partisan word in refutation. Frank Rhodes, then the president of Cornell University, but a distinguished paleontologist by training and first career, became interested in punctuated equilibrium and its links to the history of evolutionary thought. He therefore spent a sabbatical term researching the relationship of Darwin's thinking to the claims and tenets of punctuated equilibrium. He did find many genuine Darwinian resonances, while affirming our originality and concluding that "the hypothesis of punctuated equilibrium is of major importance for paleontological theory and practice" (Rhodes, 1983, p. 272).

When Gingerich (1984b, p. 116) wrote a commentary on Rhodes's article, dedicated to denying our originality and asserting once again that Darwin had said it all before, Rhodes replied with generosity and firmness (under Gingerich, 1985, p. 116):

> I do argue that punctuated equilibrium—whether true or false—is a "hypothesis of major importance" and that it has had a beneficial impact on the quality of recent paleontological studies. Gingerich asks, "Which nuances [of punctuated equilibrium] were unanticipated by Darwin?" From a long list, I suggest the following: its relationship to the genetics of stasis and the punctuation, morphological stasis and developmental constraints, evolutionary models in relation to paleoecology, stratigraphical correlation, spe-

> cies selection, mathematical models of evolutionary rates, selection of RNA molecules, phylogenetic divergence, and the evolution of communities. These topics, and many more studied from the viewpoint of punctuated equilibrium, have been the subject of recent papers . . . To suggest that there was no nuance of punctuated equilibrium which was "unanticipated by Darwin" is to make an icon of Darwin and to adopt an extravagantly Whiggish view of the history of Darwin's particular contribution—great as that was.

Some critics then followed a substitutional strategy: if one denigration fails, try another in the same form. If "Darwin said it all" fails as an optimal dismissal, then try the best available paleontological version: "G. G. Simpson said it all." Again, the search for reinterpretations and footnotes began, as this new version of denigration began to coagulate among our most committed detractors: Simpson (1944) devoted his seminal book to documenting the great variation in evolutionary rates, and punctuated equilibrium therefore has nothing new to say.

But, punctuated equilibrium was never formulated as a hypothesis about great variability in anagenetic rates (which, indeed, everyone has long acknowledged). Punctuated equilibrium presents a specific hypothesis about the location of most evolutionary change in punctuational cladogenesis, followed by pronounced stasis. Yet Simpson, as documented in *SET*, Chapter 7, pp. 562–563, denied major importance to cladogenesis at all, and held that 90 percent of evolutionary change occurred in the anagenetic mode. Moreover (see *SET*, pp. 528–531), Simpson's important hypothesis of "quantum evolution"—the idea that our detractors usually try to depict as equivalent to punctuated equilibrium—treats a vitally important, but entirely different phenomenon of different mode at a different scale: the anagenetic origin of major structural innovations, not the pacing of ordinary speciation.

Several authors, in their desire to name Simpson as the true author of punctuated equilibrium, completely misunderstood his work. Andrew Huxley, for example, misinterpreted the well-known paleontological concept of a *Stufenreihe*. Huxley quoted from Simpson's 1944 book (Huxley, 1982, p. 145): "He says (pp. 194–195): 'The pattern of step-like evolution, an appearance of successive structural steps, rather than direct sequential phyletic transitions, is a peculiarity of paleontological data more nearly universal than true rectilinearity and often mistaken for the latter,' and quotes the name *Stufenreihe* given to this mode of evolution by Abel in 1929. This is exactly equivalent to 'punctuated equilibria.'" But a *Stufenreihe* is a stratigraphic sequence displaying an evolutionary trend constructed of collateral relatives rather than direct ancestors (called, by contrast, an *Ahnenreihe*). For example, *Australopithecus robustus, Homo neanderthalensis* and *Homo sapiens* form a *Stufenreihe,* while *A. afarensis, Homo ergaster,* and *Homo sapiens* build a putative *Ahnenreihe. Stufenreihen* are necessarily discontinuous because they pile a cousin on top of an uncle on top of a grandfather, while true *Ahnenreihen* record genuine genealogical descent without breaks. In any case, the contrast bears no relationship to the concept of punctuated equilibrium (which is a hypothesis about the geometry of *Ahnenreihen*).

Mettler, Gregg and Schaffer (1988, p. 288) even grant Simpson authorship of our name! "Finally, there is the punctuated equilibrium view of Eldredge and Gould (1972), and Vrba (1983). Even though the term was coined by Simpson, these authors have given it new emphasis."

At least pathos can be balanced by bathos in our wondrously varied world. The irrepressible Beverly Halstead, labelling me with my all-time favorite epithet of "petty obnoxious infauna," while depicting Simpson as a deity watching over his loyal epigones from on high, reviewed Simpson's last book with a panegyric that left even his earlier excoriation of the British Museum in the rhetorical dust (Halstead, 1984, p. 40):

> Indeed, the original presentation of punctuated equilibrium was in anti-neo-Darwinian language but the substance was nonetheless easily accommodated within the framework given long ago by Simpson . . . It has been Simpson's overwhelming reasonableness and commonsense, as exemplified in this book, that has done so much to entrench the Modern Synthesis in the consciousness of most paleontologists, the literary pyrotechnics of Steve Gould notwithstanding. Simpson's humility before his fossils, a special kind of innocence, is perhaps one of his most endearing traits. Because he is kind and tolerant, he finds it nigh impossible to believe that some of the supporting framework of our discipline is infested with some petty obnoxious infauna. My only criticism of Simpson in his book is his apparent unwillingness to contemplate the existence of real nasties emerging from the woodwork . . . Let him [Simpson] look down from the commanding heights knowing that the citadel of neo-Darwinism still has its staunch defenders in this more combative age. We will do our best not to let him down.

THE CHARGE OF ULTERIOR MOTIVATION. When charges of dishonesty or lack of originality fail, a committed detractor can still label his opponents as unconcerned with scientific truth, but motivated by some ulterior (and nefarious) goal. Speculations about our "real" reasons have varied widely in content, but little in their shared mean spirit (see, for example, Turner, 1984; Konner, 1986; and Dennett, 1995). I will discuss only one of these peculiar speculations—the charge that punctuated equilibrium originated from my political commitments rather than from any honorable feeling about the empirical world—because, once again, the claim rests upon a canonical misquotation and exposes the apparent unwillingness or inability of our unscientific critics to read a clear text with care.

I have already discussed Halstead's version of the political charge in the great and farcical British-Museum-cum-cladism-cum-Marxism debate (see pp. 307–309). The supposed justification for this construction lies in another quotation from my writing, second in false invocation only to the "death of the Synthesis" statement discussed earlier (p. 330).

I do not see how any careful reader could have missed the narrowly focused intent of the last section in our 1977 paper, a discussion of the central and unexceptionable principle, embraced by all professional historians of science, that theories must reflect a surrounding social and cultural context. We began the section by trying to identify the cultural roots of gradualism in larger beliefs of Victorian society. We wrote (Gould and Eldredge, 1977, p. 145): "The general preference that so many of us hold for gradualism is a metaphysical stance embedded in the modern history of Western cultures: it is not a high-order empirical observation, induced from the objective study of nature . . . We mention this not to discredit Darwin in any way, but merely to point out that even the greatest scientific achievements are rooted in their cultural contexts—and to argue that gradualism was part of the cultural context, not of nature."

We couldn't then assert, with any pretense to fairness or openness to self-scrutiny, that gradualism represents cultural context, while our punctuational preferences only record unvarnished empirical truth. If all general theories embody a complex mixture of contingent context with factual adequacy, then we had to consider the cultural embeddedness of preferences for punctuational change as well. We therefore began by writing (p. 145) that "alternative conceptions of change have respectable pedigrees in philosophy." We then discussed the most obvious candidate in the history of Western thought: the Hegelian dialectic and its redefinition by Marx and Engels as a theory of revolutionary social change in human history. We cited a silly, propagandistic defense of punctuational change from the official Soviet handbook of Marxism-Leninism, in order to stress our point about the potential political employment of all general theories of change. We concluded (p. 146): "It is easy to see the explicit ideology lurking behind this general statement about the nature of change. May we not also discern the implicit ideology in our Western preference for gradualism?"

But the argument required one further step for full disclosure. We needed to say something about why we, rather than other paleontologists at other times, had developed the concept of punctuated equilibrium. We raised this point as sociological commentary about the *origin* of ideas, not as a scientific argument for the *validity* of the same ideas. An identification of cultural or ontogenetic sources says nothing about truth value, an issue that can only be settled by standard scientific procedures of observation, experiment and empirical test. So I mentioned a personal factor that probably predisposed me to openness towards, or at least an explicit awareness of, a punctuational alternative to conventional gradualistic models of change: "It may also not be irrelevant to our personal preferences that one of us learned his Marxism, literally at his daddy's knee."

I have often seen this statement quoted, always completely out of context, as supposed proof that I advanced punctuated equilibrium in order to foster a personal political agenda. I resent this absurd misreading. I spoke only about a fact of my intellectual ontogeny; I said nothing about my political beliefs (very different from my father's, by the way, and a private matter that I do not choose to discuss in this forum). I included this line within a discussion of personal and cultural reasons that might predispose certain scientists towards consideration of punctuational models—just as I had identified similar contexts behind more conventional preferences for gradualism. In the next paragraph, I stated my own personal conclusions about the general validity of punctuational change—but critics never quote these words, and only cite my father's postcranial anatomy out of context instead:

> We emphatically do not assert the "truth" of this alternate metaphysic of punctuational change. Any attempt to support the exclusive validity of such a monistic, *a priori*, grandiose notion would verge on the nonsensical. We believe that gradual change characterizes some hierarchical levels, even though we may attribute it to punctuation at a lower level—the macroevolutionary trend produced by species selection, for example. We make a simple plea for pluralism in guiding philosophies—and for the basic recognition that such philosophies, however hidden and inarticulated, do constrain all our thought. Nonetheless, we do believe that the punctuational metaphysic may

prove to map tempos of change in our world better and more often than any of its competitors—if only because systems in steady state are not only common but also so highly resistant to change.

The Most Unkindest Cut of All

If none of the foregoing charges can bear scrutiny, strategists of personal denigration still hold an old and conventional tactic in reserve: they can proclaim a despised theory both trivial and devoid of content. This charge is so distasteful to any intellectual that one might wonder why detractors don't try such a tactic more often, and right up front at the outset. But I think we can identify a solution: the "triviality caper" tends to backfire and to hoist a critic with his own petard—for if the idea you hate is so trivial, then why bother to refute it with such intensity? Leave the idea strictly alone and it will surely go away all by itself. Why fulminate against tongue piercing, goldfish swallowing, skateboarding, or any other transient fad with no possible staying power?

Nonetheless, perhaps from desperation, or from severe frustration that something regarded as personally odious doesn't seem to be fading away, this charge of triviality has been advanced against punctuated equilibrium, apparently to small effect. To cite a classic example of backfiring, Gingerich (1984a, 1984b) tried to dismiss punctuated equilibrium as meaningless and untestable by definition—and to validate gradualism *a priori* as "commitment to empiricism and dedication to the principal [sic] of testability in science" (1984a, p. 338), with stasis redefined, oxymoronically in my judgment, as "gradualism at zero rate" (1984a, p. 338). Gingerich then concludes (1984b, p. 116): "Punctuated equilibrium is unscaled, and by nature untestable. It hardly deserves recognition as a conjecture of 'major importance for paleontological theory and practice.' . . . Hypotheses that cannot be tested are of little value in science."

But how can Gingerich square this attempted dismissal with his own dedication of a decade in his career to testing punctuated equilibrium by fine-scale quantitative analysis of Tertiary mammals from the western United States (Gingerich, 1974, 1976)? These studies, which advanced a strong claim for gradualism, represent the most important empirical research published in the early phase of the punctuated equilibrium debate. Gingerich then recognized punctuated equilibrium as an interesting and testable hypothesis, for he spent enormous time and effort testing and rejecting our ideas for particular mammalian phylogenies. He then argued explicitly (1978, p. 454):

"Their [Eldredge and Gould's] view of speciation differs considerably from the traditional paleontological view of dynamic species with gradual evolutionary transitions, but it *can* be tested by study of the fossil record."

Among Darwinian fundamentalists (see my terminology in Gould, 1997b), charges of triviality have been advanced most prominently and insistently by Dawkins (1986, p. 251) who evaluates punctuated equilibrium metaphorically as "an interesting but minor wrinkle on the surface of neo-Darwinian theory"; and by Dennett (1995, p. 290) who calls punctuated equilibrium "a false-alarm revolution that was largely if not entirely in the eyes of the beholders."

But a close analysis of Dawkins's and Dennett's arguments exposes the parochiality of their judgment. They regard punctuated equilibrium as trivial because our theory doesn't speak to the restricted subset of evolutionary questions that, for them, defines an exclusive domain of interest for the entire subject. These men virtually equate evolution with the origin of intricately adaptive organic design—"organized adaptive complexity," or O.A.C. in Dawkins's terminology. They then dismiss punctuated equilibrium on the narrow criterion: "if it doesn't explain the focus of my interests, then it must be trivial." Dawkins (1984, p. 684), for example, properly notes the implications of punctuated equilibrium for validation of higher-level selection, but then writes: "Species-level selection can't explain the evolution of adaptations: eyes, ears, knee joints, spider webs, behavior patterns, everything, in short, that many of us want a theory of evolution to explain. Species selection may happen, but it doesn't seem to *do* anything much." "Everything"? Does nothing else but adaptive organismal design excite Dawkins's fancy in the entire and maximally various realm of evolutionary biology and the history of life—the "endless forms most beautiful and most wonderful" of Darwin's closing words (1859, p. 490).

But the truly curious aspect of both Dawkins's and Dennett's charge lies in their subsequent recognition, and fair discussion, of the important theoretical implication of punctuated equilibrium: the establishment of species as Darwinian individuals, and the consequent validation of species sorting and selection as a prominent process in a hierarchical theory of Darwinian evolution. In 1984, Dawkins acknowledged that this aspect of punctuated equilibrium "does, in a sense, move outside the neo-Darwinian synthesis, narrowly interpreted. This is about whether a form of natural selection operates at the level of entire lineages, as well as at the level of individual reproduction stressed by Darwin and neo-Darwinism." In his 1986 book,

Dawkins then devotes a substantial part of the chapter following his rejection of punctuated equilibrium to an evaluation of species selection. But he finishes his exploration by reimmersion in the same parochial trap of denying importance because the phenomenon doesn't explain his exclusive interest in adaptive organismal design: "To conclude the discussion of species selection, it could account for the pattern of species existing in the world at any particular time. It follows that it could also account for changing patterns of species as geological ages give way to later ages, that is, for changing patterns in the fossil record. But it is not a significant force in the evolution of the complex machinery of life . . . As I have put it before, species selection may occur but it doesn't seem to do anything much!" (Dawkins, 1986, pp. 268–269). But doesn't "the pattern of species existing in the world at any particular time" and "changing patterns in the fossil record" represent something of evolutionary importance?

At the end of his long riff against punctuated equilibrium, Dennett also pauses for breath and catches a glimmer of the concept that seems important and theoretically intriguing to many students of macroevolution (Dennett, 1995, pp. 297–298):

> The right level at which to look for evolutionary trends, he [Gould] could then claim [indeed I do], is not the level of the gene, or the organism, but the whole species or clade. Instead of looking at the loss of particular genes from gene pools, or the differential death of particular genotypes within a population, look at the differential extinction rate of whole species and the differential "birth" rate of species—the rate at which a lineage can speciate into daughter species. This is an interesting idea . . . It may be true that the best way of seeing the long-term macroevolutionary pattern is to look for differences in "lineage fecundity" instead of looking at the transformations in the individual lineages. This is a powerful proposal worth taking seriously.

I am puzzled by the discordance and inconsistency, but gratified by the outcome. Dawkins and Dennett, smart men both, seem unable to look past the parochial boundaries of their personal interest in evolution, or their feelings of jealousy towards whatever effectiveness my public questioning of their sacred cow of Darwinian fundamentalism may have enjoyed (see Gould, 1997b)—so they must brand punctuated equilibrium as trivial. But they cannot deny the logic of Darwinian argument, and they do manage to work their way to the genuine theoretical interest of punctuated equilibrium's major implication, the source of our primary excitement about the idea from the start.

The Wisdom of Agassiz's and Von Baer's Threefold History of Scientific Ideas

When I was writing *Ontogeny and Phylogeny,* I came across a wonderful, if playfully cynical, statement by the great embryologist Karl Ernst von Baer (1866, p. 63) about Louis Agassiz's view on the ontogeny of scientific theories: "Deswegen sagt Agassiz, dass wann eine neue Lehre vorgebracht würde, sie drei Stadien durchzumachen habe; zuerst sage man, sie sei nicht wahr, dann, sie sei gegen die Religion, and im dritten Stadium, sie sei längst bekannt gewesen." [Therefore, Agassiz says that when a new doctrine is presented, it must go through three stages. First, people say that it isn't true, then that it is against religion, and, in the third stage, that it has long been known.]

I won't vouch for the generality of this scenario, but Agassiz's rule certainly applies to the history of nonscientific debate about punctuated equilibrium, particularly to the aspect governed by jealousy of critics—as Eldredge and I recognized in a previous publication entitled: "Punctuated equilibrium at the third stage" (Gould and Eldredge, 1986). The first stage of empirical denial, extending roughly from our original publication in 1972 to the Chicago macroevolution meeting of 1980, featured studies of fossil sequences by paleontologists (notably Gingerich, 1974 and 1976), many of whom tried to deny that punctuated equilibrium occurred very frequently, if at all, by documenting cases of gradualism.

During the second phase, spanning the first half of the 1980s, the primary subject of this section, punctuated equilibrium, was vociferously dismissed as contrary to religion—that is, as apostate anti-Darwinian nonsense. Our theory, falsely read as a saltationist doctrine proclaiming the overthrow of the Modern Synthesis, if not of Darwinism itself, received a hefty dose of anathematization in tones usually reserved for demonizing religious heterodoxies.

The third phase then began in the mid-1980s, as documented in Chapters 3–5 of this book, and has continued ever since. The evidence became too great, and we withstood all ideological attacks without sustaining appreciable damage. Punctuated equilibrium now seemed both coherent in argument, and supported by a sufficient number of empirical studies to become a recognized evolutionary phenomenon—though at a relative frequency as yet undetermined. Such a situation must cause critics to remember the old cliché: if you can't beat 'em, join 'em (but don't grant 'em too much credit for innovation or originality). Move instead to phase three of Agassiz's continuum—"sure it's true, but we always knew this; punctuated equilibrium amounts to no more than a little wrinkle on the skin of neo-Darwinism."

As an initiating episode of the third phase, Darwinian biologists began to construct models that rendered punctuational patterns (though not always cladogenetic events of true punctuated equilibrium) by standard formulae of population genetics under certain reasonable assumptions and conditions. We have always welcomed these formulations, for we never sought the radical content of punctuated equilibrium in novel microevolutionary processes, as I have emphasized throughout this chapter—and any demonstrated mechanism for punctuational patterns evokes both our interest and satisfaction. The first two studies in this genre appeared in 1985—Newman, Cohen and Kipnis (1985) and Lande (1985). In 1986, Roger Lewin wrote a "news and views" commentary for *Science* entitled: "Punctuated equilibrium is now old hat." He ended with a gratifying comment by Joel Cohen: "In terms of the tenor of the debate, which at times has been strident, the new results will bring the various parties closer together. Cohen readily concedes that population geneticists very probably would not have applied their mathematical tools to the issue in this way had there not been such a big fuss stirred up by the paleontologists' claims. 'They deserve credit for that,' he says."

So I guess we won by Agassiz's scenario, even if personal motivations of an ungenerous nature lead our severest critics to belittle our achievement as true after all, but trivial from the outset. But why then did they ever make such a fuss?

A Coda On the Kindness and Generosity of Most Colleagues

This section, devoted to unscientific critiques by professional colleagues, centers on unhappy themes of jealousy, pettiness and meanness of spirit. But I do not wish to leave the impression that these unpleasantnesses have dominated the totality of discussion about punctuated equilibrium. Quite to the contrary, in fact—and I have already discussed the numerous tough, spirited, helpful and scientific critiques of punctuated equilibrium in Chapters 3–5 of this book. Intense and useful debate has predominated throughout the history of punctuated equilibrium. Most of our colleagues have unstintingly followed the norms and ideals of scientific discussion, and we have primarily wrestled with good argument and content, not primarily with deprecation and personal attack.

I have mentioned and cited several of these generous reactions, these fair and accurate characterizations, throughout this section—the journalistic accounts of Tudge and Gleick (pp. 317–319); the excellent textbook epitome of Curtis and Barnes (pp. 323–325); the generous assessment of punctuated equilibrium's scientific importance by Rhodes (1983); and the acknowledgment of punctuated equilibrium's prod to further exploration of formulae in population genetics by Cohen (p. 353). I also wish to emphasize that most professional colleagues have always given us generous credit, and have applauded both the debate and the interest generated by punctuated equilibrium.

I have particularly appreciated the fairness of severe critics who generally oppose punctuated equilibrium, but who freely acknowledge its legitimacy as a potentially important proposition with interesting implications, and as a testable notion that must be adjudicated in its own macroevolutionary realm. Ayala (1982) has been especially clear and gracious on this point:

> If macroevolutionary theory were deducible from microevolutionary principles, it would be possible to decide between competing macroevolutionary models simply by examining the logical implications of microevolutionary theory. But the theory of population genetics is compatible with both punctualism and gradualism; and, hence, logically it entails neither. Whether the tempo and mode of evolution occur predominantly according to the model of punctuated equilibria or according to the model of phyletic gradualism is an issue to be decided by studying macroevolutionary patterns, not by inference from microevolutionary processes. In other words, macroevolutionary theories are not reducible (at least at the present state of knowledge) to microevolution. Hence, macroevolution and microevolution are decoupled in the sense (which is epistemologically most important) that macroevolution is an autonomous field of study that must develop and test its own theories.

Such statements stand in welcome contrast to the frequent grousing of strict Darwinians who often say something like: "but we know all this, and I said so right here in the footnote to page 582 of my 1967 paper; you have stated nothing new, nothing that can alter the practice of the field." I will never forget the climactic moment of the Chicago macroevolution meeting, when John Maynard Smith rose to make such an ungenerous statement about punctuated equilibrium and macroevolutionary theory in general—and George Oster responded to him: "Yes, John, you may have had the bicycle, but you didn't ride it." In the same vein, I appreciate the comment of Marjorie Grene (quoted in Stidd, 1985), a famous philosopher who has greatly aided the clarification of evolutionary theory:

> Yet on both these counts—gradualism and neutralism—evolutionists other than paleontologists now appear unmoved. Just what we've said all along, they cry, even though, I swear, I've heard, and seen, them emphatically assert just the opposite, time and time again. No sudden changes, no non-adaptive changes, they used to exclaim, while now they ask cheerfully: why not stasis, sudden change, and neutral mutations all over the place except for an adaptive innovation here and there, now and then? We always knew it was like that. Nothing really new, no revolution here.

Finally, I am heartened by the many top-ranking biologists who have found fruitful ideas and new wrinkles in the concept of punctuated equilibrium and its macroevolutionary implications—for utility in practice remains the ultimate criterion of judgment in science. I appreciate Dan Janzen's affirmation in his article "on ecological fitting" (Janzen, 1985, p. 308): "I suddenly realize that I have blundered through the front door of the turmoil over punctuated equilibria. We don't have to dig at the fossil record; punctuated equilibria are right here in front of us, represented by most of the species that you and I have anything to do with."

I welcome the generous assessment of Kenneth Korey (1984) in the preface to his compendium of Darwin's best writing, *The Essential Darwin:*

> Unquestionably no single challenge to the synthesis has provoked more attention than the theory of punctuated equilibrium advanced by Niles Eldredge and Stephen Gould. . . . It is true that punctuated equilibrium was not a prediction of the synthesis; on the contrary, Simpson emphasized continuous, phyletic evolution as the most pervasive feature of evolution at this level . . . On the macroevolutionary front . . . punctuated equilibrium as an empirical proposition is not, perforce, in conflict with the synthesis, although if its wide province becomes established, then a more complete theoretical explanation for stasis will certainly be wanted. Species selection, in its present form, would seem to require the most profound reworking of evolutionary theory.

And I thank Paul Ehrlich (1986) for recognizing the genuine novelty and utility of punctuated equilibrium in his book, *The Machinery of Nature:*

> The jury is not in on the punctuated-equilibria controversy. That the "snapshot" of differentiation we see today seems to reveal all stages of differentiation does not necessarily signal a win for the gradualists . . . And it is not fair to swallow the punctuationist view within the gradualist orthodoxy simply because the possibility of rapid speciation has always been part of that orthodoxy. The punctuationist view is about dominant patterns, not about what is possible—and it represents a genuine challenge to one widely held tenet within evolutionary theory.

It has been a wonderful ride on Oster's bicycle, and we still have such a long road to travel.

NOTES

INTRODUCTION

1. As so much unnecessary rancor has been generated by simple verbal confusion among different meanings of this word, and not by meaningful conceptual disagreements, I should be clear that I intend only the purely descriptive definition when I write "macroevolution"—that is, a designation of evolutionary phenomenology from the origin of species on up, in contrast with evolutionary change *within* populations of a single species. In so doing, I follow Goldschmidt's own definitional preferences (1940) in the book that established his apostasy within the Modern Synthesis. Misunderstanding has arisen because, to some, the world "macroevolution" has implied a theoretical claim for distinct causes, particularly for nonstandard genetic mechanisms, that conflict with, or do not occur at, the microevolutionary level. But Goldschmidt—and I follow him here—urged a nonconfrontational definition that could stand as a neutral descriptor for a set of results that would then permit evolutionists to pose the tough question without prejudice: does macroevolutionary phenomenology demand unique macroevolutionary mechanics? Thus, in this book, "macroevolution" is descriptive higher-level phenomenology, not pugnacious anti-Darwinian interpretation.

1 WHAT EVERY PALEONTOLOGIST KNOWS

1. I may be an arrogant man, but I would never be so pompous as to use the "royal" we. I cannot separate my views on punctuated equilibrium from those of my colleague and partner in this venture from the start, Niles Eldredge. When I write "we" in this section, I mean "Eldredge and Gould."

2 THE PRIMARY CLAIMS OF PUNCTUATED EQUILIBRIUM

1. All professions maintain their parochialisms, and I trust that nonpaleontologist readers will forgive our major manifestation. We are paleontologists, so we need a name

to contrast ourselves with all you folks who study modern organisms in human or ecological time. You therefore become neontologists. We do recognize the unbalanced and parochial nature of this dichotomous division—much like my grandmother's parsing of *Homo sapiens* into the two categories of "Jews" and "non-Jews."

3 THE SCIENTIFIC DEBATE ON PUNCTUATED EQUILIBRIUM

1. Futuyma remains quite skeptical of punctuated equilibrium in general, and I would place him more among our critics than our supporters. But he does accept the empirical pattern, and he is an expert on speciation. Thus, when he developed an original way to resolve the paradox of why punctuations might correlate with events of speciation, even if processes of speciation don't accelerate the rate of evolution, he published his ideas as a constructive contribution to the general debate. Even though Futuyma disagrees with our claims for the general importance of punctuated equilibrium (while he, obviously, does not deny the phenomenon), he has granted us serious attention and has acknowledged the intellectual interest of the debate we provoked—and no one could ask for more from a good critic. Futuyma wrote (1988, p. 225), in stressing the need to integrate "synchronic" approaches as pursued by neontologists interested in evolutionary mechanisms with the "historical" themes favored by systematists and paleontologists—all (to borrow a line from elsewhere) "in order to form a more perfect union."

> We need to identify and to define rigorously questions to which both synchronic and historical evolution can make truly indispensable contributions. Some such questions have already been posed, so we now find systematists and population geneticists converging on the analysis of macromolecular sequences, geneticists publishing in *Paleobiology* (thanks to the healthy stimulus of punctuated equilibrium), systematists and students of adaptation finding a *rapprochement* in the use of phylogenetic information to test hypotheses of behavioral, physiological, and other adaptations.

4 SOURCES OF DATA FOR TESTING PUNCTUATED EQUILIBRIUM

1. As an example of the conceptual stranglehold that gradualism once imposed upon such data, the major study done before punctuated equilibrium on the evolution of these rodents presupposed anagenetic gradualism at a constant rate: "This treatment assumes that a regular increase in size continued at approximately the same rate throughout Orellan time" (Howe, 1956, p. 74). When Howe then detected accelerated change at two paleosols marking boundaries of substages within the Orellan, he assumed (without any direct evidence) that the paleosols must mark diastems, or time gaps, compressing a true

gradualism of change into the literal appearance of a small hiccup. But Heaton found no evidence for any temporal hiatus at these boundaries.

Heaton's results included both punctuated equilibrium (two stable species changing only in relative abundance) and gradualism (minor size increase within the larger species at the end of its range) in a total pattern, but he concluded (correctly, I think, in my own biased way) that punctuated equilibrium had shown greater utility in challenging previous assumptions that had stymied proper conclusions. He closed his paper by writing (1993, p. 307): "So, *Ischyromys* displays features of both 'punctuated equilibria' and 'phyletic gradualism' as defined by Eldredge and Gould (1972). But the primary revelation of this study is that what was thought to be a single gradually evolving lineage must now be seen as the replacement of one stable species by another." (In the fairness of full disclosure, Heaton did his graduate work under my direction. But I really do encourage independence and contrariness, and some of my students have documented gradualism, even in their Ph.D. dissertations, when truly (and for the only time) beholden to my "official" approval—e.g. Arnold, 1982.)

2. Interestingly, Barnosky's (1987) published version of his oral presentation refined his conclusion and tabulated a strong majority for punctuated equilibrium, even when compiled from an existing literature biased by previous traditions for ignoring stasis as non-data, and favoring apparent cases of gradualism. In his compendium for Quaternary mammals, Barnosky (1987) found punctuated equilibrium "supported twice as often as phyletic gradualism . . . The majority of species considered exhibit most of their morphological change near a speciation event, and most species seem to be discrete entities."

5 THE BROADER IMPLICATIONS OF PUNCTUATED EQUILIBRIUM FOR EVOLUTIONARY THEORY AND GENERAL NOTIONS OF CHANGE

1. I also wish to reemphasize that I assert no exclusivistic claim in this formulation. Supporters of the hierarchical theory must not repeat the parochial error of their forebears by arguing that their newly-specific, higher-level mechanisms can explain everything by reaching down, just as Darwinian traditionalists tried to develop a complete causal theory by extrapolating up. Thus, we do not challenge either the efficacy or the cardinal importance of organismal selection. As previously discussed, I fully agree with Dawkins (1986) and others that one cannot invoke a higher-level force like species selection to explain "things that organisms do"—in particular, the stunning panoply of organismic adaptations that has always motivated our sense of wonder about the natural world, and that Darwin described, in one of his most famous lines (1859, p. 3) as "that perfection of structure and coadaptation which most justly excites our admiration." But should we not regard as equally foolish, and equally vain (in both senses of the word), any proposal that insists upon explaining all "things that species and clades do" as ex-

trapolated consequences of organismic adaptation? I would not invoke species selection to explain the marvelous mechanics of beetle elytra, but the same theme of appropriate scale also leads me to equal confidence that the excellent adaptive design of beetle organisms cannot fully explain why this order so vastly predominates in species diversity, even among the most speciose of all metazoan classes—to the point of inspiring Haldane's canonical quip about God's "inordinate fondness" for these creatures (see Gould, 1993, for an exegesis of this famous anecdote).

2. One might argue that this focus only records another of Bacon's idols rather than an evident empirical reality. Bacon's *idola tribus,* or idols of the tribe, refer to mental biases deeply rooted in inherent modes of mental functioning, or human nature itself. Humans are pattern-seeking and story-telling creatures—and we prefer to tell our stories in certain modes that may reflect particular cultural traditions as well as universal preferences of thought. We shun randomness and non-directionality in favor of stories about movement in particular directions for definable reasons subject to moral judgment. We compiled the entire Bible as a grand and extended narrative in this mode, and then granted just one uncomfortable chapter to Ecclesiastes as the loyal opposition, where "time and chance happeneth to all" and "there is no new thing under the sun." Thus, our chosen focus upon trends in the paleontological record may only record their salience in piquing our interests and preferences—and not a genuinely high relative frequency among all clades in nature. This subject deserves a great deal of thought and extended study. A remarkable article by Budd and Coates, 1992, on the predominance of non-trending in the evolution of montastraeid corals may point the way to substantial reform.

3. Soon after I wrote this section, *Science* published a special issue on evolution (25 June 1999), featuring the work of Lenski's lab in a news article entitled, "Test tube evolution catches time in a bottle." The twelve populations have now been evolving for 24,000 generations. Although all have shown similar increases in cell size and fitness, the genetic bases of change have been highly varied and unpredictable. Moreover, alteration in environmental and adaptive regimes yields no common response. When, after 2000 generations of growth on glucose (with similar evolutionary responses), the 12 populations were switched to a different sugar (maltose), some populations flourished, but others grew poorly. After 1000 generations on maltose, all twelve populations did improve in fitness, but not nearly so much (and, more importantly, not nearly so consistently) as on glucose. The starting genotype for the 12 populations had been identical for the first experiment with glucose, but different (after 2000 generations of evolution on glucose for each population) for the initiation of the subsequent maltose experiment. Apparently, any departure from simple and controlled experimental conditions towards the genetic and environmental variation invariably encountered in the natural world greatly decreases the predictability, while emphasizing the contingency, of outcomes.

4. I dare not even begin to enter the deepest and most difficult of all issues raised by differences between scientific and humanistic practice: why does the history of scientific ideas, even when proceeding in a punctuational mode, marked by quirky, unpredictable

and revolutionary shifts, undeniably move to better understanding (at least as measured operationally by our technological successes)—that is, and not to mince words, to progress in knowledge—whereas no similar vector can be discerned in the history of the arts, at least in the sense that Picasso doesn't (either by any objective measure or by simple subjective consensus) trump Leonardo, and Stravinsky doesn't surpass Bach (although later ages may add new methods and styles to the arsenals of previous achievement). The naive answer—that science searches for a knowable, objective, external reality that may justly be called "true," whereas art's comparable standard of beauty must, to cite the cliché, lie in the eye of the beholder—is probably basically sound, and probably explains a great deal more of this apparent dilemma than most academic sophisticates would care to admit. (In this belief, I remain an old-fashioned, unreconstructed scientific realist—but then we all must take oaths of fealty to our chosen profession.)

But I also acknowledge that the question remains far more complicated, and far more enigmatic, than this fluffy claim of such charming naïveté would indicate. After all, we only "see" through our minds (not to mention our social organizations and their pervasive biases). And our minds are freighted with a massive cargo of all the inherent structural baggage that Kant called the *synthetic a priori*, and that modern biologists would translate as structures inherited from ancestral brains that built no adaptations *for* what we designate as "consciousness." In this light, why should we be "good" at knowing external reality? After all, our vaunted consensuses—and on this point, Kant remains as modern as the latest computer chip—may record as much about how our quirkily constructed brains must parse this "reality," as about how external nature truly "works." But enough of unanswerable questions! I only note that Kuhn himself raises this great issue in his closing thoughts on the special character of science: "It is not only the scientific community that must be special. The world of which that community is a part must also possess quite special characteristics, and we are no closer than we were at the start to knowing what these must be. That problem—What must the world be like in order that man may know it?—was not, however, created by this essay. On the contrary, it is as old as science itself, and it remains unanswered."

BIBLIOGRAPHY

Adams, Robert McC. 2000. Accelerated technological change in archaeology and ancient history. In G. M. Feinman and L. Manzanilla, eds., *Cultural Evolution: Contemporary Viewpoint.* N.Y.: Kluwer Publishers, pp. 95–118.

Ager, D. V. 1973. *The Nature of the Stratigraphic Record.* N.Y.: John Wiley.

——— 1983. Allopatric speciation—an example from the Mesozoic Brachiopoda. *Palaeontology* 26: 555–565.

Albanesi, G. L., and C. R. Barnes. 2000. Subspeciation within a punctuated equilibrium evolutionary event: phylogenetic history of the Lower-Middle Ordovician *Paristodus originalis–P. horridus* complex (Conodonta). *Jour. Paleontology* 74: 492–502.

Allaby, M. (ed.) 1985. *The Oxford Dictionary of Natural History.* N.Y.: Oxford Univ. Press.

Allen, J. C., W. M. Schaffer, and D. Rosko. 1993. Chaos reduces species extinction by amplifying local population noise. *Nature* 364: 229–232.

Alters, B. J., and W. F. McComas. 1994. Punctuated Equilibrium: the missing link in evolution education. *American Biol. Teacher* 56: 334–340.

Alvarez, L. W., W. Alvarez, F. Asaro, and H. V. Michel. 1980. Extraterrestrial cause for the Cretaceous-Tertiary extinction. *Science* 208: 1095–1108.

Anstey, R. L., and J. F. Pachut. 1995. Phylogeny, diversity, history, and speciation in Paleozoic bryozoans. In D. H. Erwin and R. L. Anstey, eds., *New Approaches to Speciation in the Fossil Record.* N.Y.: Columbia Univ. Press, pp. 239–284.

Arnold, A. J. 1982. Hierarchical structure in evolutionary theory: applications in the Foraminiferida. Ph.D. Dissertation. Dept. of Geology. Harvard University.

Asimov, I. 1989. *Asimov's Chronology of Science and Discovery.* N.Y.: Harper & Row.

Avers, C. J. 1989. *Process and Pattern in Evolution.* N.Y.: Oxford Univ. Press.

Avise, J. C. 1977. Is evolution gradual or rectangular? Evidence from living fishes. *Proc. Natl. Acad. Sci. USA* 74: 5083–5087.

Ayala, F. J. 1982. Microevolution and macroevolution. In D. S. Bendall, ed., *Evolution From Molecules to Man.* Cambridge UK: Cambridge Univ. Press, pp. 387–402.

Bahn, P. G., and J. Vertut. 1988. *Images of the Ice Age.* N.Y.: Facts on File.

Bak, P., and K. Sneppen. 1993. Punctuated equilibrium and criticality in a simple model of evolution. *Physical Rev. Letters* 71: 4083–4086.

Banner, F. T., and F. M. D. Lowry. 1985. The stratigraphical record of planktonic foraminifera and its evolutionary implications. In J. C. W. Cope and P. W. Skelton, eds., Evolutionary Case Histories From the Fossil Record. *Special Papers Palaeontology* 33: 103–116.

Barnosky, A. D. 1987. Punctuated equilibrium and phyletic gradualism: Some facts from the Quaternary mammalian record. In Genoways, H. H., ed., *Current Mammalogy, Vol 11:* 109–147.

Beatty, J. 1988. Ecology and evolutionary biology in the war and postwar years: questions and comments. *Jour. History Biol.* 21: 245–263.

Begg, C. B., and J. A. Berlin. 1988. Publication bias: a problem in interpreting medical data. *Jour. Roy. Statistical Soc.* 151: 419–463.

Bell, M. A., J. V. Baumgartner, and E. C. Olson. 1985. Patterns of temporal change in single morphological characters of a Miocene stickleback fish. *Paleobiology* 11: 258–271.

Bell, M. A. and T. R. Haglund. 1982. Fine-scale temporal variation of the Miocene stickleback *Gasterosteus doryssus. Paleobiology* 8: 282–292.

Bell, M. A., and P. Legendre. 1987. Multicharacter clustering in a sequence of fossil sticklebacks. *Syst. Zool.* 36: 52–61.

Benson, R. H. 1983. Biochemical stability and sudden change in the evolution of the deep-sea ostracode *Poseidonamicus. Paleobiology* 9: 398–413.

Bergstrom, J. and R. Levi-Setti. 1978. Phenotypic variation in the Middle Cambrian trilobite *Paradoxides davidis* Salter at Manuels, SE Newfoundland. *Geologica et Palaeontologica* 12: 1–40.

Berry, M. S. 1982. *Time, Space, and Transition in Anasazi Prehistory.* Salt Lake City: Univ. of Utah Press.

Blackburn, D. G. 1995. Saltationist and punctuated equilibrium models for the evolution of viviparity and placentation. *J. Theoret. Biol.* 174: 199–216.

Blanc, M. 1982. Les théories de l'évolution aujourd'hui. *La Recherche,* January, pp. 26–40.

Bliss, R. B., G. E. Parker, and D. T. Gish. 1980. *Fossils: Key to the Present.* San Diego CA: CLP Publishers.

Boucot, A. J. 1983. Does evolution take place in an ecological vacuum? *Jour. Paleontology* 57: 1–30.

Boulding, K. E. 1992. Punctuationalism in societal evolution. In A. Somit and S. A. Peterson, eds., *The Dynamics of Evolution: The Punctuated Equilibrium Debate in the Natural and Social Sciences.* Ithaca NY: Cornell Univ. Press, pp. 171–186.

Bown, T. M. 1979. Geology and mammalian paleontology of the Sand Creek facies,

Lower Willwood Formation (Lower Eocene), Washakie County, Wyoming. *Mem. Geol. Surv. Wyo.* 2:1–151.

Brace, C. L. 1977. *Human Evolution.* N.Y.: MacMillan.

Breton, G. 1996. Ponctualisme et gradualisme au sein d'une même lignée: réflexions sur la complexité et l'imprédictibilité des phénomènes évolutifs. *Geobios* 29: 125–130.

Brett, C. E., and G. C. Baird. 1995. Coordinated stasis and evolutionary ecology of Silurian to Middle Devonian faunas in the Appalachian Basin. In D. H. Erwin and R. L. Anstey, eds., *New Approaches to Speciation in the Fossil Record.* N.Y.: Columbia Univ. Press, pp. 285–317.

Brett, C. E., L. C. Ivany, and K. M. Schopf. 1996. Coordinated stasis: An overview. *Palaeogeol. Palaeoclimat. Palaeoecol.* 127: 1–20.

Brown, J. H., and B. A. Maurer. 1986. Body size, ecological dominance, and Cope's rule. *Nature* 324: 248–250.

Budd, A. F., and A. G. Coates. 1992. Non-progressive evolution in a clade of Cretaceous *Montastraea*-like corals. *Paleobiology* 18: 425–446.

Cain, A. J. 1979. Introduction to general discussion. *Proc. Roy. Soc. London Series B* 205: 599–604.

——— 1988. Evolution. *Jour. Evol. Biol.* 1: 185–194.

Cairns-Smith, A. G. 1971. *The Life Puzzle.* Edinburgh: Oliver & Boyd.

Carmines, E. G., and J. A. Stimson. 1989. *Issue Evolution: Race and the Transformation of American Politics.* Princeton NJ: Princeton Univ. Press.

Carson, H. L. 1975. The genetics of speciation at the diploid level. *Amer. Naturalist* 109: 83–92.

Chaisson, E. 1988. *Universe: An Evolutionary Approach to Astronomy.* Englewood Cliffs NJ: Prentice Hall.

Chaline, J. 1982. *Modalités, rythmes, mécanismes de l'évolution biologique: gradualisme phylétique ou équilibres ponctués?* Colloque International. Paris: CNRS (Centre Nationale de Recherche Scientifique), 335 pp.

Chaline, J., and B. Laurin. 1986. Phyletic gradualism in a European Plio-Pleistocene *Mimomys* lineage (Arvicolidae, Rodentia). *Paleobiology* 12: 203–216.

Chau, H. F. 1994. Scaling behavior of the punctuated equilibrium model of evolution. *Phys. Rev. E.*

Cheetham, A. H. 1986. Tempo of evolution in a Neogene bryozoan: rates of morphologic change within and across species boundaries. *Paleobiology* 12:190–202.

——— 1987. Tempo of evolution in a Neogene bryozoan: are trends in single morphologic characters misleading? *Paleobiology* 13: 286–296.

Cheetham, A. H., and J. B. C. Jackson. 1995. Process from pattern: tests for selection versus random change in punctuated bryozoan speciation. In D. H. Erwin and R. L. Anstey, eds., *New Approaches to Speciation in the Fossil Record.* N.Y.: Columbia Univ. Press, pp. 184–207.

Coope, G. R. 1979. Late Cenozoic fossil Coleoptera. *Ann. Rev. Ecol. Syst.* 10: 247–267.

——— 1980. The invasion of Northern Europe during the Pleistocene by Mediterranean species of Coleoptera. In F. Di Castri et al., eds., *Biological Invasions in Europe and the Mediterranean Basin.* Dordrecht: Kluwer, pp. 203–215.

——— 1994. The response of insect faunas to glacial-interglacial climatic fluctuations. *Philos. Trans. Roy. Soc. London B,* 344: 19–26.

Cope, J. C. W., and P. W. Skelton (eds.) 1985. Evolutionary case histories from the fossil record. *Special Papers in Palaeontology* 33: 1–203.

Courtillot, V. 1995. *La vie en catastrophes: Du hasard dans l'évolution des espèces.* Paris: Fayard.

Cronin, J. E., N. T. Boaz, C. B. Stringer, and Y. Rak. 1980. Tempo and mode in hominid evolution. *Nature* 292: 113–122.

Cronin, T. M. 1985. Speciation and stasis in marine Ostracoda: climatic modulation of evolution. *Science* 277: 60–62.

Curtis, H., and N. S. Barnes. 1985. *Invitation to Biology.* N.Y.: Worth Publishers.

Darwin, C. 1842. *The Structure and Distribution of Coral Reefs. Being the First Part of the Geology of the Voyage of the Beagle.* London: Smith, Elder.

——— 1859. *On the Origin of Species by Means of Natural Selection, or Preservation of Favored Races in the Struggle for Life.* London: Murray.

——— 1881. *The Formation of Vegetable Mould, through the Action of Worms, with Observations on their Habitats.* London: Murray.

Dawkins, R. 1986. *The Blind Watchmaker.* N.Y.: W. W. Norton.

Dennett, D. C. 1995. *Darwin's Dangerous Idea.* N.Y.: Simon & Schuster.

——— 1997. "Darwinian fundamentalism": an exchange. Letter to the Editors. *N.Y. Review of Books,* Aug. 14, 1997, pp. 64–65.

Den Tex, E. 1990. Punctuated equilibria between rival concepts of granite genesis in the late 18th, 19th, and early 20th centuries. *Geol. Jour.* 25: 215–219.

Devillers, C., and J. Chaline. 1989. *La théorie de l'évolution.* Paris: Dunod.

Dodson, E. O., and P. Dodson. 1990. *Evolution: Process and Product.* Boston MA: Prindle, Weber & Schmidt.

Dott, R. H., Jr., and D. R. Prothero. 1994. *Evolution of the Earth.* N.Y.: McGraw Hill.

Ehrlich, P. L. 1986. *The Machinery of Nature.* N.Y.: Simon & Schuster.

Eiseley, L. 1958. *Darwin's Century.* N.Y.: Doubleday.

Eldredge, N. 1971. The allopatric model and phylogeny in Paleozoic invertebrates. *Evolution* 25: 156–167.

——— 1979. Alternative approaches to evolutionary theory. In J. H. Schwartz and H. B. Rollins, eds., Models and Methodologies in Evolutionary Theory. *Bull. Carnegie Mus. Nat. Hist.* 13: 7–19.

——— 1985a. *Unfinished Synthesis: Biological Hierarchies and Modern Evolutionary Thought.* N.Y.: Oxford Univ. Press.

——— 1985b. *Time Frames.* N.Y.: Simon & Schuster.

——— 1989. *Macroevolutionary Patterns and Evolutionary Dynamics: Species, Niches and Adaptive Peaks.* N.Y.: McGraw-Hill.

——— 1995. *Reinventing Darwin: The Great Debate at The High Table of Evolutionary Theory.* N.Y.: John Wiley.
——— 1999. *The Pattern of Evolution.* N.Y.: W. H. Freeman.
Eldredge, N., and S. J. Gould. 1972. Punctuated equilibria: an alternative to phyletic gradualism. In T. J. M. Schopf, ed., *Models in Paleobiology.* San Francisco: Freeman, Cooper & Co., pp. 82–115.
——— 1977. Evolutionary models and biostratigraphic strategies. In E. G. Kauffman and J. E. Hazel, eds., *Concepts and Methods of Biostratigraphy.* Stroudsburg PA: Dowden, Hutchinson & Ross, pp. 25–40.
——— 1988. Punctuated Equilibrium prevails. *Nature* 332: 211–212.
Eldredge, N., and M. Grene. 1992. *Interactions: The Biological Context of Social Systems.* N.Y.: Columbia Univ. Press.
Elena, S. F., V. S. Cooper, and R. E. Lenksi. 1996. Punctuated Evolution caused by selection of rare beneficial mutations. *Science* 272: 1802–1804.
Emry, R. J. 1981. Additions to the mammalian fauna of the type Duchesnean, with comments on the status of the Duchesnean. *J. Paleontology* 55: 563–570.
Erdtmann, B. D. 1986. Early Ordovician eustatic cycles and their bearing on punctuations in early nematophorid (planktic) graptolite evolution. In O. H. Walliser, ed., *Global Bio-Events: A Critical Approach.* Berlin: Springer Verlag, pp. 139–152.
Erwin, D. H., and R. L. Anstey. 1995. Speciation in the Fossil Record. In D. H. Erwin and R. L. Anstry, eds., *New Approaches to Speciation in the Fossil Record.* N.Y.: Columbia Univ. Press, pp. 11–38.
Falconer, H. 1868. *Palaeontological Memoirs and Notes* (C. Murchison, ed.) 2 volumes. London: Robert Hardwicke.
Fausto-Sterling, A. 1985. *Myths of Gender: Biological Theories About Women and Men.* N.Y.: Basic Books.
Finney, S. C. 1986. Heterochrony, punctuated equilibrium, and graptolite zonal boundaries. In Hughes, C. P. and R. B. Rickards, eds., *Palaeoecology and Biostratigraphy of Graptolites,* pp. 103–113.
Fisher, R. A. 1930. *The Genetical Theory of Natural Selection.* Oxford UK: Oxford Univ. Press.
Flynn, L. J. 1986. Species longevity, stasis, and stairsteps in rhizomyid rodents. *Contributions to Geology, Univ. of Wyoming, Special Paper* 3: 273–285.
Fortey, R. A. 1985. Gradualism and punctuated equilibrium as competing and complementary theories. *Special Papers in Palaeontology* 33: 17–28.
——— 1988. Seeing is believing: gradualism and punctuated equilibria in the fossil record. *Sci. Prog.,* 72: 1–19.
Franco, A. O. 1985. La teoria del equilibrio puntuado. Una alternativa al Neodarwinismo. *Ciencias,* UNAM, Mexico City, pp. 46–59.
Fryer, G., P. H. Greenwood, and J. F. Peake. 1983. Punctuated equilibria, morphological stasis, and the paleontological documentation of speciation: a biological appraisal of a case history in an African lake. *Biol. Jour. Linnaean Soc.* 20: 195–205.
Futuyma, D. J. 1986. *Evolutionary Biology.* Sunderland MA: Sinauer.

——— 1987. On the role of species in anagenesis. *Amer. Nat.* 130: 465–473.
——— 1988a. Macroevolutionary Consequences of Speciation: Inferences from Phytophagous Insects. In J. Endler and D. Otte, eds., *Speciation and its Consequences.*
——— 1988b. *Sturm* and *Drang* and the evolutionary synthesis. *Evolution* 42: 217–226.
Gans, C. 1987. Punctuated Equilibria and political science: a neontological view. *Politics and the Life Sciences* 5: 220–244.
Geary, D. H. 1990. Patterns of evolutionary tempo and mode in the radiation of *Melanopsis* (Gastropoda: Melanopsidae). *Paleobiology* 16: 492–511.
——— 1995. The importance of gradual change in species-level transitions. In D. H. Erwin and R. L. Anstey, eds., *New Approaches to Speciation in the Fossil Record.* N.Y.: Columbia Univ. Press, pp. 67–86.
Gersick, C. J. G. 1988. Time and transition in work teams: toward a new model of group development. *Acad. Management Jour.* 31: 9–41.
——— 1991. Revolutionary change theories: a multi-level exploration of the punctuated equilibrium paradigm. *Acad. Management Rev.,* Jan., pp. 10–35.
Gilinsky, N. L. 1986. Species selection as a causal process. *Evol. Biol.* 20: 248–273.
——— 1994. Volatility and the Phanerozoic decline of background extinction intensity. *Paleobiology* 20: 445–458.
Gingerich, P. D. 1974. Stratigraphic record of early Eocene *Hyopsodus* and the geometry of mammalian phylogeny. *Nature* 248: 107–109.
——— 1976. Paleontology and phylogeny: patterns of evolution at the species level in early Tertiary mammals. *Am. Jour. Sci.* 276: 1–28.
——— 1978. Evolutionary transition from the ammonite *Subprionocyclus* to *Reedsites*—punctuated or gradual? *Evolution* 32: 454–456.
——— 1980. Evolutionary patterns in early Cenozoic mammals. *Ann. Rev. Earth Planet. Sci.* 8: 407–424.
——— 1984a. Punctuated equilibria—where is the evidence? *Syst. Zool.* 33: 335–338.
——— 1984b. Darwin's gradualism and empiricism: discussion and reply. *Nature* 309: 116.
——— (with reply by F. H. T. Rhodes) 1985. Darwin's gradualism and empiricism. *Nature* 309: 116.
——— 1987. Evolution and the fossil record: patterns, rates, and processes. *Can. Jour. Zool.* 65: 1053–1060.
——— 1989. New earliest Wasatchian mammalian fauna from the Eocene of northwestern Wyoming. *Univ. Mich. Pap. Paleontology* 28: 1–27.
Glaubrecht, M. 1995. *Der lange Atem der Schöpfung. Was Darwin gern gewusst hätte.* Berlin: Rasch und Röhring.
Glennon, L. (ed.) 1995. *Our Times.* Atlanta GA: Turner Publishing.
Godinot, M. 1985. Evolutionary implications of morphological changes in Palaeogene primates. In J. C. W. Cope and P. W. Skelton, eds., Evolutionary Case Histories From The Fossil Record. *Special Papers in Palaeontology* 33, pp. 39–47.
Gold, T. 1999. *The Deep Hot Biosphere.* N.Y.: Copernicus.

Goldschmidt, R. 1940. *The Material Basis of Evolution.* New Haven CT: Yale Univ. Press.

Golob, R., and E. Brus. 1990. *The Almanac of Science and Technology.* N.Y.: Harcourt Brace Jovanovich.

Goodfriend, G. A., and S. J. Gould. 1996. Paleontology and chronology of two evolutionary transitions by hybridization in the Bahamian land snail *Cerion. Science* 274: 1894–1897.

Gould, J. L., and C. G. Gould. 1989. *Sexual Selection.* N.Y.: W. H. Freeman, Scientific American Library.

Gould, J. L., and W. T. Keeton. 1996. *Biological Science.* N.Y.: W. W. Norton.

Gould, S. J. 1965. Is uniformitarianism necessary? *Amer. Jour. Sci.* 263: 223–228.

——— 1966. Allometry and size in ontogeny and phylogeny. *Biol. Rev.* 41: 587–640.

——— 1969. An evolutionary microcosm: Pleistocene and Recent history of the land snail *P. (Poecilozonites)* in Bermuda. *Bull. Mus. Comp. Zool.* 138: 407–532.

——— 1970a. Evolutionary paleontology and the science of form. *Earth-Sci. Rev.* 6: 77–119.

——— 1970b. Dollo on Dollo's law: irreversibility and the status of evolutionary laws. *Jour. Hist. Biol.* 3: 189–212.

——— 1971a. D'Arcy Thompson and the science of form. *New Literary Hist.* 2: 229–258.

——— 1971b. Precise but fortuitous convergence in Pleistocene land snails from Bermuda. *Jour. Paleont.* 45: 409–418.

——— 1972. Allometric fallacies and the evolution of *Gryphaea:* a new interpretation based on White's criterion of geometric similarity. In Th. Dobzhansky et al., eds., *Evolutionary Biology,* vol. 6, pp. 91–118.

——— 1974. The origin and function of "bizarre" structures: antler size and skull size in the "Irish Elk," *Megaloceros giganteus. Evolution* 28: 191–220.

——— 1977. *Ontogeny and Phylogeny.* Cambridge MA: Harvard Univ. Press.

——— 1980a. The promise of paleobiology as a nomothetic, evolutionary discipline. *Paleobiology* 6: 96–118.

——— 1980b. Is a new and general theory of evolution emerging? *Paleobiology* 6: 119–130.

——— 1980c. G. G. Simpson, Paleontology, and the Modern Synthesis. In E. Mayr and W. B. Provine, eds., *The Evolutionary Synthesis.* Cambridge MA: Harvard Univ. Press, pp. 153–172.

——— 1982a. The uses of heresy: an introduction to Richard Goldschmidt's "The Material Basis of Evolution," pp. xiii-xlii. New Haven CT and London: Yale Univ. Press.

——— 1982b. The meaning of punctuated equilibrium and its role in validating a hierarchical approach to macroevolution. In R. Milkman, ed., *Perspectives on Evolution.* Sunderland MA: Sinauer Associates, pp. 83–104.

——— 1983. The hardening of the Modern Synthesis. In: Marjorie Grene, ed., *Dimensions of Darwinism.* Cambridge UK: Cambridge Univ. Press.

——— 1984a. Covariance sets and ordered geographic variation in *Cerion* from Aruba, Bonaire, and Curaçao: a way of studying nonadaptation. *Syst. Zool.* 33: 217–237.

——— 1984b. Morphological channeling by structural constraint: convergence in styles of dwarfing and gigantism in *Cerion*, with a description of two new fossil species and a report on the discovery of the largest *Cerion. Paleobiology* 10: 172–194.

——— 1985. The paradox of the first tier: an agenda for paleobiology. *Paleobiology* 11: 2–12.

——— 1987. *Time's Arrow, Time's Cycle.* Cambridge MA: Harvard Univ. Press.

——— 1988a. The case of the creeping fox terrier clone. *Nat. Hist.* 97 (Jan.): 16–24.

——— 1988b. Trends as changes in variance: a new slant on progress and directionality in evolution (Presidential Address). *Jour. Paleont.* 62: 319–329.

——— 1989a. *Wonderful Life: The Burgess Shale and the Nature of History.* N.Y.: W. W. Norton, 347 pp.

——— 1989b. Punctuated equilibrium in fact and theory. *J. Social Biol. Struct.* 12: 117–136.

——— 1993. A special fondness for beetles. *Nat. Hist.* 102 (Jan.): 4–12.

——— 1995. *Dinosaur in a Haystack.* N.Y.: Harmony Books.

——— 1996. *Full House: The Spread of Excellence from Plato to Darwin.* N.Y.: Harmony Books.

——— 1997a. Cope's rule as psychological artifact. *Nature* 385: 199–200.

——— 1997b. Darwinian Fundamentalism, part 1. *The New York Review of Books,* June 12, pp. 34–37. Evolution: The Pleasures of Pluralism, part 2. *The New York Review of Books,* June 26, pp. 47–52.

——— 1997c. The paradox of the visibly irrelevant. *Nat. Hist.* 106 (Dec.): 12–18, 60–66.

——— 1998. *Leonardo's Mountain of Clams and The Diet of Worms.* N.Y.: Harmony Books.

Gould, S. J., and N. Eldredge. 1971. Speciation and punctuated equilibria: an alternative to phyletic gradualism. G. S. A. Ann. Meeting, Washington, DC, *Abstracts with Programs,* pp. 584–585.

——— 1977. Punctuated equilibria: the tempo and mode of evolution reconsidered. *Paleobiology:* 3: 115–151.

——— 1983. Darwin's gradualism. *Systematic Zool.* 32: 444–445.

——— 1986. Punctuated equilibrium at the third stage. *Systematic Zool.* 35: 143–148.

——— 1993. Punctuated equilibrium comes of age. *Nature* 366: 223–227.

Gould, S. J., and R. C. Lewontin. 1979. The spandrels of San Marco and the Panglossian paradigm: a critique of the adaptationist programme. *Proc. R. Soc. Lond. B* 205: 581–598.

Gould, S. J., and E. A. Lloyd. 1999. Individuality and adaptation across levels of selection: how shall we name and generalize the unit of Darwinism? *Proc. Natl. Acad. Sci. USA* 96: 11904–11909.

Gould, S. J., D. M. Raup, J. J. Sepkoski, Jr., T. J. M. Schopf, and D. S. Simberloff. 1977.

The shape of evolution: a comparison of real and random clades. *Paleobiology* 3: 23–40.

Gould, S. J., and S. Vrba, 1982. Exaptation—a missing term in the science of form. *Paleobiology* 8:4–15.

Grant, V. 1983. The Synthetic Theory strikes back. *Biol. Zentralblatt* 102: 149–158.

Greiner, G. O. G. 1974. Environmental factors controlling the distribution of Recent benthic Foraminifera. *Breviora Mus. Comp. Zool. Harvard Univ.* Number 420.

Grine, F. E. 1993. Australopithecine taxonomy and phylogeny. In R. L. Ciochon and J. Fleagle, eds., *The Human Evolution Source Book.* Englewood Cliffs NJ: Prentice Hall, pp. 145–175.

Haldane, J. B. S. 1932. *The Causes of Evolution.* London: Longmans Green.

Hallam, A. 1968. Morphology, palaeoecology and evolution of the genus *Gryphaea* in the British Lias. *Phil. Trans. Roy. Soc. London* 254: 91–128.

——— 1978. How rare is phyletic gradualism and what is its evolutionary significance? Evidence from Jurassic bivalves. *Paleobiology* 4:16–25.

——— 1990. Biotic and abiotic factors in the evolution of early marine molluscs. In R. M. Ross and W. D. Allmon, eds., *Causes of Evolution: A Paleontological Perspective.* Chicago IL: Univ. of Chicago Press, pp. 249–269.

Halstead, B. 1984. Neo-Darwinism rules. *New Scientist,* May 3, p. 40.

——— 1985. The Evolution debate continues. *Modern Geology* 9: 317–326.

Hansen, T. A. 1978. Larval dispersal and species longevity in Lower Tertiary gastropods. *Science* 199: 885–887.

——— 1980. Influence of larval dispersal and geographic distribution on species longevity in neogastropods. *Paleobiology* 6: 193–207.

Hanson, N. R. 1961. *Patterns of Discovery.* Cambridge UK: Cambridge Univ. Press.

Heaton, T. H. 1993. The Ologocene rodent *Ischyromys* of the Great Plains: replacement mistaken for anagenesis. *Jour. Paleontology* 67: 297–308.

——— 1996. Ischyromyidae. In *The Terrestrial Eocene-Oligocene Transition in North America,* pp. 373–398.

Hoffman, A. 1989. *Arguments on Evolution.* N.Y.: Oxford.

Howe, J. A. 1956. The Oligocene rodent *Ischyromys* in relationship to the paleosols of the Brule Formation. MS. Thesis, Univ. of Nebraska, 89 pp.

Huxley, A. 1982. Address of the President. *Proc. Roy. Soc. London Series B* 214: 137–152.

Imbrie, J. 1957. The species problem with fossil animals. In E. Mayr, ed., *The Species Problem.* Am. Assoc. Adv. Sci. Pub. No. 50, pp. 125–153.

Ivany, L. C. 1996. Coordinated stasis or coordinated turnover? Exploring intrinsic *vs.* extrinsic controls on pattern. *Palaeogeog. Palaeoclimat. Palaeoecol.* 127: 1–18.

Ivany, L. C., and K. M. Schopf, eds., 1996. New Perspectives on Faunal Stability in the Fossil Record. *Special Issue of Palaeogeog. Palaeoclimatol. Palaeoecol.* volume 127, 359 pp.

Jablonski, D. 1997. Body-size evolution in Cretaceous molluscs and the status of Cope's rule. *Nature* 385: 250–252.

——— 1999. The future of the fossil record. *Science* 284: 2114–2116.

Jablonski, D., S. Lidgard, and P. D. Taylor. 1997. Comparative ecology of bryozoan radiations: origin of novelties in cyclostomes and cheilostomes. *Palaios* 12: 505–523.

Jackson, J. B. C., and A. H. Cheetham. 1990. Evolutionary significance of morphospecies: a test with Cheilostome Bryozoa. *Science* 248: 579–582.

——— 1994. Phylogeny reconstruction and the tempo of speciation in cheilostome Bryozoa. *Paleobiology* 20: 407–423.

——— 1999. Tempo and mode of speciation in the sea. *Trends Ecol. Evol.* 14: 72–77.

Jacobs, K., and L. Godfrey. 1982. Cerebral leaps and bounds: a punctuational perspective on hominid cranial capacity increase. *Man and His Origins* 21: 77–87.

Janzen, D. 1977. What are dandelions and aphids? *Amer. Nat.* 111: 586–589.

——— 1985. On ecological fitting. *Oikos* 45: 308–310.

Johanson, D., and M. Edey. 1981. *Lucy.* N.Y.: Simon & Schuster.

Johanson, D., and B. Edgar. 1996. *From Lucy to Language.* N.Y.: Simon & Schuster.

Johnson, A. L. A. 1985. The rate of evolutionary change in European Jurassic scallops. In J. C. W. Cope and P. W. Skelton, eds., *Evolutionary Case Histories From The Fossil Record. Special Papers in Palaeontology* 33: 91–102.

Johnson, J. G. 1975. Allopatric speciation in fossil brachiopods. *Jour. Paleontol.* 49: 646–661.

——— 1982. Occurrence of phyletic gradualism and punctuated equilibria through geological time. *Jour. Palaeontol.* 56: 1329–1331.

Jones, D. S., and S. J. Gould. 1999. Direct measurement of age in fossil *Gryphaea:* the solution to a classic problem in heterochrony. *Paleobiology* 25: 158–187.

Kammer, T. W., T. K. Baumiller, and W. I. Ausich. 1997. Species longevity as a function of niche breadth: Evidence from fossil crinoids. *Geology* 25: 219–222.

Kauffman, S. A. 1993. *The Origins of Order: Self-Organization and Selection in Evolution.* Oxford: Oxford Univ. Press.

Kelley, P. H. 1983. Evolutionary patterns of eight Chesapeake group molluscs: evidence for the model of puntuated equilibria. *Jour. Paleontol.* 57: 581–598.

——— 1984. Multivariate analysis of evolutionary patterns of seven Miocene Chesapeake Group molluscs. *Jour. Paleontol* 58: 1235–1250.

Kerr, R. A. 1994. Between extinctions, evolutionary stasis. *Science* 266: 29.

——— 1995. Did Darwin get it all right? *Science* 267: 1421–1422.

Kilgour, F. G. 1998. *The Evolution of the Book.* N.Y.: Oxford Univ. Press.

Kimbel, W. H., D. C. Johanson, and Y. Rak. 1994. The first skull and other new discoveries of *Australopithecus afarensis* at Hadar, Ethiopia. *Nature* 368: 449–451.

Kimura, M. 1968. Evolutionary rate at the molecular level. *Nature* 217: 624–626.

Konner, M. 1986. Revolutionary biology. *The Sciences,* p. 608.

Korey, K. 1984. *The Essential Darwin.* Boston MA: Little, Brown.

Kraus, D. 1983. *Concepts in Modern Biology.* N.Y.: Globe Books.

Krishtalka, L., and R. K. Stuckey. 1985. Revision of the Wind River faunas, early Eocene of central Wyoming. Part 7. Revision of *Diacodexis* (Mammalia, Artiodactyla). *Ann. Carnegie Mus.* 54: 413–486.

——— 1986. Early Eocene artiodactyls from the San Juan Basin, New Mexico, and the

Piceance Basin, Colorado. In K. M. Flanagan and J. A. Lillegraven, eds., *Vertebrates, Phylogeny and Paleontology. Univ. Wyo. Contrib. Geol. Spec. Pap. 3:* 183–197.

Kucera, M., and B. A. Malmgren. 1998. Differences between evolution of mean form and evolution of new morphotypes: an example from Late Cretaceous planktonic foraminifera. *Paleobiology* 24: 49–63.

Kuhn, T. S. 1962. *The Structure of Scientific Revolutions.* Chicago IL: Univ. of Chicago Press.

Lampl, M., J. D. Veldhuis, and M. L. Johnson. 1992. Saltation and stasis: a model of human growth. *Science* 258: 801–803.

Lande, R. 1976. Natural selection and random genetic drift in phenotypic evolution. *Evolution* 30: 314–334.

——— 1986. The dynamics of peak shifts and the pattern of morphological evolution. *Paleobiology* 12: 343–354.

Lawless, J. V. 1998. Punctuated equilibrium and paleohydrology. *Proc. New Zealand Geothermal Workshop* 10: 165–169.

Leakey, M. G., F. Spoor, F. H. Brown, P. N. Gathogo, C. Klarie, L. N. Leakey, and I. McDougall. 2001. New hominin genus from eastern Africa shows diverse middle Pliocene lineages. *Nature* 410: 433–440.

Lemen, C. A., and P. W. Freeman. 1989. Testing macroevolutionary hypotheses with cladistic analysis: evidence against rectangular evolution. *Evolution* 43: 1538–1554.

Lenski, R. E., and M. Travisano. 1994. Dynamics of adaptation and diversification: a 10,000-generation experiment with bacterial populations. *Proc. Natl. Acad. Sci. USA* 91: 6808–6814.

Lerner, I. M. 1954. *Genetic Homeostasis.* N.Y.: John Wiley.

Leroi-Gourhan, A. 1967. *Treasures of Prehistoric Art.* N.Y.: H. N. Abrams.

Levin, H. L. 1991. *The Earth Through Time.* Fort Worth TX: W. B. Saunders.

Levine, D. 1991. Punctuated Equilibrium: the modernization of the proletarian family in the age of ascendant capitalism. *International Labor and Working Class History* No. 39.

Levinton, J. 1988. *Genetics, Paleontology, and Macroevolution.* Cambridge UK: Cambridge Univ. Press.

Lewin, R. 1986. Punctuated Equilibrium is now old hat. *Science* 231: 672–673.

Lich, D. K. 1990. *Cosomys primus:* a case for stasis. *Paleobiology* 16: 384–395.

Lieberman, B. S., C. E. Brett, and N. Eldredge. 1994. Patterns and processes of stasis in two species lineages of brachiopods from the Middle Devonian of New York State. *Amer. Mus. Nat. Hist. Novitates* Number 3114.

——— 1995. A study of stasis and change in two species lineages from the Middle Devonian of New York state. Paleobiology 21: 15–27.

Lieberman, B. S. and S. Dudgeon. 1996. An evaluation of stabilizing selection as a mechanism for stasis. *Palaeogeog. Palaeoclimat. Palaeoecol.* 127: 229–238.

Lister, A. M. 1993a. “Gradualistic” evolution: its interpretation in Quaternary large mammal species. *Quarternary International* 19: 77–84.

——— 1993b. Mammoths in miniature. *Nature* 362: 288–289.

——— 1996. Dwarfing in island elephants and deer: processes in relation to time and isolation. *Symp. Zool. Soc. Lond.* 69: 277–292.

Lloyd, E. A., and S. J. Gould. 1993. Species selection on variability. *Proc. Natl. Acad. Sci. USA* 90: 595–599.

Loch, Christoph H. 1999. A punctuated equilibrium model of technology diffusion. Abstract for Dynamics of Computation Group Meeting, Xerox Palo Alto Research Center.

Losos, J. B., K. I. Warheit, and T. W. Schoener. 1997. Adaptive differentiation following experimental island colonization in *Anolis* lizards. *Nature* 387: 70–73.

Lyne, J., and H. F. Howe. 1986. "Punctuated equilibria": rhetorical dynamics of a scientific controversy. *Quart. Jour. Speech* 72: 132–147.

MacFadden, B. J. 1986. Fossil horses from "eohippus" *(Hyracotherium)* to *Equus:* scaling laws and the evolution of body size. *Paleobiology* 12: 355–369.

MacFadden, B. J., N. Solounias, and T. E. Cerling. 1999. Ancient diets, ecology, and extinction of 5-million-year-old horses from Florida. *Science* 283: 824–827.

MacGillavry, H. J. 1968. Modes of evolution mainly among marine invertebrates. *Bijdragen tot de Dierkunde* 38: 69–74.

MacLeod, N. 1991. Punctuated anagenesis and the importance of stratigraphy to paleobiology. *Paleobiology* 17: 167–188.

Maddux, J. 1994. Punctuated equilibrium by computer. *Nature* 371: 197.

Malmgren, B. A., W. A. Berggren, and G. P. Lohman. 1983. Evidence for punctuated gradualism in the Late Neogene *Globorotalia tumida* lineage of planktonic foraminifera. *Paleobiology* 9: 377–389.

Malmgren, B. A., and J. P. Kennett. 1981. Phyletic gradualism in a Late Cenozoic lineage: DSDP Site 284, Southwest Pacific. *Paleobiology* 7: 230–240.

Marshall, C. R. 1994. Confidence intervals on stratigraphic ranges: partial relaxation of the assumption of randomly distributed fossil horizons. *Paleobiology* 20: 459–469.

——— 1995. Distinguishing between sudden and gradual extinction in the fossil record: Predicting the position of the Cretaceous-Tertiary iridium anomaly using the ammonite fossil record on Seymour Island, Antarctica. *Geology* 23: 731–734.

Mayden, R. L. 1986. Speciose and depauperate phylads and tests of punctuated *vs.* gradual evolution: fact or artifact? *Syst. Zool.* 35: 591–602.

Maynard Smith, J. 1983. The genetics of stasis and punctuation. *Ann. Rev. Genetics* 17: 11–25.

Maynard Smith, J., R. Burian, S. Kauffman, P. Alberch, J. Campbell, B. Goodwin, R. Lande, D. Raup, and L. Wolpert. 1985. Developmental constraints and evolution. *Quart. Rev. Biol.* 60: 265–287.

Mayr, E. 1954. Change of genetic environment and evolution. In J. S. Huxley, A. C. Hardy, and E. B. Ford, eds., *Evolution As A Process.* London: G. Allen & Unwin., pp. 157–180.

——— 1963. *Animal species and evolution.* Cambridge MA: Harvard Univ. Press.

——— 1982. Adaptation and selection. *Biologisches Zentralblatt* 101: 161–174.

——— 1992. Speciational evolution and punctuated equilibria. In A. Somit, and S. A. Peterson, eds., *The Dynamics of Evolution.* Ithaca NY: Cornell Univ. Press, pp. 21–53.

Mazur, A. 1992. Periods and question marks in the punctuated evolution of human social behavior. In A. Somit and S. A. Peterson, eds., *The Dynamics of Evolution: The Puntuated Equilibrium Debate in the Natural and Social Sciences.* Ithaca NY: Cornell Univ. Press, pp. 221–234.

McAlester, A. L. 1962. Some comments on the species problem. *Jour. Paleontology* 36: 1377–1381.

McHenry, H. M. 1994. Hominid dualism. *Science* 265.

McKinney, F. K., S. Lidgard, J. J. Sepkoski Jr., and P. D. Taylor. 1998. Decoupled temporal patterns of evolutionary and ecology in two Post-Paleozoic clades. *Science* 281: 807–809.

McKinney, M. L. 1988. Classifying heterochrony: allometry, size, and time. In M. L. McKinney, ed., *Heterochrony in Evolution: A Multidisciplinary Approach.* N.Y.: Plenum Press, pp. 17–34.

McKinney, M. L., and W. D. Allmon. 1995. Metapopulations and disturbance: from patch dynamics to biodiversity dynamics. In: D. H. Erwin and R. L. Anstrey, eds., *New Approaches to Speciation in the Fossil Record.* N.Y.: Columbia Univ. Press, pp. 123–183.

McKinney, M. L., and D. S. Jones. 1983. Oligopycoid echinoids and the biostratigraphy of the Ocala limestone of peninsular Florida. *Southeastern Geology* 23: 21–30.

McKinney, M. L., and K. J. McNamara. 1991. *Heterochrony: The Evolution of Ontogeny.* N.Y.: Plenum Press.

McNamara, K. J. 1997. *Shapes of Time.* Baltimore MD: Johns Hopkins Univ. Press.

McShea, D. W. 1994. Mechanisms of large-scale evolutionary trends. *Evolution* 48: 1747–1763.

Mettler, L. E., T. G. Gregg, and H. E. Schaffer. 1988. *Population Genetics and Evolution.* Englewood Cliffs NJ: Prentice Hall.

Meyer, C. J. A. 1878. Micrasters in the English chalk: two or more species? *Geol. Mag.* 5: 115–117.

Michaux, B. 1987. An analysis of allozymic characters of four species of New Zealand *Amalda* (Gastropoda: Olividae: Ancillinae). *New Zealand Jour. Zool.* 14: 359–366.

——— 1989. Morphological variation of species through time. *Biol. Jour. Linnaean Soc.* 38: 239–255.

Miller, A. I. 1998. Biotic Transitions in Global Marine Diversity. *Science* 281: 1157–1160.

Mindel, D. P., J. W. Sites, Jr., D. Grauer. 1989. Speciational evolution: a phylogenetic test with allozymes in *Sceloporus* (Reptilia). *Cladistics* 5: 49–61.

Mokyr, J. 1990. Punctuated equilibrium and technological progress. *Amer. Econ. Rev.* 80: 350–354.

Moore, R. C., C. G. Lalicker, and A. G. Fischer. 1952. *Invertebrate Fossils.* N.Y.: McGraw-Hill.

Moore, R. C., and L. R. Landon. 1943. Evolution and classification of Paleozoic crinoids. *Geol. Soc. Amer. Special Paper* 46: 1–153.

Moretti, F. 1996. *Modern Epic: The World System From Goethe to Garcia Marquez.* N.Y.: Verso.

Morris, P. J. 1996. Testing patterns and causes of stability in the fossil record, with an example from the Pliocene Lusso Beds of Zaire. *Palaeogeog. Palaeoclimatol. Palaeoecol.* 127: 313–337.

Morris, P. J., L. C. Ivany, K. M. Schopf, and C. E. Brett. 1995. The challenge of paleoecological stasis: reassessing sources of evolutionary stability. *Proc. Natl. Acad. Sci. USA* 92: 11269–11273.

Nehm, R. H., and D. H. Geary. 1994. A gradual morphological transition during a rapid speciation event in marginellid gastropods (Neogene, Dominican Republic). *Jour. Paleontology* 68: 787–795.

Newell, N. D. 1949. Phyletic size increase—an important trend illustrated by fossil invertebrates. *Evolution* 3: 103–124.

Newman, C. M., J. E. Cohen, and C. Kipnis. 1985. Neo-Darwinian evolution implies punctuated equilibria. *Nature* 315: 400–401.

Nichols, D. J. 1982. Phyletic gradualism or punctuated equilibria: the evidence from fossil pollen of the Juglandaceae. *Palynology* 6: 288–289.

Nield, E. W., and V. C. T. Tucker. 1985. *Palaeontology: An Introduction.* Oxford UK: Pergamon Press.

Olson, E. C. 1952. The evolution of a Permian vertebrate chronofauna. *Evolution* 6: 181–196.

O'Neill, R. V., D. L. DeAngelis, J. B. Waide, T. H. F. Allen. 1986. *A Hierarchical Concept of Ecosystems.* Princeton NJ: Princeton Univ. Press.

Papadopoulos, D., D. Schneider, J. Meier-Eiss, W. Arber, R. E. Lenski, and M. Blot. 1999. Genonic evolution during a 10,000-generation experiment with bacteria. *Proc. Natl. Acad. Sci. USA* 96: 3807–3812.

Parsons, P. A. 1993. Stress, extinctions and evolutionary change: from living organisms to fossils. *Biol. Rev.* 68: 313–333.

Paul, C. R. C. 1985. The adequacy of the fossil record reconsidered. In J. C. W. Cope and P. W. Skelton, eds., *Evolutionary Case Histories From The Fossil Record. Special Papers in Palaeontology* 33: 7–15.

Penny, D. 1983. Charles Darwin, gradualism and punctuated equilibrium. *Systematic Zool.* 32: 72–74.

——— 1985. Two hypotheses on Darwin's gradualism. *Systematic Zool.* 34: 201–205.

Price, P. W. 1996. *Biological Evolution.* Fort Worth TX: W. B. Saunders.

Prothero, D. R., and T. H. Heaton. 1996. Faunal stability during the Early Oligocene climatic crash. *Palaeogeog. Palaeoclimatol. Palaeoecol.* 127: 257–283.

Prothero, D. R., and N. Shubin. 1989. The evolution of Oligocene horses. In D. R.

Prothero and R. M. Schoch, eds., *The Evolution of Perissodactyls.* Oxford UK: Oxford Univ. Press, pp. 142–175.

Provine, W. B. 1986. *Sewall Wright and Evolutionary Biology.* Chicago IL: Univ. of Chicago Press.

Raff, R. A. 1996. *The Shape of Life.* Chicago IL: Univ. of Chicago Press.

Rand, D. A., and H. B. Wilson. 1993. Evolutionary catastrophes, punctuated equilibria and gradualism in ecosystem evolution. *Proc. R. Soc. Lond* 253: 137–141.

Rand, D. A., H. B. Wilson, and J. M. McGlade. 1993. Dynamics and evolution: evolutionary stable attractors, invasion exponents and phenotype dynamics. *Warwick Preprints* 53/92.

Raup, D. M. 1975. Taxonomic survivorship curves and Van Valen's Law. *Paleobiology* 1: 82–96.

——— 1985. Mathematical models of cladogenesis. *Paleobiology* 11: 42–52.

——— 1992. Large-body impact and extinction in the Phanerozoic. *Paleobiology* 18: 80–88.

Raup, D. M., and S. J. Gould. 1974. Stochastic simulation and evolution of morphology—towards a nomothetic paleontology. *Syst. Zool.* 23: 305–322.

Raup, D. M., S. J. Gould, T. J. M. Schopf, and D. S. Simberloff. 1973. Stochastic models of phylogeny and the evolution of diversity. *Jour. Geology* 81: 525–542.

Ray, T. S. 1992. *Artificial Life.* In C. G. Langton, ed., Santa Fe Institute Studies in the Sciences of Complexity, *Proc. Vol. X.* Redwood City CA: Addison Wesley, pp. 371–408.

Rensberger, B. 1986. *How the World Works: A Guide to Science's Greatest Discoveries.* N.Y.: William Morrow.

Rensch, B. 1947. *Neuere Probleme der Abstammungslehre.* Stuttgart: F. Enke.

——— 1960. *Evolution Above the Species Level.* N.Y.: Columbia Univ. Press.

Reyment, R. A. 1975. Analysis of a generic level transition in Cretaceous ammonites. *Evolution* 28: 665–676.

——— 1982. Analysis of trans-specific evolution in Cretaceous ostracodes. *Paleobiology* 8: 293–306.

Reznick, D. N., F. H. Shaw, F. H. Rodd, and R. G. Shaw. 1997. Evolution of the rate of evolution in natural populations of guppies *(Poecilia reticulata). Science* 275: 1934–1937.

Rhodes, F. H. T. 1983. Gradualism, punctuated equilibrium, and the *Origin of Species. Nature* 305: 269–272.

Ridley, M. 1993. *Evolution.* Boston MA: Blackwell.

Rightmire, G. P. 1981. Patterns in the evolution of *Homo erectus. Paleobiology* 7: 241–246.

——— 1986. Stasis in *Homo erectus* defended. *Paleobiology* 12: 324–325.

Roberts, J. 1981. Control mechanisms of Carboniferous brachiopod zones in eastern Australia. *Lethaia* 14: 123–134.

Robison, R. A. 1975. Species diversity among agnostoid trilobites. *Fossils and Strata* 4: 219–226.

Roe, D. 1980. The handaxe makers. In A. Sherratt, ed., *The Cambridge Encyclopedia of Archaeology.* N.Y.: Columbia Univ. Press, pp. 71–78.

Ross, R. M. 1990. The evolution and biogeography of Neogene Micronesian ostracodes: the role of sea level, geography, and dispersal. Ph.D. Dissertation, Harvard University.

Rubinstein, E. 1995. Punctuated equilibrium in scientific publishing. *Science* 268: 1415.

Ruse, M. 1992. Is the theory of punctuated equilibria a new paradigm? In A. Somit and S. A. Peterson, eds., *The Dynamics of Evolution: The Punctuated Equilibrium Debate in the Natural and Social Sciences.* Ithaca N.Y.: Cornell Univ. Press, pp. 139–167.

Sacher, G. A. 1966. Dimensional analysis of factors governing longevity in mammals. *Proc. Int. Congr. Gerontology,* Vienna, p. 14.

Sadler, P. M. 1981. Sediment accumulation rates and the completeness of stratigraphic sections. *Jour. Geol.* 89: 569–584.

Salvatori, N. 1984. Paleontologi a confronto sui tempi e i modi dell'evoluzione. *Airone,* December, p. 34.

Savage, C. H., and G. F. F. Lombard. 1983. *Sons of the Machine: Case Studies of Social Change in the Workplace.* Cambridge MA: MIT Press.

Schankler, D. M. 1981. Faunal zonation of the Willwood Formation in the central Bighorn Basin, Wyoming. In P. D. Gingerich, ed., Early Cenozoic Paleontology and Stratigraphy of the Bighorn Basin, Wyoming. *Univ. Mich. Pap. Paleontology* 24: 99–114.

Schindel, D. E. 1982. Resolution analysis: a new approach to the gaps in the fossil record. *Paleobiology* 8: 340–353.

Schoch, R. M. 1984. Possible mechanisms for punctuated patterns in the fossil record. *GSA Abstracts with Programs* 16: 62.

Schoonover, L. M. 1941. A stratigraphic study of the molluscs of the Calvert and Choptank Formations of Southern Maryland. *Bull. Amer. Paleontol.* 25: 169–299.

Schwartz, J. H. 1999. *Sudden Origins: Fossils, Genes, and the Emergence of Species.* N.Y.: John Wiley and Sons.

Scudo, F. M. 1985. Darwin, Darwinian theories and punctuated equilibria. *Systematic Zool.* 34: 239–242.

Selander, R. K., S. Y. Yang, R. C. Lewontin, and W. E. Johnson. 1970. Genetic variation in the horseshoe crab (*Limulus polyphemus*), a phylogenetic "relic." *Evolution* 24: 402–419.

Sepkoski, J. J. 1982. A compendium of fossil marine families. *Milwaukee Publ. Mus. Contrib. Biol. Geol.* 51: 1–25.

——— 1988. Alpha, beta, or gamma—where does all the diversity go? *Paleobiology* 14: 221–234.

——— 1991. Population biology models in macroevolution. In N. L. Gilinsky and P. W. Signor, eds., *Analytical Paleontology.* Paleont. Soc. Short Courses in Paleontology No. 4. Knoxville TN, pp. 136–156.

——— 1997. Biodiversity: past, present, and future. *Jour. Paleontol.* 71: 533–539.

Sequieros, L. 1981. La evolucion biologica, teoria en crisis. *Razon y Fe,* December, pp. 586–593.

Shalizi, C. R. 1998. Scientific models: claiming and validating. *Santa Fe Institute Bulletin,* pp. 9–10.

Shapiro, E. A. 1978. Natural selection in a Miocene pectinid: a test of the punctuated equilibria model. *Geol Soc. Am. Abstracts Annual Meeting* 10: 490.

Shattuck, G. B. 1904. Geological and paleontological relations with a review of earlier investigations. *Maryland Geological Survey, Miocene Volume.*

Shaw, A. B. 1969. Adam and Eve, paleontology and the non-objective arts. *Jour. Paleontology* 43: 1085–1098.

Sheldon, P. R. 1987. Parallel gradualistic evolution of Ordovician trilobites. *Nature* 330: 561–563.

——— 1996. Plus ça change—a model for stasis and evolution in different environments. *Palaeogeog. Palaeoclimatol. Palaeoecol.* 127: 209–227.

Simpson, G. G. 1944. *Tempo and Mode in Evolution.* N.Y.: Columbia Univ. Press.

Smith, A. B. and C. R. C. Paul. 1985. Variation in the irregular echinoid *Discoides* during the early Cenomanian. *Special Papers in Palaeontology* 33: 29–37.

Smith, C. G. 1994. Tempo and mode in deep-sea benthic ecology: punctuated equilibrium revisited. *Palaios* 9: 3–13.

Smocovitis, V. B. 1996. *Unifying Biology.* Princeton NJ: Princeton Univ. Press.

Sneppen, K., P. Bak, H. Flyvbjerg, and M. H. Jensen. 1994. Evolution as a self-organized critical phenomenon. *Preprint.*

Sober, E., and D. S. Wilson. 1998. *Unto Others: The Evolution and Psychology of Unselfish Behavior.* Cambridge MA: Harvard Univ. Press.

Somit, A., and S. A. Peterson. 1992. *The Dynamics of Evolution: The Punctuated Equilibrium Debate in the Natural and Social Sciences.* Ithaca N.Y.: Cornell Univ. Press.

Sorhannus, U. 1990. Punctuated morphological change in a Neogene diatom lineage: "local" evolution or migration? *Historical Biol.* 3: 241–247.

Stanley, S. M. 1973. An explanation for Cope's rule. *Evolution* 27: 1–26.

——— 1975. A theory of evolution above the species level. *Proc. Natl. Acad. Sci. USA* 72: 646–650.

——— 1979. *Macroevolution: Pattern and Process.* San Francisco CA: W. H. Freeman.

——— 1982. Macroevolution and the fossil record. *Evolution* 36: 460–473.

Stanley, S. M., and X. Yang. 1987. Approximate evolutionary stasis for bivalve morphology over millions of years: a multivariate, multilineage study. *Paleobiology* 13: 113–139.

Stebbins, G. L., and F. J. Ayala. 1981a. Is a new evolutionary synthesis necessary? *Science* 213: 967–971.

——— 1981b. The evolution of Darwinism. *Scientific American:* 72–82.

Stidd, B. M. 1985. Are punctuationists wrong about the Modern Synthesis? *Philosophy of Science* 52: 98–109.

Swisher, C. C., W. J. Rink, S. C. Anton, H. P. Schwarcz, G. H. Curtis, A. Suprijo, and Widiasmoro. 1996. Latest *Homo erectus* of Java. *Science* 274: 1870–1874.

Sylvester-Bradley, P. C. (ed.) 1956. *The Species Concept in Paleontology.* London: Systematics Assoc. Publication No. 2, 145 pp.

Tattersall, I., and J. Schwartz. 2000. *Extinct Humans.* Boulder CO: Westview Press.

Tax, S. (ed.) 1960. *Evolution After Darwin.* 3 vols. Chicago: Univ. of Chicago Press. Vol. 1: *The Evolution of Life;* vol. 2: *The Evolution of Man;* vol. 3: *Issues in Evolution,* edited with C. Callender.

Thain, M., and M. Hickman. 1990. *The Penguin Dictionary of Biology,* New Edition. N.Y.: Penguin Books.

Thompson, D'Arcy W. 1917. *On Growth and Form.* Cambridge UK: Cambridge Univ. Press.

——— 1942. *On Growth and Form.* 2nd Edition. Cambridge UK: Cambridge Univ. Press.

Thompson, P. 1983. Tempo and mode in evolution: punctuated equilibria and the modern synthetic theory. *Philosophy of Science* 50: 432–452.

Thurow, L. C. 1996. *The Future of Capitalism.* N.Y.: Penguin Books.

Traub, J. 1995. Shake them bones. *New Yorker,* March 13, p. 60.

Trueman, A. E. 1922. The use of *Gryphaea* in the correlation of the Lower Lias. *Geol. Mag.* 49: 256–268.

Turner, J. R. G. 1984. Why we need evolution by jerks. *New Scientist,* February 9.

——— 1986. The genetics of adaptive radiation: a neo-Darwinian theory of punctuational evolution. In D. M. Raup and D. Jablonski, eds., *Patterns and Processes in the History of Life.* Berlin: Springer Verlag.

Valdecasas, A. G., and D. V. Herreros. 1982. La teoria de la evolucion: los terminos de controversia. *Revista Universidad Complutense Madrid,* pp. 153–158.

Valentine, J. W. 1990. The macroevolution of clade shape. In R. M. Ross and W. D. Allmon, eds., *Causes of Evolution: A Paleontological Perspective.* Chicago IL: Univ. of Chicago Press, pp. 128–150.

Vermeij, G. J. 1977. The Mesozoic marine revolution: evidence from snails, predators, and grazers. *Paleobiology* 3: 245–258.

——— 1987. *Evolution and Escalation: An Ecological History of Life.* Princeton NJ: Princeton Univ. Press.

Vogel, K. 1983. *Macht die biologische Evolution Sprünge?* Wiesbaden: Franz Steiner.

Von Baer, K. E. 1828. *Über Entwickelungsgeschichte der Thiere.* Königsberg: Bornträger.

——— 1866. Über Prof. Nic. Wagner's Entdeckung von Larven die sich fortpflanzen, und über die Pädogenesis überhaupt. *Bull. Acad. Imp. Sciences St. Petersburg* 9: 63–137.

Vrba, E. S. 1980. Evolution, species and fossils: how does life evolve? *South African Jour. Sci.* 76: 61–84.

——— 1984a. Evolutionary pattern and process in the sister-group Alcelaphini-Aepycerotini (Mammalia: Bovidae). In N. Eldredge and S. M. Stanley, eds., *Living Fossils.* N.Y.: Springer-Verlag, pp. 62–79.

——— 1984b. What is species selection? *Systematic Zool.* 33: 318–328.

——— (ed.) 1985a. *Species and Speciation.* Pretoria: Transvaal Museum.

——— 1985b. Environment and evolution: alternative causes of temporal distribution of evolutionary events. *South Afr. Jour. Sci.* 81: 229–236.

Vrba, E., and S. J. Gould. 1986. The hierarchical expansion of sorting and selection: Sorting and selection cannot be equated. *Paleobiology* 12: 217–228.

Wagner, P. J. 1995. Testing evolutionary constraint hypotheses with early Paleozoic gastropods. *Paleobiology* 21: 248–272.

——— 1996. Contrasting the underlying pattern of active trends in morphologic evolution. *Evolution* 50: 990–1007.

——— 1999. The utility of fossil data in phylogenetic analyses: a likelihood example using Ordovician-Silurian species of the Lophospiridae (Gastropoda: Murchisoniina). *Am. Malacol. Bull.* 15: 1–31.

Wagner, P. J., and D. H. Erwin. 1995. Phylogenetic patterns as tests of speciation models. In D. H. Erwin and R. L. Anstey, eds., *New Approaches to Speciation in the Fossil Record.* N.Y.: Columbia Univ. Press, pp. 87–122.

Wake, D. B., G. Roth, and M. H. Wake. 1983. On the problem of stasis in organismal evolution. *J. Theor. Biol.* 101: 211–224.

Wei, K. 1994. Allometric heterochrony in the Pliocene-Pleistocene planktic foraminiferal clade *Globoconella. Paleobiology* 20: 66–84.

Wei, K. Y., and J. P. Kennett. 1988. Phyletic gradualism and punctuated equilibrium in the late Neogene planktonic foraminiferal clade *Globoconella. Paleobiology* 14: 345–363.

Weiss, H., and R. S. Bradley. 2001. What drives societal collapse? *Science* 291: 609–610.

Weller, J. M. 1961. The species problem. *Jour. Paleontology* 35: 1181–1192.

Wessels, N. K., and J. L. Hopson. 1988. *Biology.* N.Y.: Random House.

West, R. M. 1979. Apparent prolonged evolutionary stasis in the middle Eocene hoofed mammal *Hyopsodus. Paleobiology* 5: 252–260.

Westoll, T. S. 1949. On the evolution of the Dipnoi. In G. L. Jepsen, E. Mayr, and G. G. Simpson, eds., *Genetics, Paleontology and Evolution.* Princeton NJ: Princeton Univ. Press.

Whatley, R. C. 1985. Evolution of the ostracods *Bradleya* and *Poseidonamicus* in the deep-sea Cainozoic of the south-west Pacific. *Special Papers in Palaeontology* 33: 103–116.

White, T. D., and J. M. Harris. 1977. Suid evolution and correlation of African localities. *Science* 198: 13–21.

Wiggins, V. D. 1986. Two punctuated equilibrium dinocyst events in the Upper Miocene of the Bering Sea. *AASP Contrib. Series* 17: 159–167.

Williams, G. C. 1966. *Adaptation and Natural Selection.* Oxford UK: Oxford Univ. Press.

——— 1992. *Natural Selection: Domains, Levels and Challenges.* N.Y.: Oxford Univ. Press.

Williamson, P. G. 1981. Paleontological documentation of speciation in Cenozoic molluscs from Turkana Basin. *Nature* 252: 298–300.

——— 1985. Punctuated equilibrium, morphological stasis and the paleontological documentation of speciation: A reply to Fryer, Greenwood and Peake's critique of the Turkana Basin mollusk sequence. *Biol. Jour. Linn. Soc.* 26: 307–324.

——— 1987. Selection or constraint?: A proposal on the mechanism of stasis. In K. S. W. Campbell and M. F. Day, eds., *Rates of Evolution.* London: Allen & Unwin, pp. 129–142.

Wilson, D. S., and E. Sober. 1998. Reviving the superorganism. *Jour. Theor. Biol.* 136: 337–356.

Wollin, Andrew. 1996. A hierarchy-based approach to punctuated equilibrium: an alternative to thermodynamic self-organization in explaining complexity. Abstract, INFORMS National Meeting, Atlanta GA.

Wolpoff, M. H. 1984. Evolution in *Homo erectus:* the question of stasis. *Paleobiology* 10: 389–406.

Wray, G. A. 1995. Punctuated evolution of embryos. *Science* 267: 1115–1116.

Wynne-Edwards, V. C. 1962. *Animal Dispersion in Relation to Social Behavior.* Edinburgh: Oliver & Boyd.

Yoon, C. K. 1996. Bacteria seen to evolve in spurts. *New York Times,* June 25.

Ziegler, A. M. 1966. The Silurian brachiopod *Eocoelia hemisphaerica* (J. de C. Sowerby) and related species. *Palaeontology* 9: 523–543.

ILLUSTRATION CREDITS

Figure	Source
1.1	Drawing by Laszlo Meszoly.
1.2	From author's collection.
2.1a	Reprinted with permission from *Nature,* vol. 293, no. 5832, Oct. 8, 1981, p. 4, fig. 8. Copyright ©1981 Macmillian Magazines Limited.
2.1b	Reprinted with permission from *Nature,* vol. 293, no. 5832, Oct. 8, 1981, p. 5, fig. 4. Copyright ©1981 Macmillian Magazines Limited.
2.2	From Goodfriend and Gould, 1996, p. 1896, fig. 3. Copyright ©1996 American Association for the Advancement of Science.
2.3	From Moore, Lalicker, and Fischer, 1952, p. 33, fig. 1.14. Copyright ©1952 by The McGraw-Hill Companies. Reproduced with permission of The McGraw-Hill Companies.
2.4	From Simpson, 1944.
2.5	From Moore, Lalicker, and Fischer, 1952, p. 33, fig. 1.15. Copyright ©1952 by The McGraw-Hill Companies. Reproduced with permission of The McGraw-Hill Companies.
3.1	From Michaux, 1989.
3.2	Drawing by Laszlo Meszoly.
3.3	From Gould and Eldredge, 1986, p. 142.
3.4	From *Proc. Natl. Acad. Sci.,* USA, 91 (1994), p. 6811, fig. 5. Copyright ©1994 National Academy of Sciences, USA. Used by permission.
3.5a	From Glenn L. Jepson, Ernst Mayr, and George Gaylord Simpson (eds.), *Genetics, Paleontology, and Evolution* (Princeton: Princeton University Press, 1949). Copyright ©1949 by Princeton University Press. Reprinted by permission of Princeton University Press.
3.5b	From Raup, David M. and Steven M. Stanley, *Principles of Paleontology* (San Francisco: W. H. Freeman and Co. 1971), p. 266, fig. 10-10a.

3.6 From Erwin and Anstey, 1995. Copyright ©1995 by Columbia University Press. Reproduced with permission of Columbia University Press in the format Textbook via Copyright Clearance Center.

4.1 From Smith and Paul, 1985, p. 35, fig. 4. Reprinted by permission of the Paleontological Association.

4.2 From Cronin, 1985, p. 61, fig. 1b.

4.3 From Kucera and Malmgren, 1998, p. 56, fig. 6.

4.4 From Kucera and Malmgren, 1998, p. 57, fig. 7.

4.5 From Cheetham, 1986, p. 196, fig. 5.

4.6 From H. M. McHenry, "Tempo and mode in human evolution," *Proc. Natl. Acad. Sci.,* USA 91, 1994, p. 6781, fig. 1b. Copyright ©1994 National Academy of Sciences, USA. Used by permission.

4.7 From Heaton, 1993, p. 302, fig. 11.

4.8 From Heaton, 1993, p. 299, fig. 3.

4.9 From Heaton, 1993, p. 301, fig. 9.

4.10 From Kelley, 1984, p. 1247, fig. 11.

4.11 From Stanley and Yang, 1987, p. 124, fig. 8.

4.12 From Stanley and Yang, 1987, p. 132, fig. 16.

4.13 From Prothero and Heaton, 1996, p. 262, fig. 2. Copyright ©1996, with permission from Elsevier Science.

4.14 From Erwin and Anstey, 1995, p. 68, fig. 3.1. Copyright ©1995 by Columbia University Press. Reproduced with permission of Columbia University Press in the format Textbook via Copyright Clearance Center.

4.15 From Sheldon, 1996, p. 214, fig. 1. Copyright ©1996, with permission from Elsevier Science.

5.1 Adapted from illustrations by David Starwood, from "The Evolution of Life on the Earth" by Stephen Jay Gould. *Scientific American,* October 1994, p. 86. Copyright ©1994 by *Scientific American.* All rights reserved.

5.2 Adapted from "Universal Phylogenetic Tree in Rooted Form." Copyright ©1994 by Carl R. Woese. *Microbiological Reviews,* 58, 1994, pp. 1–9. Adapted with permission of the author.

5.3 From Gould, 1996, p. 150, fig. 21. Copyright ©1996. Used by permission of Crown Publishers.

5.4 Adapted from Gould, 1988b. Copyright ©1988 by Stephen Jay Gould. Adapted with permission of *Journal of Paleontology.*

5.5 From "The Evolution of the Horse" by W. D. Matthew. Appeared in *Quarterly Review of Biology* 1926. Neg. no. 123823. Courtesy Department of Library Services, American Museum of Natural History.

5.6 From Lenski and Travisano, 1994, p. 6810, fig. 2. Copyright ©1994 National Academy of Sciences, USA. Used by permission.

5.7 From Blackburn, 1995, p. 203, fig. 1. Used by permission of Academic Press, London.

5.8 From Kilgour, 1998. Copyright ©1998 by Frederick Kilgour. Used by permission of Oxford University Press, Inc.

A.1 "Poorly punctuated equilibrium," *American Scientist,* May-June 1997, p. 225. Used by permission of Mark Heath.

A.2 Adapted from Bliss, Parker, and Gish, 1980. Drawing by Laszlo Meszoly.

A.3 From Price, 1996. Copyright ©1996. Reprinted with permission of Brooks/Cole, an imprint of Wadsworth Group, a division of Thompson Learning. Fax 800-730-2215.

INDEX